ENCYCLOPAEDIA OF
BIOTECHNOLOGY

ENCYCLOPAEDIA OF

BIOTECHNOLOGY

Bethany J. Mathews

ANMOL PUBLICATIONS PVT. LTD.

NEW DELHI - 110 002 (INDIA)

ANMOL PUBLICATIONS PVT. LTD.
H.O.: 4374/4B, Ansari Road, Daryaganj
New Delhi - 110 002
Ph.: 23278000, 23261597
B.O.: No. 1015, Ist Main Road, BSK III Stage
III Phase, III Block
Bangalore - 560 085 (India)
Visit us at: www.anmolpublications.com

Encyclopaedia of Biotechnology

First Published, 2006

ISBN 81-261-2906-9

PRINTED IN INDIA

Printed at Mehra Offset Press, Delhi.

Preface

For a long time there has been a need for a volume in biotechnology defining terms in a clear and lucid manner. The normal reader is deluged continually with complex terms in biotechnology, which are incomprehensible to a layman. This volume attempts to cover broadly the whole dominion of biotechnology. Many words from fields closely related to biotechnology are included because of their wide usage in biotechnology literature.

The Encyclopaedia of Biotechnology covers an enormous range of technical terms from both pure and applied Biotechnology; this superb reference goes beyond basic definitions to provide helpful explanations and examples. All aspects of biotechnology and genetic engineering are presented in accurate, easy-to-understand terms. This text provides concise definitions of terms for persons unfamiliar to biotechnology, and clarifies new terms and how they are being used for those who are already somewhat conversant in the area. It explore the most up-to-date research in the field-from gene cloning, to the creation of genetically engineered plants and animals, to the ability to derive genetic fingerprints from microscopic samples of tissue. The explanations of entries range from brief definitions of a few lines to expository discussions of various lengths. In terms of its broader objectives, this volume covers the whole range of terms related with biotechnology and includes essential entries from such fields as genetic engineering, DNA recombination, enzyme technology, biochemical technology, fermenter technology, somatic cell cultures, biophysics, biomedical sciences. . Both succinct and wide-ranging, this volume is an ideal handy reference work and great introduction for students and non-scientists alike.

The aim of this volume is to provide a concise repository of number of major concepts in this field. In preparation of this book, the author has freely consulted large number of books and journals so no authenticity is claimed. Author is especially thankful to M/S Anmol Publications for shaping this book in its final form. Suggestions for further improvement of this book are not only welcome but also greatly appreciated.

Author

A

ABC transporters Members of a large superfamily of membrane transport proteins that hydrolyze ATP and transfer a diverse array of small molecules across a membrane.

A DNA A right-handed DNA double helix that has 11 base pairs per turn. DNA exists in this form when partially dehydrated.

A kinase (cyclic AMP-dependent protein kinase) Enzyme that phosphorylates target proteins in response to a rise in intracellular cyclic AMP. First identified in skeletal muscle as part of the pathway of regulation of glycogen breakdown in response to adrenaline.

Abiotic Arising from non-living organisms.

Abiotic stress Outside (nonliving) factors which can cause harmful effects to plants, such as soil conditions, drought, extreme temperatures.

Abscisic acid Abscisic acid is a single compound unlike the auxins, gibberellins, and cytokinins. It was called "abscisin II" originally because it was thought to play a major role in abscission of fruits. At about the same time another group was calling it "dormin" because they thought it had a major role in bud dormancy. The name abscisic acid (ABA) was coined by a compromise between the two groups. Though ABA generally is thought to play mostly inhibitory roles, it has many promoting functions as well.

In 1963, abscisic acid was first identified and characterized by Frederick Addicott and his associates. They were studying compounds responsible for the abscission of fruits (cotton). Two compounds were isolated and called abscisin I and abscisin II. Abscisin II is presently called abscisic acid (ABA). Two other groups at about the same time discovered the same compound. One group headed by Philip Wareing was studying bud dormancy in woody plants. The other group led by Van Steveninck was studying abscission of flowers and fruits from lupine. Plant physiologists agreed to call the compound abscisic acid.

ABA is a naturally occurring compound in plants. It is a sesquiterpenoid (15-carbon) which is partially produced via the mevalonic pathway in chloroplasts and other plastids. Because it is sythesized partially in the chloroplasts, it makes sense that biosynthesis primarily occurs in the leaves. The production of ABA is accentuated by stresses such as water loss and freezing temperatures. It is believed that biosynthesis occurs indirectly through the production of carotenoids. Carotenoids are pigments produced by the chloroplast which have 40 carbons. Breakdown of these carotenoids occurs by the following mechanism:

Violaxanthin is a carotenoid which has forty carbons.

It is isomerized and then split via an isomerase reaction followed by an oxidation reaction. One molecule of xanthonin is produced from one molecule of violaxanthonin and it is uncertain what happens to the remaining biproduct.

The one molecule of xanthonin produced is unstable and spontaneously changed to ABA aldehyde. Further oxidation results in ABA.

Activation of the molecule can occur by two methods. In the first method, an ABA-glucose ester can form by attachment of glucose to ABA. In the second method, oxidation of ABA can occur to form phaseic acid and dihyhdrophaseic acid.

The transport of ABA can occur in both xylem and phloem tissues. It can also be translocated through paranchyma cells. The movement of abscisic acid in plants does not exhibit polarity like auxins. ABA is capable of moving both up and down the stem.

The following are some of the phyysiological responses known to be associated with abscisic acid.

Stimulates the closure of stomata (water stress brings about an increase in ABA synthesis).

Inhibits shoot growth but will not have as much affect on roots or may even promote growth of roots.

Induces seeds to synthesize storage proteins.

Inhibits the affect of gibberellins on stimulating de novo synthesis of a-amylase.

Has some effect on induction and maintanance of dormancy.

Induces gene transcription especially for proteinase inhibitors in response to wounding which may explain an apparent role in pathogen defense.

Absciss; abscissa The horizontal axis of a graph.

Absorption In general: the process of absorbing; taking up of water and nutrients by assimilation or imbibition. The taking up by capillary, osmotic, chemical or solvent action, such as the taking up of a gas by a solid or liquid, or taking up of a liquid by a solid.

In biology: the movement of a fluid or a dissolved substance across a cell membrane.

In plants: water and mineral salts are absorbed from the soil by roots.

In animals: solubulized food material is absorbed into the circulatory system through cells lining the alimentary canal.

Abstraction reaction A reaction that removes an atom from a structure.

Acaricide A pesticide used to kill or control mites or ticks.

Accessory bud Lateral bud occurring at the base of a terminal bud or at the side of an axillary bud.

Acclimatization The adaptation of a living organism (plant, animal or micro-organism) to a changed environment that subjects it to physiological stress. Acclimatization should not be confused with adaptation.

Acellular Describing tissues or organisms that are not made up of separate cells but often have more than one nucleus.

Acentric chromosome Chromosome fragment lacking a centromere.

Acetyl Chemical group derived from acetic acid. Acetyl groups are important in metabolism and often added as a covalent modification of proteins.

Acetyl co-enzyme A A compound formed in the mitochondria when an acetyl group (CH_3CO-) - derived from breakdown of fats, proteins, or carbohydrates - combines with the thiol group (-SH) of co-enzyme A.

Acetylcholine Neurotransmitter that functions at cholinergic chemical synapses, found both in the brain and in the peripheral nervous system. It is the neurotransmitter at vertebrate neuromuscular junctions.

Acetylcholine receptor Lon channel that opens in response to acetylcholine binding, thereby converting a chemical signal into an electrical one. Best understood example of a ligand-gated channel. Sometimes called the *nicotinic* acetylcholine receptor to distinguish it from a *muscarinic* receptor, which is a G-protein-linked cell-surface receptor.

Acid In the Brψnsted definition, an acid is a chemical species that can donate a proton to another species (a base). In the Lewis definition, an acid is a chemical species that can accept (and share) a pair of electrons from another species. Hydrochloric acid is a Brψnsted acid; the proton it donates is a Lewis acid. A neutral Lewis acid and a neutral Lewis base can commonly form a dipolar bond.

Acquired Developed in response to the environment, not inherited, such as a character trait (acquired characteristic) resulting from environmental effect(s).

Acquired immune deficiency syndrome (AIDS) Clinically human immunodeficiency virus (HIV) infection first was recognized in 1981 in homosexual men in New York City who presented with evidence of a profound acquired immune deficiency syndrome (AIDS). We now appreciate that HIV infection is a worldwide health problem that affects millions of men and women. HIV has the capability to affect every organ system in the body by direct damage by the virus or by rendering the host susceptible to opportunistic infections. This article discusses some of the more common HIV-related illnesses that are seen frequently in the ED.

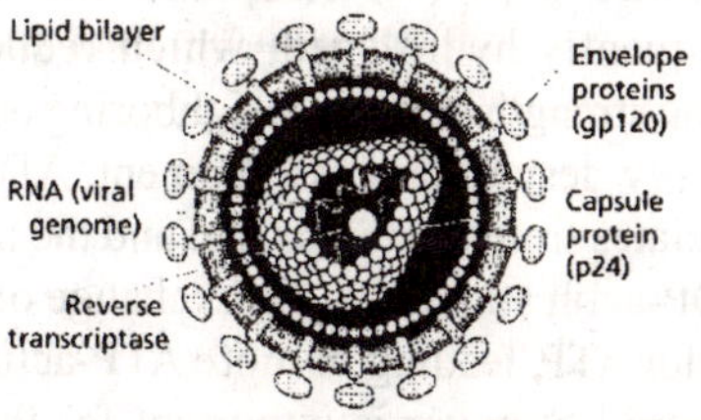

Fig. Acquired immune deficiency syndrome (AIDS)

HIV is a Lentivirus, a subgroup of retroviruses. This family of viruses is known for latency, persistent viremia, infection of the nervous system, and weak host immune responses. HIV has high affinity for CD4 T lymphocytes and monocytes. HIV binds to CD4 cells and becomes internalized. The virus replicates itself by generating a DNA copy by reverse transcriptase. Viral DNA becomes incorporated into the host DNA, enabling further replication.

HIV is transmitted primarily through sexual contact (over 70%); worldwide, it is more common in heterosexual men and women than in homosexual men. Although the majority of initial HIV-related AIDS cases in the United States were homosexual men, more recently the majority of new cases of HIV infection are in the heterosexual population. Parenteral transmission occurs largely among intravenous drug users; transmission by contaminated blood products is rare in the United States, though this remains a serious problem in developing countries. Since the introduction of universal precaution practices, infection of health care workers through parenteral exposure remains rare. Children are infected primarily by perinatal transmission.

More than 887,000 persons have AIDS, and an estimated 1-2 million persons have asymptomatic HIV infection.

Since the AIDS epidemic began, more than 20 million deaths have been attributed to AIDS. The current estimate of worldwide disease prevalence is more than 38 million HIV infections. Ninety-five percent of these cases are in developing countries, generally in sub-Saharan Africa and Southeast Asia.

The course of HIV infection is characterized primarily by latency. Unfortunately, profound immune suppression eventually develops and the illness appears to be almost uniformly lethal. More than 500,000 persons have died of AIDS in the United States.

Progression from HIV infection to AIDS occurs at a median of 11 years after infection. In the recent past, most patients would not survive more than 1-2 years following diagnosis of AIDS. However, since the introduction of highly active antiretroviral therapy (HAART) and prophylaxis against opportunistic pathogens, death rates from AIDS have begun to decline significantly.

Most HIV infections still occur in men; however, the frequency of infection in women is increasing, especially in developing countries. In the United States, fewer than 18% of all HIV cases are in women, whereas worldwide an estimated 50% of all HIV patients are women.

Most AIDS cases occur in adults aged 25-44 years. Children represent fewer than 1% of AIDS cases in the United States. Internationally, children younger than 15 years are estimated to account for close to 10% of all HIV cases.

Acridine dyes A class of positively charged polycyclic molecules that intercalate into DNA and induce frameshift mutations.

Acrocentric A chromosome that has its centromere near the end.

Acropetal 1. Developing or blooming in succession towards the apex, such as leaves or flowers developing acropetally.

2. Transport or movement of substances towards the apex, such as the movement of water through the plant.

The opposite tendency is termed basipetal.

Acrosomal process Long, thin, actin-containing spike produced from the head of certain sperm when they make contact with the egg. Seen in sea urchins and other marine invertebrates whose eggs are surrounded by a thick gelatinous coat.

Acrosome Region at the head end of a sperm cell that contains a sac of hydrolytic enzymes used to digest the protective coating of the egg.

Actin Actin is a globular structural protein that polymerizes in a helically fashion to form actin filaments (or microfilaments). These form the cytoskeleton - a three-dimensional network inside an eukaryotic cell. Actin filaments provide mechanical support for the cell, determine the cell shape, enable cell movements (through lamellipodia, filopodia, or pseudopodia); and participate in certain cell junctions, in cytoplasmic streaming and in contraction of the cell during cytokinesis. In muscle cells they play an essential role, along with myosin, in muscle contraction. In the cytosol, actin is predominantly bound to ATP, but can also bind to ADP. An ATP-actin complex polymerizes faster and dissociates slower than an ADP-actin complex. Actin is one of the most abundant proteins in many eukaryotic cells, with concentrations of over 100 μM. It is also one of the most highly conserved proteins, differing by no more than 5% in species as diverse as algae and humans.

The individual subunits of actin are known as globular actin (G-actin), while the filamentous polymer composed of G-actin subunits (a microfilament), is called F-actin. The microfilaments are the thinnest component of the cytoskeleton, measuring only 7 nm in diameter. Much like the microtubules, actin filaments are polar, with a fast growing plus (+) or barbed end and a slow growing minus (-) or pointed end. The terms barbed and pointed end come from the arrow-like appearance of microfilaments decorated with the motor domain of myosin as seen in electronmicrographs. Filaments elongate approximately 10 times faster at the plus (+) end than the minus (-) end. This phenomenon is known as the treadmill effect. The process of actin polymerization, nucleation, starts with the association of three G-actin monomers into a trimer. ATP-actin then binds the plus (+) end, and the ATP is subsequently hydrolyzed, which reduces the binding strength between neighboring units and generally destabilizes the filament. ADP-actin dissociates from the minus end and the increase in ADP-actin stimulates the exchange of bound ADP for ATP, leading to more ATP-actin units. This rapid turnover is important for the cell's movement. End-capping proteins such as CapZ prevent the addition or loss of monomers at the filament end where actin turnover is unfavourable like in the muscle apparatus.

The protein cofilin binds to ADP-actin units and promotes their dissociation from the minus end and prevents their reassembly. The protein profilin reverses this effect by stimulating the exchange of bound ADP for ATP. In addition, ATP-actin units bound to profilin will dissociate from cofilin and are then free to polymerize. Another important component in filament production is the Arp2/3 complex, which nucleates new actin filaments while bound to existing filaments, thus creating a branched network. All of these three proteins are regulated by cell signaling mechanism.

Actin filaments are assembled in two general types of structures: bundles and networks. Actin-binding proteins dictate the formation of either structure since they cross-link actin filaments. Actin filaments have the appearance of a double-stranded helix.

In non-muscle actin bundles, the filaments are held together such that they are parallel to each other by actin-bundling proteins and/or cationic species. Bundles play a role in many cellular processes such as cell division (cytokinesis) and cell movement. For example, in vertebrates, the actin-bundling protein villin is almost entirely responsible for causing bundle formations in the microvilli of intestinal cells.

Actin, together with myosin filaments, form actomyosin, which provides the mechanism for muscle contraction. Muscular contraction uses ATP for energy. The ATP allows, through hydrolysis, the myosin head to extend up and bind with the actin filament. The myosin head then releases after moving the actin filament in a relaxing or contracting movement by usage of ADP.

In contractile bundles, the actin-bundling protein actinin separates each filament by 40 nm. This increase in distance allows the motor protein myosin to interact with the filament, enabling deformation or contraction. In the first case, one end of myosin is bound to the plasma membrane while the other end walks towards the plus end of the actin filament. This pulls the membrane into a different shape relative to the cell cortex. For contraction, the myosin molecule is usually bound to two separate filaments and both ends simultaneously walk towards their filament's plus end, sliding the actin filaments over each other. This results in the shortening, or contraction, of the actin bundle (but not the filament). This mechanism is responsible for muscle contraction and cytokinesis, the division of one cell into two.

Actin networks, along with many actin-binding proteins (such as the Arp2/3 complex and filamin) form a complex network at the cortical regions of the cell. Recent studies have also suggested that actin network on the cell cortex serve as barriers for molecular diffusion within the plasmic membrane.

Actin is one of the most highly conserved proteins, with 80.2% sequence conservation at the gene level between Homo sapiens and Saccharomyces cerevisiae, and 95% conservation of the primary structure of the protein product.

Although most yeasts have only a single actin gene, higher eukaryotes generally express several isoforms of actin encoded by a family of related genes. Mammals have at least six actins, which are divided into three classes (alpha, beta and gamma) according to their isoelectric point. Alpha actins are generally found in muscle, whereas beta and gamma isoforms are prominent in non-muscle cells. Although there are small differences in sequence and properties between the isoforms, all actins assemble into microfilaments and are essentially identical in the majority of tests performed in vitro.

The typical actin gene has an approximately 100 nucleotide 5' UTR, a 1200 nucleotide translated region, and a 200 nucleotide 3' UTR. The majority of actin genes are interrupted by introns, with up to 6 introns in any of 19 well-characterised locations. The high conservation of the family makes actin the favoured model for studies comparing the introns-early and introns-late models of intron evolution.

All non-spherical prokaryotes appear to possess genes such as MreB which encode homologues of actin; these genes are required for the cell's shape to be maintained. The plasmid-derived gene ParM encodes an actin-like protein whose polymerised form is dynamically unstable, and appears to partition the plasmid DNA into the daughter cells during cell division by a mechanism analogous to that employed by microtubules in eukaryotic mitosis.

Actin was first observed experimentally in 1887 by W.D. Halliburton, who extracted a protein from muscle which 'coagulated' preparations of myosin, and which he dubbed "myosin-ferment". However, Halliburton was unable to further characterise his findings and the discovery of actin is generally credited instead to Brïnσ F. Straub, a young biochemist working in Albert Szent-Gyorgyi's laboratory at the Institute of Medical Chemistry at the University of Szeged, Hungary.In 1942 Straub developed a novel technique for extracting muscle protein that allowed him to isolate substantial amounts of relatively pure actin. Straub's method is essentially the same as that used in laboratories today. Szent-Gyorgyi had previously described the more viscous form of myosin produced by slow muscle extractions as 'activated' myosin, and since Straub's protein produced the activating effect, it was dubbed "actin'. The hostilities of World War II meant that Szent-Gyorgyi and Straub were unable to publish the work in Western scientific journals; it became well-known in the West only in 1945, when it was published as a supplement to the Acta Physiologica Scandinavica.

Straub continued to work on actin and in 1950 reported that actin contains bound ATP and that, during polymerisation of the protein into microfilaments, the nucleotide is hydrolysed to ADP and inorganic phosphate (which remain bound in the microfilament). Straub suggested that the transformation of ATP-bound actin to ADP-bound actin played a role in muscular contraction. In fact this is only true in smooth muscle, and was not experimentally supported until 2001.

The crystal structure of G-actin was solved in 1990 by Kabsch and colleagues. In the same year a model for F-actin was proposed by Holmes and colleagues. The model was derived by fitting a helix of G-actin structures according to low-resolution fibre diffraction data from the filament. Several models of the filament have been proposed since. However there is still no x-ray structure of F-actin.

Actin-binding protein Protein that associates with either actin monomers or actin filaments in cells and modifies their properties. Examples include myosin, a-actinin, and profilin.

Actin filament (microfilament) Helical protein filament formed by the polymerization of globular actin molecules. A major constituent of the cytoskeleton of all eucaryotic cells and part of the con- tractile apparatus of skeletal muscle.

Action potential A spontaneous self-propagating change in membrane potential that travels as a 'wave' along electrically excitable cell membranes found in neurons and muscle cells. Action potentials are triggered when the cell's membrane potential depolarizes (i.e. becomes more positive with respect to the internal side) beyond a threshold value of usually -40mV. Voltage sensitive ion channels (Na and K channels) open and close in fast succession causing first in influx of sodium ions (making the inside more positive = depolarization) followed by potassium efflux (making the outside more positive and inside more negative = hyperpolarization). The potassium efflux brings the membrane potential below the threshold and thus to rest. A new action potential can only be triggered with a new stimulus which comes from neurotransmitter activated ion channels at locations of synaptic interaction between neurons or neurons and muscle cells.

Activated charcoal; activated carbon Charcoal which has been treated to remove hydrocarbons and to increase its adsorptive properties. It acts by condensing and holding a gas or solute onto its surface; thus inhibitory substances in nutrient medium may be adsorbed to charcoal included in the medium. Rooting factors such as phenolamines present as contaminants in charcoal may stimulate growth *in vitro*. Its addition to rooting medium may stimulate root initiation in some plant species. Activated charcoal may differ in origin and in composition.

Activation energy Distinct states correspond to minima of a potential energy surface in a configuration space. In this classical picture, the activation energy for transforming state A into state B is the maximum increase in energy (relative to the ground state of A) encountered on a minimum-energy path from A to B. Energy here refers to potential energy; an analogous definition based on free energy can be constructed. When tunneling is considered, lower energy paths become possible, but an activation energy can be associated with the reaction (at a given temperature) via the relationship between temperature and reaction rate.

Activator 1. A substance or physical agent that stimulates transcription of a specific gene or operon.

2. A compound that, by binding to an allosteric site on an enzyme, enables the active site of the enzyme to bind to the substrate.

Active collection In PGR: Defined in the International Undertaking on Plant Genetic as a collection which complements a base collection and is a collection from which seed samples are drawn for distribution, exchange and other purposes such as multiplication and evaluation.

Active immunity A type of acquired immunity whereby resistance to a disease is built up by either having the disease or receiving a vaccine to it.

Active ingredient The individual element of a given compound or mixture responsible for that compound's primary characteristics.

Active site 1. A site on the surface of a catalyst at which activity occurs.

2. The site on the surface of an enzyme molecule that binds the substrate molecule.

Active transport Movement of a molecule across a membrane or other barrier driven by energy other than that stored in the concentration gradient or electrochemical gradient of the transported molecule.

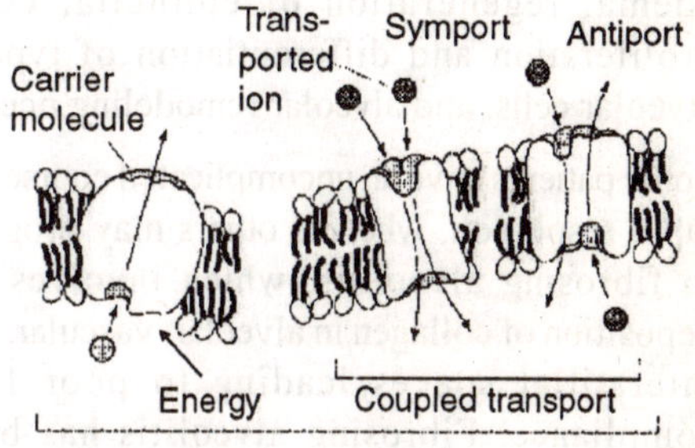

Fig. Active transport

Acute respiratory distress syndrome (ARDS) Acute respiratory distress syndrome (ARDS) was first described in 1967 by Ashbaugh, who described a syndrome of severe respiratory failure associated with pulmonary infiltrates, similar to infant hyaline membrane disease.

Put simply, ARDS is the presence of pulmonary edema in the absence of volume overload or depressed left ventricular function. This is distinguished from acute lung injury, which is similar to ARDS with the exception of a PaO_2/FIO_2 ratio of less than 300 mm Hg.

ARDS occurs in children as well as adults. The condition originates from a number of insults involving damage to the alveolocapillary membrane with subsequent fluid accumulation within the airspaces of the lung. Histologically, these changes have been termed diffuse alveolar damage.

The development of ARDS starts with damage to the alveolar epithelium and vascular endothelium resulting in increased permeability to plasma and inflammatory cells into the interstitium and alveolar space. Damage to the surfactant-producing type II cells and the presence of protein-rich fluid in the alveolar space disrupts the production and function of pulmonary surfactant leading to microatelectasis and impaired gas exchange.

Ultimately, regional variations in pulmonary perfusion, ventilation/perfusion (V/Q) mismatch with shunting of blood through unventilated alveoli, and increased alveolar-arterial oxygen gradient occur. Eventually, resorption of alveolar edema, regeneration of epithelial cells, proliferation and differentiation of type II alveolar cells, and alveolar remodeling occur.

Some patients have an uncomplicated course and rapid resolution, whereas others may progress to fibrosing alveolitis, which involves the deposition of collagen in alveolar, vascular, and interstitial spaces leading to poor lung compliance. Fibrosing alveolitis has been reported histologically as early as 5-7 days.

The mortality rate is approximately 30-40%. Deaths usually result from multisystem organ failure rather than lung failure alone. Patients may present early in the course of the disease without symptoms or signs.Mild tachypnea may be the only manifestation.Severe respiratory distress eventually ensues.

Rales are frequently heard on auscultation of the lungs. However, examination findings may be surprisingly minimal when contrasted with chest radiographic findings.Signs of volume overload, such as a third heart sound (S_3) on auscultation of the heart or jugular venous distention, should be noticeably absent.

A number of clinical conditions are associated with development of ARDS.

Sepsis and the systemic inflammatory response syndrome (SIRS) are the most common predisposing factors associated with development of ARDS. These conditions may result from the indirect toxic effects of neutrophil-derived inflammatory mediators in the lungs.

Severe traumatic injury (especially multiple fractures), severe head injury, and pulmonary contusion are strongly associated with development of ARDS. Long bone fractures may give rise to ARDS through fat embolism. After severe head injury, ARDS is thought to result from a sudden discharge of the sympathetic nervous system, leading to acute pulmonary hypertension and injury to the pulmonary capillary bed. Pulmonary contusions cause ARDS through direct trauma to the lung.

Multiple transfusions are another important risk factor for ARDS, independent of the reason for transfusion or the coexistence of trauma. The incidence of ARDS increases with the number of units transfused. Preexisting liver disease or coagulation abnormalities further contribute to this risk.

Patients who have nearly drowned can develop ARDS. Development of ARDS is slightly more common with salt-water aspiration than with fresh-water aspiration. Infiltrates and hypoxia develop within 12-24 hours of the initial accident. Patients who are symptomatic after 6 hours of observation generally do well. Aspiration is particularly damaging to lung tissue, leading to an osmotic gradient that favors movement of water into airspaces of the lung. Aspiration may be visible with chest radiography, although the chest radiograph may be normal early in the course of the disease.

Smoke inhalation causes lung tissue damage from direct heat, toxic chemicals, and particulate matter carried into the lower lung. Patients with smoke inhalation initially may be asymptomatic. Patients with airway burns and/or exposure to carbon monoxide or toxic fumes should be monitored closely for development of ARDS, even if symptoms initially are absent.

Overdoses of narcotics (eg, heroin), salicylates, tricyclic antidepressants, and other sedatives have been associated with development of

ARDS. (Overdoses of tricyclic antidepressants are the most common.) This risk is independent of the risk from concurrent aspiration.

Adaptation A process in biological evolution as the result of natural selection where a species becomes better adjusted to the living conditions of its environment (temperature, food sources, and predators). In adaptation, individuals that have the most offspring contribute more of their genetic makeup to the next generation. Beneficial traits are favored in this editing process and the next generation of a species or population (for instance a smaller geographical subset of a species) is better adapted to new environmental conditions that caused certain individuals to have fewer offspring. Thus genes that are not optimal for certain conditions will become rarer as a result. Adaptation, as evolution in general, is studied at the level of a population of interbreeding individuals.

Adaptation traits The complex of traits related to reproduction and survival of the individual in a particular production environment. Adaptation traits contribute to individual fitness; they are the traits subjected to selection during the evolution of animal genetic resources. By definition, these traits are also important to the ability of the animal genetic resource to be sustained in the production environment.

Adaptive mutation Bacteria and yeast cells in the stationary (non-growing) phase, have a way of producing (or keeping) only the mutations most appropiate for their making use of new substrates for growth.

Adaptive radiation The evolution of new forms, sub-species or species from one species of plant or animal in order to exploit new habitats or food sources.

Adaptor 1. A synthetic double-stranded oligonucleotide that has a blunt end, while the other end has a nucleotide extension that can base pair with a cohesive end created by cleavage of a DNA molecule with a specific type II restriction endonuclease. The blunt end of the adaptor can be ligated to the ends of a target DNA molecule and the construct can be cloned into a vector by using the cohesive ends of the adaptor.

2. A synthetic single-stranded oligonucleotide that, after self-hybridization, produces a molecule with cohesive ends and an internal restriction endonuclease site. When the adaptor is inserted into a cloning vector by means of the cohesive ends, the internal sequence provides a new restriction endonuclease site.

Addendum In formulation of tissue culture media: an item or a constituent substance to be added.

Additive allelic effects Effects of alleles at a locus, where the heterozygote is exactly intermediate between the two homozygotes.

Adenilate cyclase The enzyme that catalyses the formation of cyclic AMP.

Adenine ($C_5H_5N_5$) (symbol: A) Adenine is one of the two purine nucleobases used in forming nucleotides of the nucleic acids DNA and RNA. In DNA, adenine binds to thymine or uracil via two hydrogen bonds to assist in stabilizing the nucleic acid structures. In RNA, adenine binds to uracil, which is used in the cytoplasm for protein synthesis.

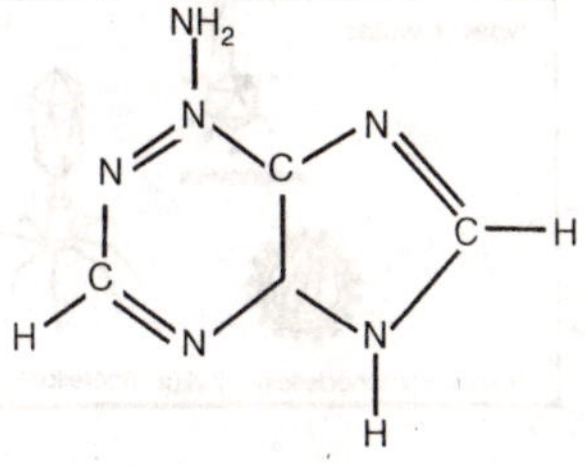

Fig. Adenine

Adenine forms adenosine, a nucleoside, when attached to ribose, and deoxyadenosine when attached to deoxyribose; it forms adenosine triphosphate (ATP), a nucleotide, when three phosphate groups are added to adenosine. Adenosine triphosphate is used in cellular metabolism as one of the basic methods of transferring chemical energy between chemical reactions.

In older literature, adenine was sometimes called Vitamin B4. However it is no longer considered a true vitamin or part of Vitamin B.

Some think that, at the origin of life on Earth, the first adenine was formed by the polymerizing of five hydrogen cyanide (HCN) molecules.

Adenosine deaminase deficiency A lack of an enzyme called adenosine deaminase. Its main characteristic is corresponding lack of immune defences against infection. Sufferers die shortly after birth unless kept in a germ-free environment ('bubbles').

Adenovirus Adenoviruses are viruses of the family Adenoviridae. They infect both humans and animals. Adenoviruses were first isolated in human adenoids (tonsils), from which the name is derived.

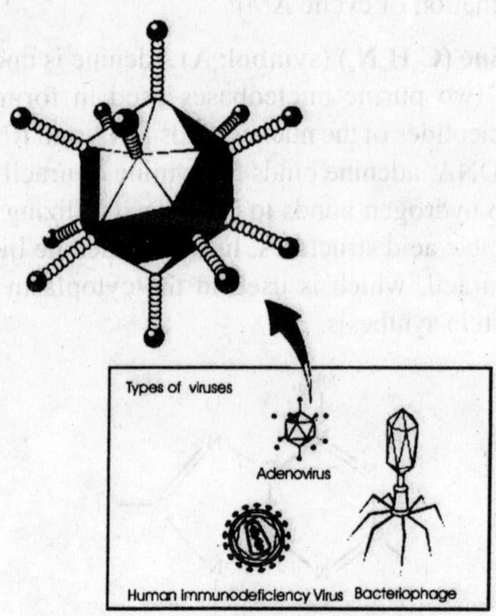

Fig. Adenovirus

Adenoviruses are classifed as group I under the Baltimore classification scheme. They're medium-sized (60-90 nm), nonenveloped icosahedral viruses containing double-stranded DNA. The virion also has a unique "spike" or fibre associated with each penton base of the capsid (see picture below) that aids in attachment to the host cell via the coxsackie-adenovirus receptor on the surface of the host cell. There are 51 immunologically distinct types (6 subgenera: A through F) that can cause human infections ranging from respiratory disease (mainly subgenera B and C), and conjunctivitis (subgenera B and D), to gastroenteritis (subgenera F serotypes 40 and 41). Adenoviruses are unusually stable to chemical or physical agents and adverse pH conditions, allowing for prolonged survival outside of the body and water. Adenoviruses are primarily spread via respiratory droplets, however they can also be spread by fecal routes as well.

The adenovirus genome is linear, non-segmented double stranded (ds) DNA which is around 30-38Kbp. This allows the virus to theoretically carry 30 to 40 genes. Although this is significantly larger than other viruses in its Baltimore group it is still a very simple virus and is heavily reliant on the host cell for survival and replication. An interesting feature of this viral genome is that it has a terminal 55 kDA protein associated with each of the 5' ends of the linear dsDNA, these are used as primers in viral replication and ensure that the ends of the virus' linear genome are adequately replicated.

Adenoviruses possess a linear dsDNA genome and are able to replicate in the nucleus of mammalian cells using the host's replication machinery.

The adenovirus life cycle is separated, by the DNA replication process, into two phases: an early and a late phase. In both phases a primary transcript is generated which is alternatively spliced to generate monocistronic mRNAs compatible with the host's ribosome, allowing for the products to be translated.

The early genes are responsible for expressing mainly non-structural, regulatory proteins. These goal of these proteins is three-fold: to alter the expression of host proteins that are necessary for DNA synthesis; to activate other virus genes (such as the virus-encoded DNA polymerase); and to avoid premature death of the infected cell by the host-immune defenses (blockage of apoptosis, blockage of interferon activity, and blockage of MHC class I translocation and expression).

Some adenoviruses under specialised conditions can transform cells using their early gene products. E1a has been found to immortalise primary cells in vitro allowing E1b to assist and stably transform the cells. Nevertheless, they are reliant upon each other to successfully transform the host cell and form tumours.

DNA replication separates the early and late phases. Once the early genes have liberated adequate virus genes, replication machinery and replication substrates, replication of the adenovirus genome can occur. A terminal protein that is covalently bound to the 5' end of the adenovirus genome acts as a primer for replication. The viral DNA polymerase then uses a strand displacement mechanism, as opposed to the conventional Okazaki fragments used in mammalian DNA replication, to replicate the genome.

The late phase of the adenovirus life cycle is focused on producing sufficient quantities of structural protein to pack all the genetic material produced by DNA replication. Once the viral components have successfully been replicated the virus is assembled into its protein shells and released from the cell as a result of virally induced cell lysis.

ADEPT (antibody-directed enzyme pro-drug therapy) A way to target a drug to a specific tissue. The drug is administered as an inactive pro-drug, and then converted into an active drug by an enzyme administered with a second injection. The enzyme is coupled to an antibody that concentrates it in the target tissue. When the enzyme arrives at the target tissue, the pro-drug is activated to form the active drug, while elsewhere it remains inactive.

Adhesion The attraction of dissimilar molecules for each other. A sticking together of unlike substances, such as soil and water.

Adipose tissue Tissue, found throughout the body, in which fat is stored as an energy reserve. Excessive expansion leads to obesity.

Adjuvant Insoluble material that increases the formation and persistence of antibodies when injected with an antigen.

Adsorbent A substance to which compounds adhere. In tissue culture, an adsorbent is added to the culture medium to adsorb compounds released by cultured cells or tissues, thus minimizing any adverse effect on the subsequent growth in culture. A common adsorbent in tissue culture is *activated charcoal*.

Adsorption The formation of a layer of gas, liquid or solid on the surface of a solid.

Adult cloning The creation of identical copies of an adult animal by nuclear transfer from differentiated adult tissue.

Adult stem cell A specialized cell that is needed for growth, wound healing and tissue regeneration. Adult stem cells are found in all tissues and organs of animals and plants.

Advanced Applied to an organism or a part thereof, implying considerable development from the ancestral stage or from the explant stage.

Adventitious A structure arising at sites other than the usual ones, e.g., shoots from roots or leaves, and embryos from any cell other than a zygote.

Adventitious presence The unintended presence of genetic material or whole seeds from another variety, crop, or weed, in a seed or grain shipment.

Aerate To supply with or mix with air or gas. The process is aeration.

Aerobe A micro-organism that grows in the presence of oxygen.

Aerobic bacteria Bacteria that can live in the presence of oxygen.

Aerobic respiration A type of respiration in which foodstuffs are completely oxidized to carbon dioxide and water, with the release of chemical energy, in a process requiring atmospheric oxygen.

Affinity chromatography A method for separating molecules by exploiting their ability to bind specifically to other molecules. There are several types of biological affinity chromatography. A biological molecule can be immobilized and a smaller molecule to which it is to bind can be stuck to it, or the smaller ligand can be immobilized and the macromolecule stuck to it. A variant is to use an antibody as the immobilized molecule and use it to "capture" its antigen: this is often called immuno-affinity chromatography. A variation is pseudo-affinity chromatography, in which a compound which is like a biological ligand is immobilized on a solid material, and enzymes or other proteins are bound to it. Other techniques include metal affinity chromatography, where a metal ion is immobilized on a solid support: metal ions bind tightly and specifically to many biomolecules. The metal ion is bound to a chelator or chelating group, a chemical group that binds specifically and extremely tightly to that metal.

Affinity constant The reciprocal of the dissociation constant; a measure of the binding energy of a ligand in a receptor.

Affinity tag; purification tag An amino acid sequence that has been engineered into a protein to make its purification easier. These can work in a number of ways. The tag could be another protein, which binds to some other material very tightly and thus allowing the protein to be purified by affinity chromatography. The tag could be a short amino acid sequence, which is recognized by an antibody. The antibody would then bind to the protein whereas it would not have done so before. One such short peptide, called FLAG, has been designed so that it is particularly easy to make antibodies against it. The tag could be a few amino acids, which are then used as a chemical tag on the protein. For example, a string of positively charged amino acids will bind very strongly to a negatively charged filter: this could be used as the basis of a separation system. Some amino acids bind metals very strongly, especially in pairs: this chemical property can be exploited by using a filter with metal atoms chemically linked onto it to pull a protein out of a mixture of proteins.

Aflatoxin Toxic compounds, produced by moulds (fungi) of the *Aspergillus flavus* group, that bind to DNA and prevent replication and transcription. Aflatoxins can cause acute liver damage and cancer. Animals may be poisoned by eating stored food or feed contaminated with the mould.

Agar (*agar-agar*) A polysaccharide solidifying agent used in nutrient media preparations and obtained from certain types of red algae (*Rhodophyta*). Both the type of agar and its concentration can affect the growth and appearance of cultured explants.

Agarose gel electrophoresis A matrix composed of a highly purified form of agar that is used to separate larger DNA and RNA molecules ranging 20,000 nucleotides.

Aggregate 1. A clump or mass formed by gathering or collecting units.

2. A body of loosely associated cells, such as a friable callus or cell suspension.

3. Coarse inert material, such as gravel, that is mixed with soil to increase its porosity.

4. A serological reaction (aggregation) in which the antibody and antigen react and precipitate out of solution.

Agonist A drug, hormone or transmitter substance that forms a complex with a receptor site that is capable of triggering an active response from a cell.

Agricultural biotechnology A range of tools, including traditional breeding techniques, that alter living organisms, or parts of organisms, to make or modify products; improve plants or animals; or develop microorganisms for specific agricultural uses. Modern biotechnology today includes the tools of genetic engineering.

Agrobacterium tumefaciens At the turn of the century *Agrobacterium tumefaciens* was identified as the causal agent in crown gall disease in dicotyledonous plants. Since then, thorough research has been done on this bacterium's mechanism of tumor induction; in addition, *Argorbacterium* is used in numerous research projects as a means with which to introduce new genes into the genomes of a number of plants.

Agrobacterium tumefaciens str. C58 has an unusual chromosomal organization - it has a 2 Mb linear and a 2.8 Mb circular chromosome as well as a 206.479 kbp Ti (tumor-inducing) plasmid. The genes that cause gall formation in plants are located for the most part on the the Ti plasmid. Interestingly, if *Agrobacterium* is grown near its maximum temperature of about 30°C, then the plasmid is lost as well as the pathogenicity of the bacterium. The bacterium itself is still functional and can thrive in culture (Deacon). The genome of *Agrobacterium vitis* is currnetly being sequenced. It causes gall disease in grapes.

Agrobacterium tumefaciens is a Gram-negative, non-sporeforming, rod-shaped bacterium. *Agrobacterium* strains use different carbohydrates and are classified into three main biovars. The differences among biovars are mainly determined by the genes on the circular chromosome. *A. tumefaciens* is known for calling the formation of galls on plants that it infects. When a wound opens on the plant tissue, the motile cells of *A. tumefaciens* move into the tissue by chemotaxis as a response to the release of sugars and other components normally in the roots. While *A. tumefaciens* cells without Ti plasmids recognize and move towards plant wounds, the strains containing the Ti plasmids respond even more strongly because they recognize phenolic compounds such as acetosyringone that come out of the wound.

Agrobacterium tumefaciens can generally be found on and around root surfaces known as the rhizosphere. There it seems to use nutrients that leak from the root tissue. It will infect the tissue at wound sites formed from transplanting seedlings, burrowing animals or bugs, etc (Deacon). *Agrobacterium radiobacter* grows on various explosives such as nitroglycerine - they use this as their sole source of nitrogen. It removes two nitro groups from nitroglycerine by an NADH-dependent oxidoreductase, but can not use the carbon in nitroglycerine for growth because it cannot remove the third nitro group to release glycerol (White).

Agrobacterium tumefaciens is most widely known for causing crown gall disease that affects many dicotyledonous (broad-leaved) plants; another strain called biovar 3 causes crown gall disease in grapevines. The disease causes the formation of tumor-like swellings called galls that can generally be found on the crown of the plant just above the soil. Crown gall disease does not usually seriously harm older plants; however, it may reduce the value of a plant in a nursery.

Agrobacterium tumefaciens causes crown gall disease by first transferring part of its DNA into an opening in the plant. The DNA is attches itself to the plants DNA through histones. The DNA then integrates itself into the plant's genome and causes the formation of the gall. Because of how smoothly *Agrobacterium tumefaciens* is able to transfer DNA it has become a helpful tool for scientists to use.

Bananas have potential to become the world's first edible vaccine due to *Agrobacterium*. An edible vaccine doesn't need sterile syringes, costly refrigeration, or multiple injections. According to the World Health Organization (WHO), more than 2 million children die worldwide each year from diarrhea that can be prevented easily with vaccines. Thus, researchers lead by Dr. Charles Arntzen are looking into making the food vaccines to prevent diarrhea caused by *Escherichia coli* and *Vibrio cholara* bacteria. First, the gene for an immune system-stimulating disease protein would have to be spliced into the bacterium. The bacterium is then allowed to contaminate banana cells, inserting the Ti plasmid (containing the spliced gene) into their chromosomes. The next step is to grow the cells until they are mature fruit-bearing plants. The fruits, when eaten, would transfer the immunity to a particular disease into the consumer's body. This would be a relatively inexpensive and feasible method to distribute certain vaccines to people, especially in developing countries.

This has already been shown to work using potatoes and a modified *E. coli* protein that is known to cause severe diarrhea. When mice ate the raw engineered potatoes, they developed the antibodies to the *E. coli* toxin. Clinical tests on humans eating genetically engineered potatoes also showed that they started producing antibodies against Norwalk virus, which causes acute bouts of diarrhea. For now, all clinical trials involve genetically modified potatoes or tomatoes, both of which can easily be freeze-dried, transported, and reconstituted. These can be more easily dosage regulated and quality controlled, unlike a banana tree growing in the middle of a village, for example. But the ideal of fresh banana vaccines is still being developed and perfected.

Agrobacterium tumefaciens-mediated transformation The process of DNA transfer from Agrobacterium tumefaciens to plants, that occurs naturally during crown gall disease, and can be used as a method of transformation.

Agrobiodiversity; agricultural biological diversity That component of biodiversity that is relevant to food and agriculture production. The term agrobiodiversity encompasses within-species, species and ecosystem diversity.

Agronomic(s) The science of soil management and crop production.

Airlift fermenter A cylindrical fermentation vessel in which the cells are mixed by air introduced at the base of the vessel and that rises through the column of culture medium. The cell suspension circulates around the column as a consequence of the gradient of air bubbles in different parts of the reactor.

Albinism Hereditary absence of pigment in an organism. Albino animals have no colour in their skin, hair and eyes. The term is also used for absence of chlorophyll in plants.

Albino 1. An organism lacking pigmentation, due to genetic factors. The condition is *albinism*.

2. A conspicuous plastome (plastid) mutant involving loss of chlorophyll.

Albumin A human protein that makes up part of blood plasma. It helps to regulate the pressure and volume of plasma in the body. It is also responsible for transporting fatty acids in blood.

Alcohol In chemistry, an alcohol is any organic compound in which a hydroxyl group (*-OH*) is bound to a carbon atom, which in turn is bound to other hydrogen and/or carbon atoms. The general formula for a simple acyclic alcohol is $C_nH_{2n+1}OH$.

In general usage, alcohol refers almost always to ethanol, also known as grain alcohol, a colorless, volatile liquid, formed by the fermentation of sugars. It also often refers to any beverage that contains ethanol (see *alcoholic beverage*). This sense underlies the term alcoholism (addiction to alcohol). As a drug, ethanol is known to have a depressing effect that decreases the responses of the central nervous system. Other forms of alcohol are usually described with a clarifying adjective, as in *isopropyl alcohol* or by the suffix *-ol*, as in *isopropanol*.

The word dates to the 16th century when it was used to refer to any chemical substance arrived at by sublimation, which is also the source of *kohl*, attested already in Akkadian as referring to eye makeup.

The functional group of an alcohol is a hydroxyl group bonded to an sp^3 hybridized carbon. It can therefore be regarded as a derivative of water, with an alkyl group replacing one of the hydrogens. If an aryl group is present rather than an alkyl, the compound is generally called a phenol rather than an alcohol. Also, if the hydroxyl group is bonded to one of the sp^2 hybridized carbons of an alkenyl group, the compound is referred to as an enol. The oxygen in an alcohol has a bond angle of around 109° (c.f. 104.5° in water), and two nonbonded electron pairs. The O-H bond in methanol (CH_3OH) is around 96 picometres long.

There are three major subsets of alcohols-'primary' (1°), 'secondary' (2°) and 'tertiary' (3°), based upon the number of carbons the C-OH carbon (shown in red) is bonded to. Methanol is the simplest 'primary' alcohol. The simplest secondary alcohol is isopropanol (propan-2-ol), and a simple tertiary alcohol is *tert*-butanol (2-methylpropan-2-ol).

The simplest and most commonly used alcohols are methanol and ethanol (common names methyl alcohol and ethyl alcohol, respectively), which have the structures shown above.

Methanol was formerly obtained by the distillation of wood, and was called "wood alcohol". It is now a cheap commodity chemical produced by the high pressure reaction of carbon monoxide with hydrogen. In common usage, "alcohol" often refers simply to ethanol or "grain alcohol". Methylated spirits ("Meths"), also called "surgical spirits", is a form of ethanol rendered undrinkable by the addition of methanol. Aside from its major use in alcoholic beverages, ethanol is also used (though highly controlled) as an industrial solvent and raw material.

Alcohols are in wide use in industry and science as reagents, solvents, and fuels. Ethanol and methanol can be made to burn more cleanly than gasoline or diesel. Because of its low toxicity and ability to dissolve non-polar substances, ethanol is often used as a solvent in medical drugs, perfumes, and vegetable essences such as vanilla. In organic synthesis, alcohols frequently serve as versatile intermediates.

Ethanol is also commonly used in beverages after fermentation to promote flavor or induce a euphoric intoxication commonly known as "drunkenness" or "being drunk". The use of ethanol for this purpose is illegal in some countries, mainly Islamic. In such instances of consumption, alcohol is a drug, with potential for overdose, toxic poisoning, and physiological and/or psychological dependency (known as alcoholism) if taken in excessive quantities. Alcoholism has become one of the most common drug addictions (perhaps behind caffeine) in the world. The physiological dependency caused by alcoholism means that the user experiences physical withdrawal (in the form of a headache known as a "hangover," extremely high anxiety known as "the shakes," and restlessness or trouble sleeping) upon cessation or decrease of use.

Because of such particular uses, historically, ethanol has been regulated by taxation. Those who manufacture it for other purposes often avoid this expense by "denaturing" it in a manner that renders it unfit for drinking.

Many alcohols can be created by fermentation of fruits or grains with yeast, but only ethanol is commercially produced this way, chiefly for fuel and drink. Other alcohols are generally produced by synthetic routes from natural gas, petroleum, or coal feed stocks, for example via acid catalyzed hydration of alkenes.

In the IUPAC system, the name of the alkane chain loses the terminal "e" and adds "ol", e.g. "methanol" and "ethanol". When necessary, the position of the hydroxyl group is indicated by a number between the alkane name and the "ol": propan-1-ol for $CH_3CH_2CH_2OH$, propan-2-ol for $CH_3CH(OH)CH_3$. Sometimes, the position number is written before the IUPAC name: 1-propanol and 2-propanol. If a higher priority group is present (such as an aldehyde, ketone or carboxylic acid), then it is necessary to use the prefix "hydroxy", for example: 1-hydroxy-2-propanone (CH_3COCH_2OH).

The word "alcohol" almost certainly comes from the Arabic language (the "al-" prefix being the Arabic definite article); however, the precise origin is unclear. It was introduced into Europe, together with the art of distillation and the substance itself, around the 12th century by various European authors who translated and popularized the discoveries of Islamic alchemists.

The hydroxyl group generally makes the alcohol molecule polar. Those groups can form hydrogen bonds to one another and to other compounds. Two opposing solubility trends in alcohols are: the tendency of the polar OH to promote solubility in water, and of the carbon chain to resist it. Thus, methanol, ethanol, and propanol are miscible in water because the hydroxyl group wins out over the short carbon chain. Butanol, with a four-carbon chain, is moderately soluble because of a balance between the two trends. Alcohols of five or more carbons (Pentanol and higher) are effectively insoluble because of the hydrocarbon chain's dominance.

Because of hydrogen bonding, alcohols tend to have higher boiling points than comparable hydrocarbons and ethers. All simple alcohols are miscible in organic solvents. This hydrogen bonding means that alcohols can be used as protic solvents.

The lone pairs of electrons on the oxygen of the hydroxyl group also makes alcohols nucleophiles.

Alcohols, like water, can show either acidic or basic properties at the O-H group. With a pK_a of around 16-19 they are generally slightly weaker acids than water, but they are still able to react with strong bases such as sodium hydride or reactive metals such as sodium. The salts that result are called alkoxides, with the general formula RO^-M^+.

Alcohols conjugated to aromatic rings have a lower pKa (around 10). Electron-withdrawing groups also work to make alcohols more acidic. For example, Para-nitro phenol has a pKa of 7.15.

Meanwhile the oxygen atom has lone pairs of nonbonded electrons that render it weakly basic in the presence of strong acids such as sulfuric acid.

The word "alcohol" almost certainly comes from the Arabic language (the "al-" prefix being the Arabic definite article); however, the precise origin is unclear. It was introduced into Europe, together with the art of distillation and the substance itself, around the 12th century by various European authors who translated and popularized the discoveries of Islamic alchemists.

The hydroxyl group generally makes the alcohol molecule polar. Those groups can form hydrogen bonds to one another and to other compounds. Two opposing solubility trends in alcohols are: the tendency of the polar OH to promote solubility in water, and of the carbon chain to resist it. Thus, methanol, ethanol, and propanol are miscible in water because the hydroxyl group wins out over the short carbon chain. Butanol, with a four-carbon chain, is moderately soluble because of a balance between the two trends. Alcohols of five or more carbons (Pentanol and higher) are effectively insoluble because of the hydrocarbon chain's dominance.

Because of hydrogen bonding, alcohols tend to have higher boiling points than comparable hydrocarbons and ethers. All simple alcohols are miscible in organic solvents. This hydrogen bonding means that alcohols can be used as protic solvents.

The lone pairs of electrons on the oxygen of the hydroxyl group also makes alcohols nucleophiles.Alcohols, like water, can show either acidic or basic properties at the O-H group. With a pK_a of around 16-19 they are generally slightly weaker acids than water, but they are still able to react with strong bases such as sodium hydride or reactive metals such as sodium. The salts that result are called alkoxides, with the general formula $RO^- M^+$.

Alcohols conjugated to aromatic rings have a lower pKa (around 10). Electron-withdrawing groups also work to make alcohols more acidic. For example, Para-nitro phenol has a pKa of 7.15.

Meanwhile the oxygen atom has lone pairs of nonbonded electrons that render it weakly basic in the presence of strong acids such as sulfuric acid. For example, with methanol:

Alcohols can also undergo oxidation to give aldehydes, ketones or carboxylic acids, or they can be dehydrated to alkenes. They can react to form ester compounds, and they can (if activated first) undergo nucleophilic substitution reactions. Alcohols often have an odor described as 'biting' that 'hangs' in the nasal passages. Ethanol in the form of alcoholic beverages has been consumed by humans since pre-historic times, for a variety of hygienic, dietary, medicinal, religious, and recreational reasons. While infrequent consumption of ethanol in small quantities may be harmless or even beneficial, larger doses result in a state known as drunkenness or intoxication and, depending on the dose and regularity of use, can cause acute respiratory failure or death and with chronic use has medical repercussions.

Other alcohols are substantially more poisonous than ethanol, partly because they take much longer to be metabolized, and often their metabolism produces even more toxic substances. Methanol, or *wood alcohol*, for instance, is oxidized by alcohol dehydrogenase enzymes in the liver to the poisonous formaldehyde, which can cause blindness or death.

An effective treatment to prevent formaldehyde toxicity after methanol ingestion is to administer ethanol. This will bind to alcohol dehydrogenase, preventing methanol from binding and thus acting as a substrate. Any formaldehyde will be converted to formic acid and excreted before it causes damage.

The OH group is not a good leaving group in nucleophilic substitution reactions, so neutral alcohols do not react in such reactions. However if the oxygen is first protonated to give $R\text{-}OH_2^+$, the leaving group (water) is much more stable, and nucleophilic substitution can take place. For instance, tertiary alcohols react with hydrochloric acid to produce tertiary alkyl halides, where the hydroxyl group is replaced by a chlorine atom. If primary or secondary alcohols are to be reacted with hydrochloric acid, an activator such as zinc chloride is needed. Alternatively the conversion may be performed directly using thionyl chloride.

Alcohol dehydrogenase The enzyme, present in most tissues of the body, that catalyses the oxidation of alcohol to acetaldehyde. There are several genetically determined forms of the enzyme, whose pattern differs among individuals.

Aleurone The outermost layer of the endosperm in a seed.

Algal biomass Single-celled plants, such as *Chlorella* spp. and *Spirulina* spp., are grown commercially in ponds to make feed materials. Chlorella is grown commercially to make into fish food: it is fed to zooplankton, and these in turn are harvested as feed for fish farms. This is a means of converting sunlight into food in a way more convenient and controllable than normal farming.

Alginate Polysaccharide gelling agent.

Alkane A saturated, acyclic hydrocarbon structure; usually quite inert.

Alkylating agents Chemicals that transfer alkyl (methyl, ethyl, etc.) groups to the bases in DNA.

Allele One of two or more alternative forms of a gene. A person may have two copies of the same allele which would be called homozygous or two different forms which is heterozygous. Different alleles arise from changes in the base sequence of that gene through mutations. For example, the gene for eye colour has different alleles resulting in blue or brown eyes.

Allele frequency The number of copies of an allele in a population, expressed as a proportion of the total number of copies of all alleles at a locus in a population.

Allele-specific amplification (ASA) The use of polymerase chain reaction (PCR) at a sufficiently high stringency that only a primer with exactly the same sequence as the target DNA will be amplified. A powerful means of genotyping for single-locus disorders that have been characterized at the molecular level.

Allelic exclusion A phenomenon whereby only one functional allele of an antibody gene can be assembled in a given B lymphocyte. The "allele" on the other homologous chromosome in a diploid mammalian cell cannot undergo a functional re-arrangement, which would result in the production of two different antibodies by a single plasma cell.

Allelopathy The phenomenon by which the secretion of chemicals, such as phenolic and terpenoid compounds, by a plant inhibits the growth or reproduction of other plant species with which it is competing.

Allergen A substance, usually a protein, that can cause an allergy or allergic reaction in the body.

Allergenicity Refers to the potential of a substance to cause allergies.

Allergy A hypersensitivity (characterised by a damaging immune response) to an allergen or substance in a person's environment such as dust, pollen, fur or a chemical. Can manifest as hayfever, skin rash, asthma, food allergies etc. Referred to as an allergic reaction.

Allogamy Cross fertilization in plants.

Allometric When the growth rate of one part of an organism differs from that of another part or of the rest of the body.

Allopatric speciation Speciation occurring at least in part because of geographic isolation.

Allopolyploid A polyploid organism (usually a plant) having multiple sets of chromosomes derived from different species. Hybrids are usually sterile, because they do not have sets of homologous chromosomes and therefore pairing cannot take place. However, if doubling of the chromosome number occurs in a hybrid derived from two diploid (2n) species, the resulting tetraploid (4n) is a fertile plant, since it contains two sets of homologous chromosome and pairing may occur; this tetraploid is an allotetraploid.

Allosteric enzyme An enzyme that has two structurally distinct forms, one of which is active and the other inactive. Active forms of allosteric enzymes tend to catalyse the initial step in a pathway leading to the synthesis of molecules. The end product of this synthesis can act as a feedback inhibitor, converting the enzyme to the inactive form, thus controlling the amount of product synthesized.

Allosteric regulation A catalysis-regulating process in which the binding of a small effector molecule to one site on an enzyme affects the activity at another site.

Allosteric transition A reversible interaction of a small molecule with a protein molecule, resulting in a change in the shape of the protein and a consequent alteration of the interaction of that protein with a third molecule.

Allotetraploid An organism with four genomes derived from hybridization of different species. Usually, in forms that become established, two of the four genomes are from one species and two are from another species.

Allozygote A diploid individual that is homozygous at a locus in which the two genes are not identical by descent from a common ancestor.

Alphalactalbumin Protein component of milk.

Alternative mRNA splicing The inclusion or exclusion of different exons to form different mRNA transcripts.

Alu sequences A family of 300-bp sequences occurring nearly a million times in the human genome.

Alzheimer's disease A disease characterized by, among other things, progressive loss of memory. The development of Alzheimer's disease is thought to be associated, in part, with possessing certain alleles of the gene that encodes apolipoprotein E.

Ambient temperature Air temperature at a given time and place; not radiant temperature.

Amide A molecule containing an amine bonded to a carboxyl group; the resulting bond has substantial double-bond character. Also termed a peptide; amide bonds link amino acids in proteins.

Amine A molecule containing N with a single bond to C and two other single bonds to H or C (but not an amide); the amine group or moiety.

Amino acid An acid containing the group NH_2. In particular, any of 20 basic building blocks of proteins with a free amino (NH2) and a free carboxyl (COOH) group, and having the basic formula NH_2 - CR - COOH. According to the side group R, they are subdivided into: polar or hydrophilic (serine, threonine, tyrosine, asparagine and glutamine); non-polar or hydrophobic (glycine, alanine, valine, leucine, isoleucine, proline, phenylalanine, tryptophan and cysteine); acidic (aspartic acid and glutamic acid) and basics (lysine, arginine, hystidine). The sequence of amino acids determines the shape, properties and the biological role of a protein. Plants and many micro-organisms can synthesize amino acids from simple inorganic compounds, but animals are unable to synthesize some of them, called essential amino acids, so they must be present in the diet.

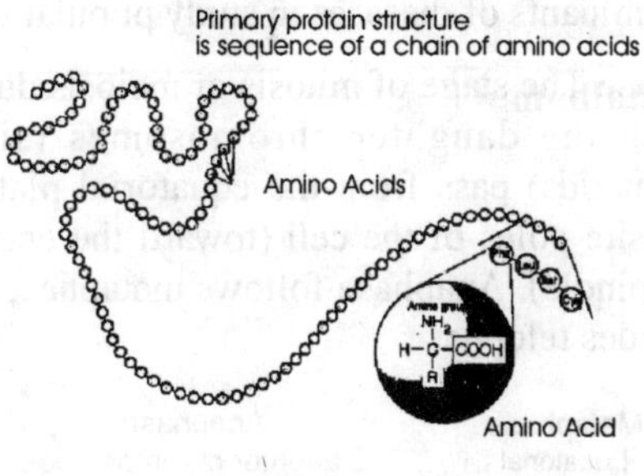

Fig. Amino acid

Aminoacyl site; A site One of two sites on ribosomes to which the incoming aminoacyl tRNA binds.

Aminoacyl tRNA synthetase Enzyme that attaches each amino acid to its specific tRNA molecule.

Amitosis Cell division (cytokinesis), including nuclear division through constriction of the nucleus, without chromosome differentiation as in mitosis. The maintenance of genetic integrity and diploidy during amitosis is uncertain. This process occurs in the endosperm of flowering plants.

Ammonia A colourless gas made up of one nitrogen and three hydrogen atoms. It is highly alkaline in water based solutions.

Amniocentesis A procedure in which a small amount of amniotic fluid (the fluid surrounding a baby in the womb) is drawn out using a needle inserted in to the abdomen. This fluid contains cells from the baby and can be used to test for chromosomal and genetic disorders, as well as certain birth defects before the baby is born.

Amniotic fluid Liquid contents of the amniotic sac of higher vertebrates, containing cells of the embryo (not of the mother). Both fluid and cells are used for diagnosis of genetic abnormalities in the embryo or foetus.

Amorph; null mutation A mutation that obliterates gene function.

Amphidiploid A species or type of plant derived from doubling the chromosomes in the F1 hybrid of two species; an allopolyploid. In an amphidiploid the two species are known, whereas in other allopolyploids they may not be known.

Amphimixis True sexual reproduction involving the fusion of male and female gametes and the formation of a zygote.

Ampicillin (beta-lactamase) An antibiotic derived from penicillin that prevents bacterial growth by interfering with cell wall synthesis.

Amplification 1. Treatment (e.g., use of chloramphenicol) designed to increase the proportion of plasmid DNA relative to that of bacterial (host) DNA.

2. Replication of a gene library in bulk.

3. Duplication of gene(s) within a chromosomal segment.

4. Creation of many copies of a segment of DNA by the polymerase chain reaction (PCR)

Amplified fragment length polymorphism (AFLP) A type of DNA marker, generated by digestion of genomic DNA with two restriction enzymes to create many DNA fragments, ligation of specific sequences of DNA (called adaptors) to the ends of these fragments, amplification of the fragments via PCR (using a set of primers with sequences corresponding to the adapters, plus various random combinations of three additional bases at the end), and visualization of fragments via gel electrophoresis. The PCR will amplify any fragment whose sequence happens to start with any of the three-base sequences in the set of primers. AFLPs have the important advantage that many markers can be generated with relatively little effort. They are a very useful means of quantifying the extent of genetic diversity within and between populations. Their major disadvantage is that they are not specific to a particular locus and, because they are scored as the presence or absence of a band, heterozygotes cannot be distinguished from homozygotes, i.e., they are inherited in a dominant fashion.

Amplify To increase the number of copies of a DNA sequence, either *in vivo* by inserting into a cloning vector that replicates within a host cell, or *in vitro* by polymerase chain reaction (PCR).

Amylase A group of enzymes that degrade starch, glycogen and other polysaccharides, producing a mixture of glucose and maltose. Plants have both a- and b-amylase; animal have only a-amylase.

Amylolytic The capability of breaking down starch into sugars.

Amylopectin A polysaccharide comprising highly branched chains of glucose molecules. The water-insoluble portion of starch.

Amylose A polysaccharide consisting of linear chains of 100 to 1000 glucose molecules. The water-soluble portion of starch.

Anabolic pathway A pathway by which a metabolite is synthesized; a biosynthetic pathway.

Anabolism Biosynthesis of molecules in cells and part of metabolism.

Anadromous Fish that return from oceans to fresh water to spawn (e.g. salmon).

Anaemia A condition that is due to a reduced number of red blood cells or reduced amounts of haemoglobin within them. This results in reduced oxygen carrying capacity and reduced aerobic activity in body cells.

Anaerobic An environment or condition in which molecular oxygen is not available for chemical, physical or biological processes.

Anaerobic digestion Digestion of materials in the absence of oxygen.

Anaerobic respiration Respiration in which foodstuffs are partially oxidized, with the release of chemical energy, in a process not involving atmospheric oxygen, such as alcoholic fermentation, in which one of the end products is ethanol.

Analogous Features of organisms or molecules that are superficially or functionally similar but have evolved in a different way or contain different compounds.

Analytical epidemiology Epidemiological investigations specifically aimed at studying the determinants of diseases in study populations.

Anaphase The stage of mitosis or meiosis during which the daughter chromosomes (sister chromatids) pass from the equatorial plate to opposite poles of the cell (toward the ends of the spindle). Anaphase follows metaphase and precedes telophase.

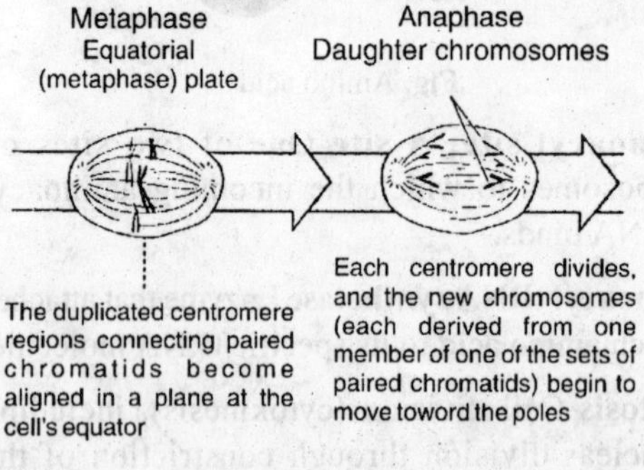

Fig. Anaphase

Anaphylaxis A generalised inflammatory immune reaction to a foreign protein in a sensitised individual which may be severe enough to be life-threatening.

Anchor Embedded marker enabling authors to link to a specific part of a Web document.

Anchor gene A gene that has been positioned on both the physical map and the linkage map of a chromosome.

Androgenesis Male parthenogenesis, i.e., the development of a haploid embryo from a male nucleus. The maternal nucleus is eliminated or inactivated subsequent to fertilization of the ovum, and the haploid individual (referred to as androgenetic) contains in its cells the genome of the male gamete only. Androgenesis is detected by cytological staining.

Aneuploid An organism or cell having a chromosome number that is not an exact multiple of the monoploid *(x)* with one chromosome being present in greater (e.g., trisomic $2n + 1$) or lesser (e.g., monosomic $2n - 1$) number than the normal diploid number.

Angelman's syndrome A genetic disorder caused by abnormal function of the gene UBE3A. It causes developmental problems, speech impairment and seizures.

Angioedema Swelling of the skin or mucous membrane and underlying tissue.

Angiosperm A division of the plant kingdom that includes all flowering plants, i.e. vascular plants in which double fertilization occurs resulting in development of fruit containing seeds. Divided into two major groups, monocotyledons and dicotyledons

Animal cell immobilization Entrapment of animal cells in some solid material in order to produce some natural product or genetically engineered protein. Animal cells have the advantage that they already produce many proteins of pharmacological interest, and that genetically engineered proteins are produced by them with the post-translation modifications normal to animals. However, because animal cells are much more fragile than bacterial ones, they cannot tolerate a commercial fermentation process. Typical materials are hollow fibre membrane bioreactors, or porous carriers made of polysaccharide, protein, plastic or ceramic materials with microscopic holes inside which the cells grow.

Animal genome (gene) bank A planned and managed repository containing animal genetic resources. Repositories include the environment in which the genetic resource has developed, or is now normally found *(in situ)* or facilities elsewhere *(ex situ - in vivo* or *in vitro).* For *in vitro, ex situ* genome bank facilities, germplasm is stored in the form of one or more of the following: semen, ova, embryos and tissue samples.

Animal model A laboratory animal with a specific disease that researchers can experiment with to find out more about that disease and how it occurs in humans. Animal models are used to learn more about the causes of a disease, its diagnosis in humans, and to investigate or trial new treatments or preventative actions. Animal models of disease may occur naturally in an animal population, or may be created using techniques such as genetic engineering, or by exposing animals to environments that induce that disease to develop.

Annealing The process of heating (de-naturing step) and slowly cooling (re-naturing step) double-stranded DNA to allow the formation of hybrid DNA or complementary strands of DNA or of DNA and RNA.

Annual 1 Taking one year, or occurring at intervals of one year.

2. In botany, a plant that completes its life cycle within one year. During this time the plant germinates, grows, flowers, produces seeds, and dies. A DNA marker detectable by virtue of variation in its sequence, irrespective of whether or not it actually occurs in or near a coding sequence. Microsatellites are typical anonymous DNA markers.

Anovulation Lack of ovulation caused by a medical condition, drug, or menopause.

Antagonism An interaction between two organisms (e.g., moulds or bacteria) in which the growth of one is inhibited by the other.

Antagonist A compound that inhibits the effect of an agonist in such a way that the combined biological effect of the two becomes smaller than the sum of their individual effects.

Anther Microsporangium bearing microspores which develop into pollen (microgametophytes). The upper part of a stamen, containing pollen sacs within which are numerous pollen grains.

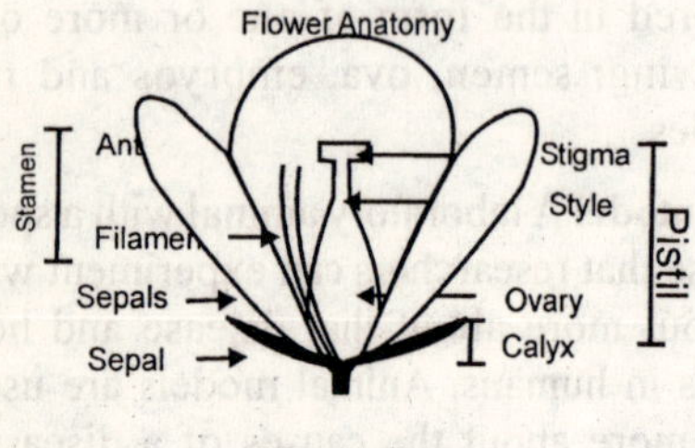

Fig. Anther

Anther culture The aseptic culture of anthers for the production of haploid plants from microspores.

Anthesis The flowering period or efflorescence. Anthesis is the time of full bloom, which lasts till fruit set.

Anthocyanin Water-soluble blue, purple and red flavonoid pigments found in vacuoles of cells.

Anthropocentrism Anthropocentrism is a view that regards humans as the central element of the universe. Proponents of this believe that we should only protect and replenish the environment so that it serves human purposes such as producing food and drugs; and that the fate of animals and plants are not morally significant except in terms of sustaining human well being.

Antiauxin A chemical that interferes with the auxin response. Antiauxin may or may not involve prevention of auxin transport or movement in plants. Some antiauxins are said to promote morphogenesis *in vitro,* such as 2,3,5-triiod̆obenzoate (TIBA; f.w. 499.81), or 2,4,5-trichlorophenoxyacetate (2,4,5-T; m.w. 255.49), which stimulate the growth of some cultures.

Antibiotic A class of natural and synthetic compounds that inhibit the growth of or kill some micro-organisms. Antibiotics such as penicillin are often used to control (to some extent kill) contaminating organisms. However, resistance to particular antibiotics can be acquired through mutations. Some contaminating organisms are only suppressed or their metabolism slowed to an insignificant level.

Anti idiotype antibodies Antibodies which recognize the binding sites of other antibodies. Their binding sites are complementary to the binding sites of another immunoglobulin. When an animal becomes immune to something, it not only acquires antibodies against that something, it also acquires antibodies against those antibodies. This forms a network of antibodies which can all bind to each other to various degrees, helping to regulate the immune response. Some allergic responses are in part due to the breakdown of this sort of regulation.

Anti oncogene A gene whose product prevents the normal growth of tissue.

Anti terminator A type of protein which enables RNA polymerase to ignore certain transcriptional stop or termination signals and read through them to produce longer mRNA transcripts.

Antibiotic resistance The tendency for a bacteria to develop a resistance to one or more antibiotics. This can occur when antibiotics are used repetitively or incorrectly. Most targeted bacteria are killed by a dose of antibiotics, but some are not. These remaining bacteria have a genetic resistance to the antibiotic, and this trait can be passed on to their offspring. For this reason there is a push to reduce the unnecessary use of antibiotics, to reduce the risk of them being rendered ineffective in the treatment of certain diseases.

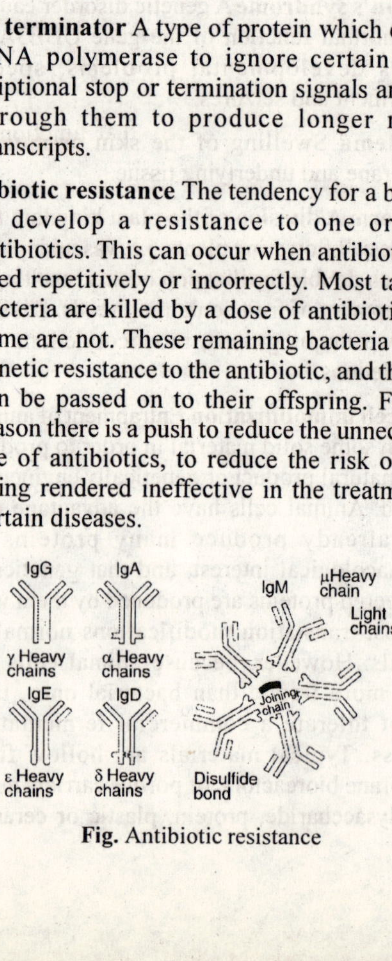

Fig. Antibiotic resistance

Antibody An antibody is a protein used by the immune system to identify and neutralize foreign objects like bacteria and viruses. Each antibody recognizes a specific antigen unique to its target. Production of antibodies is referred to as the humoral immune system.

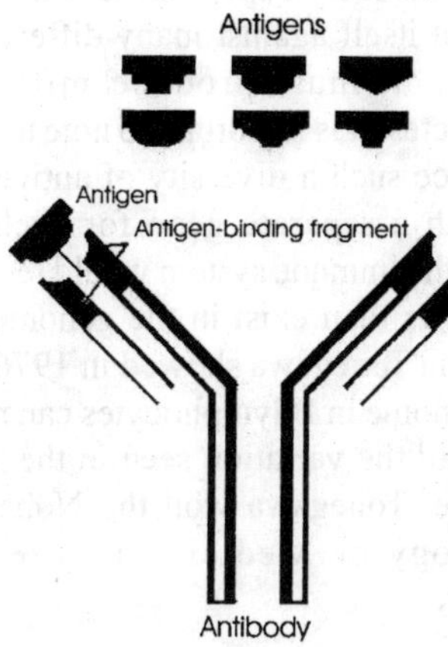

Fig. Antibody

Immunoglobulins are glycoproteins in the immunoglobulin superfamily that function as antibodies. The terms *antibody* and *immunoglobulin* are often used interchangeably. They are found in the blood and tissue fluids, as well as many secretions. In structure, they are globulins (in the γ-region of protein electrophoresis). They are synthesized and secreted by plasma cells that are derived from the B cells of the immune system. B cells are activated upon binding to their specific antigen and differentiate into plasma cells. In some cases, the interaction of the B cell with a T helper cell is also necessary.

Immunoglobulins are heavy plasma proteins, often with added sugar chains (see glycosylation) on N-terminal (all antibodies) and occasionally O-terminal (IgA1 and IgD) amino acid residues.

The basic unit of each antibody is a monomer. An antibody can be monomeric, dimeric, trimeric, tetrameric, pentameric, etc. The monomer is a "Y"-shape molecule that consists of two identical heavy chains and two identical light chains connected by disulfide bonds.

There are five types of heavy chain: γ, δ, α, μ and ε. They define classes of immunoglobulins. Heavy chains α and γ have approximately 450 amino acids, while μ and ε have approximately 550 amino acids. Each heavy chain has a constant region, which is the same by all immunoglobulins of the same class, and a variable region, which differs between immunoglobulins of different B cells, but is the same for all immunoglobulins produced by the same B cell. Heavy chains γ, α and δ have the constant region composed of three domains but have a hinge region; the constant region of heavy chains μ and ε is composed of four domains. The variable domain of any heavy chain is composed of one domain. These domains are about 110 amino acids long. There are also some amino acids between constant domains.

There are only two types of light chain: λ and κ. In humans, they are similar, but only one type is present in each antibody. Each light chain has two successive domains: one constant and one variable domain. The approximate length of a light chain is from 211 to 217 amino acids.

The "Y"-shaped monomer has two heavy and two light chains. Together this gives six to eight constant domains and four variable domains. Each half of the forked end of the "Y" is called an Fab fragment. It is composed of one constant and one variable domain of each the heavy and the light chain, which together shape the antigen binding site at the amino terminal end of the monomer. The two variable domains bind their specific antigens.

The enzyme papain cleaves a monomer into two *Fab* (*f*ragment *a*ntigen *b*inding) fragments and an *Fc* (*f*ragment *c*rystallizable) fragment. The enzyme pepsin cleaves below hinge region, so a f(ab)2 fragment and a fc fragment is formed.

Together, the antibodies in an organism can bind a wide variety of foreign antigens. Somatic recombination events generate this diversity.

This is when genes are selected (variable (V), diversity (D) and joining (J) for heavy chains, and only V and J for light chains) to form countless combinations. The main reason that the human immune system is capable of binding so many antigens is the variable region of the heavy chain. To be specific, it is the area where these V, D and J genes are found - otherwise known as the complementarity determining region 3 (CDR3).

The Fc fragment, the stem of the "Y," is composed of two heavy chains that each contribute two to three constant domains (depending on the class of the antibody). Fc binds to various cell receptors and complement proteins. In this way, it mediates different physiological effects of antibodies (opsonization, cell lysis, mast cell, basophil and eosinophil degranulation and other processes).

The variable regions of the heavy and light chains can be fused together to form a single chain variable fragment (scFv), which retains the original specificity of the parent immunoglobulin.

A crude estimation of immunoglobulin levels can be made by protein electrophoresis. Here the plasma proteins are separated into albumin, alpha-globulins (1 and 2), beta-globulins (1 and 2) and gamma-globulins according to weight. Immunoglobulins are all in the gamma region. In myeloma and some other disease states, a very high concentration of one particular immunoglobulin will show up as a *monoclonal* band.

According to differences in their heavy chain constant domains, immunoglobulins are grouped into five classes, or isotypes: *IgG*, *IgA*, *IgM*, *IgD*, and *IgE*. (The isotypes are also defined with light chains, but they do not define classes, so they are often neglected.) Other immune cells partner with antibodies to eliminate pathogens depending on which IgG, IgA, IgM, IgD, and IgE constant binding domain receptors it can express on its surface.

The antibodies that a single B lymphocyte produces can differ in their heavy chain, and the B cell often expresses different classes of antibodies at the same time. However, they are identical in their specificity for antigen, conferred by their variable region. To achieve the large number of specificities the body needs to protect itself against many different foreign antigens, it must produce millions of B lymphoyctes. It is important to note that, in order to produce such a diversity of antigen binding sites with a separate gene for each possible antigen, the immune system would require many more genes than exist in the genome. Instead, as Susumu Tonegawa showed in 1976, portions of the genome in B lymphocytes can recombine to form all the variation seen in the antibodies and more. Tonegawa won the Nobel Prize in Physiology or Medicine in 1987 for his discovery.

IgG is a monomeric immunoglobulin, built of two heavy chains γ and two light chains. Each molecule has two antigen binding sites. This is the most abundant immunoglobulin and is approximately equally distributed in blood and in tissue liquids. This is the only isotype that can pass through the placenta, thereby providing protection to the fetus in its first weeks of life before its own immune system has developed. It can bind to many kinds of pathogens, for example viruses, bacteria, and fungi, and protects the body against them by complement activation (classic pathway), opsonization for phagocytosis and neutralisation of their toxins. There are 4 subclasses: IgG1 (66%), IgG2 (23%), IgG3 (7%) and IgG4 (4%). -IgG1, IgG3 and IgG4 cross the placenta easily. -IgG3 is the most effective complement activator, followed by IgG1 and then IgG2. IgG4 does not activate complement. - IgG1 and IgG3 bind with high affinity to Fc receptors on phagocytic cells. IgG4 has intermediate affinity and IgG2 affinity is extremely low.

IgG (along with IgA below) gluten specific antibodies are one of the two main antibodies which are measured in patients with coeliac(gb)/ celiac(am) disease.

IgA represents about 15% to 20% of immunoglobulins in the blood, although it is primarily secreted across the mucosal tract into the stomach and intestines. It is also found in maternal milk, tears and saliva. This immunoglobulin helps to fight against pathogens that contact the body surface, are ingested, or are inhaled. It does not activate complement, and opsonises only weakly. Its heavy chains are of the type α. It exists in two forms, IgA1 (90%) and IgA2 (10%) that differ in the structure. IgA1 is composed like other proteins, however in IgA2 the heavy and light chains are not linked with disulfide but with noncovalent bonds. Though IgA2 is less in serum, it accounts for major secretory antibody.

The IgA found in secretions have a special form. They are dimeric molecules, linked by two additional chains. One of these is the J chain (from *j*oin), which is a polypeptide of molecular mass 1,5 kD, rich with cysteine and structurally completely different from other immunoglobulin chains. This chain is formed in the antibodies secreting cells. The dimeric form of IgA in the outer secretions has also a polypeptide of the same molecular mass (1,5 kD) that is called the secretory chain and is produced by the epithelial cells. It is also possible to find trimeric and even tetrameric IgA.

Decreased or absent IgA, termed *selective IgA deficiency*, can be a clinically significant immunodeficiency.

IgM forms polymers where multiple immunoglobulins are covalently linked together with disulfide bonds, normally as a pentamer or occasionally as a hexamer. It has a large molecular mass of approximately 900 kD (in its pentamer form). The J chain is attached to most pentamers, while hexamers do not possess the J chain due to space constraints in the complex. Because each monomer has two antigen binding sites, an IgM has 10 of them, however it cannot bind 10 antigens at the same time because they hinder each other. Because it is a large molecule, it cannot diffuse well, and is found in the interstitium only in very low quantities. IgM is primarily found in serum; however, because of the J chain, it is also important as a secretory immunoglobulin. Due to its polymeric nature, IgM possesses high avidity, and is particularly effective at complement activation. It is also a so-called "natural antibody": it is found in the serum without any evidence of prior contact with antigen.

In germline cells, the gene segment encoding the μ constant region of the heavy chain is positioned first among other constant region gene segments. For this reason, IgM is the first immunoglobulin expressed by mature B cells.

IgD makes up about 1% of proteins in the plasma membranes of mature naive B-lymphocytes (coexpressed with IgM) and is also found in serum in very small amounts. It is monomeric and incorporates the δ heavy chain in its structure. IgD's function is currently unknown, as mice lacking IgD seem to retain normal immune responses (implying redundancy if not lack of function), and IgD ceases to be expressed in activated B-lymphocytes. It may function as a regulatory antigen receptor.

IgE is a monomeric immunoglobulin with the heavy chain ε. It contains a high proportion of carbohydrates. Its molecular mass is 190 kD. It can be found on the surface of the plasma membrane of basophils and mast cells of connective tissue. IgE plays a role in immediate hypersensitivity and the defense against parasites such as worms. The IgE antibodies are present also in outer excretions. They do not activate complement. Only IgE is heat-labile.

When a macrophage ingests a pathogen, it attaches parts of the pathogen's proteins to a class II MHC protein. This complex is moved to the outside of the cell membrane, where it can be recognized by a T lymphocyte, which compares it to similar structures on the cell membrane of a B lymphocyte. If it finds a matching pair, the T lymphocyte activates the B lymphocyte, which starts producing antibodies. A B lymphocyte can produce antibodies only against the structure it presents on its surface.

Antibodies exist freely in the bloodstream or bound to cell membranes. They are part of the humoral immune system. Antibodies exist in clonal lines that are specific to only one antigen, e.g., a virus hull protein. In binding to such antigens, they can cause agglutination and precipitation of antibody-antigen products primed for phagocytosis by macrophages and other cells, block viral receptors, and stimulate other immune responses, such as the complement pathway.

Antibodies that recognize viruses can block these directly by their sheer size. The virus will be unable to dock to a cell and infect it, hindered by the antibody. They can also agglutinate them so the phagocytes can capture them. Antibodies that recognize bacteria mark them for ingestion by phagocytes, a process called opsonization. Together with the plasma component complement, antibodies can kill bacteria directly. They neutralize toxins by binding with them.

It is important to note that antibodies cannot attack pathogens within cells, and certain viruses "hide" inside cells (as part of the lysogenic cycle) for long periods of time to avoid them. This is the reason for the chronic nature of many minor skin diseases (such as cold sores); any given outbreak is quickly suppressed by the immune system, but the infection is never truly eradicated because some cells retain viruses that will resume the apparent symptoms later.

Detection of particular antibodies is a very common form of medical diagnostics. Serology depends on these methods. Autoimmune disorders can often be traced to antibodies that bind the body's own epitopes; many can be detected through blood tests. Antibodies directed against RBC surface antigens in immune mediated hemolytic anemia can be detected with the Coombs test. The Coombs test is also used for antibody screening in blood transfusion prepration and also for antibody screening in antenatal women.

"Designed" monoclonal antibody therapy is already being employed in a number of diseases (including rheumatoid arthritis) and in some forms of cancer. Presently, many antibody-related therapies are undergoing extensive clinical trials for use in practice.

In biochemistry, antibodies are used for immunological identification of proteins, using the Western blot method. A similar technique is used in ELISPOT and ELISA assays, in which detection antibodies are used to detect cell secretions such as cytokines or antibodies. Antibodies are also used to separate proteins (and anything bound to them) from the other molecules in a cell lysate.

These purified antibodies are often produced by injecting the antigen into a small mammal, such as a mouse or rabbit. Sometimes, in order to obtain large quantity of antibodies, goats, sheep, or horses are used. Blood isolated from these animals contains *polyclonal antibodies* — multiple antibodies that stick to the same antigen. The serum (blood from which blood-clotting proteins and red-blood cells were removed), also known as the antiserum, because it now contains the desired antibodies, is commonly purified with Protein A/G purification or antigen affinity chromatography. If the lymphocytes that produce the antibodies can be isolated and immortalized, then a *monoclonal antibody* can be obtained.

Antibody binding site The part of an antibody that binds to the antigenic determinant.

Antibody class The class to which an antibody belongs, depending on the type of heavy chain present. In mammals, there are five classes of antibodies: IgA, IgD, IgE, IgG, and IgM.

Antibody-mediated (humoral) immune response The synthesis of antibodies by B cells in response to an encounter of the cells of the immune system with a foreign immunogen.

Antibody structure Antibodies have a well-defined structure. Each antibody has two identical "light" chains and two identical "heavy" chains. Each chain comprises a constant region, i.e., a region that is the same between antibodies of the same class and sub-class, and a variable region that differs between. The antigen-binding region or binding site - *complementarity determining region* - is in the variable region. The antibody can be cut by proteases into several fragments, known as Fab, Fab', and Fc.

Anticlinal The plane of cell wall orientation or cell division perpendicular (at right angles) to the surface of an organ.

Anticoagulant Substance that prevents blood from clotting.

Anticoding strand Is also called the antisense or template strand, and is the single strand of the DNA double helix that becomes transcribed.

Anticodon A sequence of three bases in a molecule of transfer RNA (tRNA) that binds to a complementary codon in messenger RNA (mRNA). Each anticodon designates a specific amino acid to be added to a growing polypeptide.

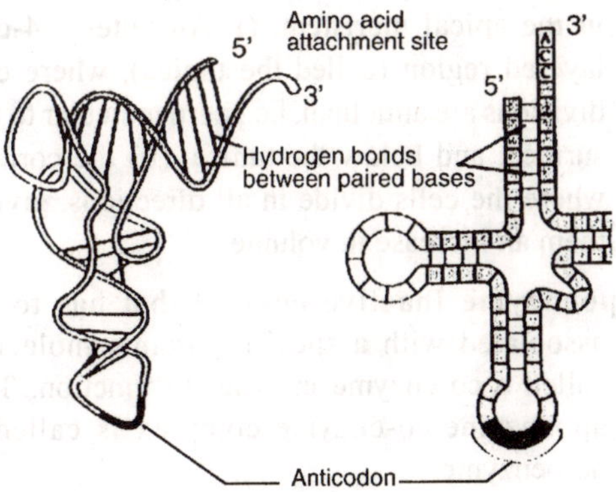

Fig. Anticodon

Antigen; immunogen A compound that elicits an immune response by stimulating the production of antibodies. The antigen, usually a protein, when introduced into a vertebrate organism is bound by the antibody or a T cell receptor.

Antigenic determinant The individual surface feature of an antigen, that elicits the production of a specific antibody in the course of the immune response. Each antigenic determinant, typically a few amino acids in size, causes the synthesis of a different antibody and thus exposure to a single antigen may result in the expression of a number of antibodies.

Antigenic switching The altering of a micro-organism's surface antigens through genetic re-arrangement, to elude detection by the host's immune system.

Antihaemophilic globulin Blood globulin that reduces the clotting time of haemophilic blood.

Antihemophilic factors A family of whole-blood proteins that initiate blood clotting. Some of these proteins, such as factor VIII, can be used to treat hemophilia.

Anti-infective Describes agents that fight infections.

Antimicrobial agent Any chemical or biological agent that harms the growth of micro-organisms.

Antinutrient An undesirable substance in food that can inhibit nutrient metabolism or absorption.

Antioxidant A molecule that protects cells from oxidative damage of oxygen and free radical molecules that are chemically unstable and cause random reactions damaging proteins, nucleic acids, and cell membranes. Examples of dietary antioxidants are vitamins C, E, and K, and diverse plant products such as lycopene, a nutraceutical found in tomatoes.

Antioxidant solution Pre-treatment solution (e.g., Vitamin C; citric acid) that retards senescence and browning of tissue. It is employed to incubate explants prior to surface sterilization.

Antiparallel orientation The normal arrangement of the two strands of a DNA molecule, and of other nucleic-acid duplexes (DNA-RNA, RNA-RNA), in which the two strands are oriented in opposite directions so that the 5′-phosphate end of one strand is aligned with the 3′-hydroxyl end of the complementary strand.

Antisense A piece of DNA producing a mirror image ("antisense") messenger RNA that is opposite in sequence to one directing protein synthesis. Antisense technology is used to selectively turn off production of certain proteins.

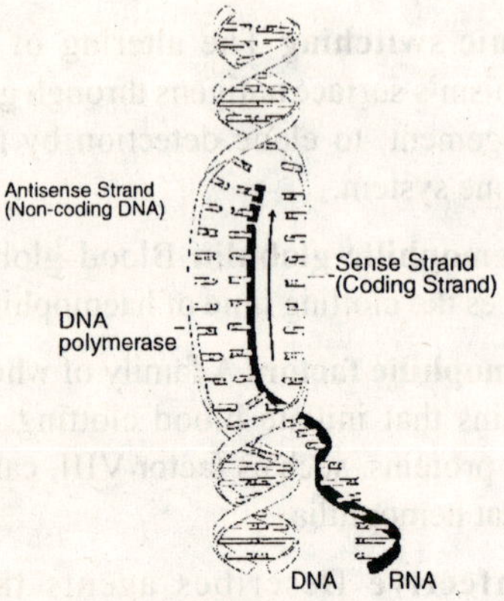

Fig. Antisense

Antisense gene A gene that produces a transcript (mRNA) that is complementary to the pre-mRNA or mRNA of a normal gene (usually constructed by inverting the coding region relative to the promoter).

Antisense RNA A complementary RNA sequence that binds to a naturally occurring (sense) mRNA molecule, thus blocking its translation.

Antisense therapy The *in vivo* treatment of a genetic disease by blocking translation of a protein with a DNA or an RNA sequence that is complementary to a specific mRNA.

Antiseptic Any substance that kills or inhibits the growth of disease-causing micro-organism (a micro-organism capable of causing sepsis), but is essentially non-toxic to cells of the body.

Antiserum Blood serum containing specific antibodies against an antigen. Antisera are used to confer passive immunity to many diseases.

Antitranspirant A compound designed to reduce transpiration when sprayed or painted on leaves of newly transplanted trees, shrubs or vines, or used as a dip for cuttings in lieu of misting; may interfere with photosynthesis and respiration if the coating is too thick or unbroken.

Aperiodic Refers to the lack of symmetry in molecular structures or functions. An important insight into the mechanism of biological structures is their aperiodic composition and distribution of atoms causing the extraordinary complexity of cells.

Apex The tip, point or angular summit. The tip of a leaf; the portion of a root or shoot containing apical and primary meristems. Usually used to designate the apical tip of the meristem.

Apical cell A meristematic initial in the apical meristem of shoots or roots of plants. As this cell divides, new tissues are formed.

Apical dominance The phenomenon of inhibition of growth of lateral (axillary) buds in a plant by the presence of the terminal (apical) bud on the branch, due to auxins produced by the apical bud.

Apical meristem A region of the tip of each shoot and root of a plant in which cell division is continually occurring to produce new stem and root tissue, respectively. Two regions are visible in the apical meristem: (i) An outer 1-4-cell layered region (called the tunica), where cell divisions are anticlinal, i.e., perpendicular to the surface; and below the tunica, (ii) the corpus, where the cells divide in all directions, giving them an increase in volume.

Apoenzyme Inactive enzyme that has to be associated with a specific organic molecule called a co-enzyme in order to function. The apoenzyme/co-enzyme complex is called a holoenzyme.

Apolipoprotein E (Apo E) Certain alleles of the gene that encodes the protein apolipoprotein E have been associated with the development of heart disease and Alzheimer's disease.

Apomixis The asexual production of diploid offspring without the fusion of gametes. The embryo develops by mitotic division of the maternal or paternal gamete, or, in the case of plants, by mitotic division of a diploid cell of the ovule.

Apoptosis The process of cell death, which occurs naturally as a part of normal development, maintenance and renewal of tissue in an organism. Apoptosis differs from necrosis, in which cell death is caused by a toxic substance.

Aquaculture Growing of water plants and animals, rather than harvesting them from wherever they happen to grow in rivers or seas. Usually aquaculture uses fresh water; when it uses sea water it can be called mariculture. It is considered to be a part of biotechnology (although peripheral) because it is a new commercial development, and because it often involves growing organisms in large volumes of water, which has similarities to growing large volumes of yeast or bacteria. Biotechnology also provides clean, well-aerated water for the animals to grow in; food, such as krill or powdered synthetic food; and food additives, such as astaxanthins to ensure that fish and prawns have the right colour. Aquaculture has also been used to mass-produce macro- and micro-algae for chemicals, vitamins and pigments. For both animals and plants, biotechnologists have been using genetic methods to produce triploid and tetraploid organisms, and hybrid algae through plant cell fusions. Triploid trout, for example, are sterile, and can be used for biocontrol of weeds without the threat of their being able to breed themselves.

Arabidopsis A genus of flowering plants in the Cruciferae. *A. thaliana* is used in research as a model plant because it has a small genome (5 pairs of chromosomes; 2n = 10) and can be cultured easily, with a generation time of two months.

Archaea A prokaryotic form of life that forms a domain in the tree of life. There are three domains: bacteria, archaea, and eukarya. Bacteria are also prokaryotic organisms. Eukaryotes include animals, plants, fungi, and protozoan and have very different cell structures, bigger and with internal membrane bound structures (organelles). While bacteria and archaea look similar in structure, they have very different metabolic and genetic activity. One defining physiological characteristic of archaea is their ability to live in extreme environments. They are often called extremophiles and unlike bacteria and eukarya depend on either high salt, high or low temperature, high pressure, or high or low pH.

Aromatic A term used to describe cyclic pi-bonded structures of special stability.

ARS (autonomous replicating sequence) Any eukaryotic DNA sequence that initiates and supports chromosomal replication; they have been isolated in yeast cells. Also called autonomous(ly) replicating segment.

Arthropods The largest animal kingdom, branch. It includes insects, arachnids and crustaceans. This group can be identified by their segmented bodies, and paired, jointed antennae, wings or legs.

Artificial inembryonation Non-surgical transfer of embryo(s) to a recipient female. As *in vitro* embryo technology develops, artificial inembryonation will gradually replace artificial insemination.

Artificial insemination The placement of sperm inside the female reproductive tract to improve the chances of fertilisation and pregnancy occurring. Artificial insemination is also called *intrauterine insemination.*

Artificial seed Encapsulated or coated somatic embryos (embryoids) that are planted and treated like seed.

Artificial selection The practice of choosing individuals from a population for reproduction, usually because these individuals possess one or more desirable traits.

Ascorbic acid; vitamin C ($C_6H_8O_6$) A water-soluble vitamin present naturally in some plants, and also synthetically produced. Aside from its role as a vitamin, it is used as an antioxidant in plant tissue culture; and included in disinfection solutions.

Ascospore One of the spores contained in the ascus of certain fungi.

Ascus (pl: asci) Reproductive sac in the sexual stage of a type of fungi (*Ascomycetes*) in which ascospores are produced.

Aseptic Asepsis or sterile. The state of being free of contaminating organisms (bacteria, fungi, algae and all micro-organisms except viruses) but not necessarily free of internal symbionts.

Asexual Any type of reproduction not involving meiosis or the union of gametes.

Asexual embryogenesis The sequence of events whereby embryos develop from somatic cells.

Asexual propagation Vegetative, somatic, non-sexual reproduction of a plant without fertilization.

Asexual reproduction Reproduction that does not involve the formation and union of gametes from the different sexes or mating types. It occurs mainly in lower animals, micro-organisms and plants. In plants, asexual reproduction is by vegetative propagation (e.g., bulbs, tubers, corms) and by formation of spores.

Asparagine ($C_4H_8N_2O_3$) One of the 20 essential amino acids. It is occasionally included in plant tissue culture media as a source of reduced nitrogen.

Assay 1. To test or evaluate.

2. The procedure for measuring the quantity of a given substance in a sample (chemically or by other means).

3. The substance to be analysed.

Assembler In recent popular usage, any nanomachine, usually assumed to offer magical, universal capabilities in an atom-sized package. In the author's usage, any programmable nanomechanical system able to perform a wide range of mechanosynthetic operations.

Assisted hatching Sometimes used during in vitro fertilization, this procedure removes the outer coating of the mother's egg so it can implant properly in the uterus.

Assisted reproductive technologies Assisted reproductive technologies (ART) refer to advanced fertility techniques such as *in vitro* fertilisation (IVF) which are used to bring eggs and sperm together to help achieve pregnancy.

Assortative mating Mating in which the partners are chosen on the basis of phenotypic similarity.

Asthma Breathing difficulty caused by narrowing of the air passages that is reversible.

Asynapsis The failure or partial failure in the pairing of homologous chromosomes during the meiotic prophase.

Atherosclerosis A degenerative disease of arteries in which there is thickening caused by an accumulation of material (plaque) beneath the inner lining, eventually restricting blood flow. The material characteristically contains cholesterol and macrophage cells.

Atom (atomistic) The smallest unit of matter as recognized by chemical properties of molecules. Atoms are composed of protons, neutrons and electrons. The latter provide all properties described by molecular interactions and chemical reactions that are essential processes in biology.

Atomic force microscope A device in which the deflection of a sharp stylus mounted on a soft spring is monitored as the stylus is moved across a surface. If the deflection is kept constant by moving the surface up and down by measured increments, the result (under favorable conditions) is an atomic-resolution topographic map of the surface. Also termed a scanning force microscope.

ATP (adenosine triphosphate) A nucleotide of fundamental importance as a carrier of chemical energy in all living organisms. It consists of adenosine with three phosphate groups, linked together linearly. The phosphates are attached to adenosine through its ribose (sugar) portion. Upon hydrolysis, bonds yield either one molecule of ADP (adenosine diphosphate) and an inorganic phosphate, or one molecule of AMP (adenosine monophosphate) and pyrophosphate; in both cases releasing energy that is used to power biological processes. ATP is regenerated by rephosphorilation of AMP and ADP, using chemical energy derived from the oxidation of food.

Attenuated vectors Micro-organisms or pieces of DNA such as plasmids which have certain infective components removed by recombinant DNA technology to destroy their virulence, thereby making them safer to use.

Attenuated Weakened; with reference to vaccines, made from pathogenic organisms that have been treated so as to render them avirulent.

Attenuator A nucleotide sequence in the 5′ region of a prokaryotic gene (or in its RNA) that causes premature termination of transcription, possibly by forming a secondary structure.

Australia consumer's association (ACA) Are an independent organisation who represent and act in Australian consumers' interests. They lobby and campaign on behalf of consumers to promote their rights, to influence government policy, and to ensure consumer issues have a high profile in the public arena.

Authentic protein A recombinant protein that has all the properties - including any post-translational modifications - of its naturally occurring counterpart.

Auto immunity A disorder in the body's defence mechanism in which an immune response is elicited against its own (self) tissues.

Auto-immune disease Disorder in which the immune systems of affected individuals produce antibodies against molecules that are normally produced by those individuals (called self antigens).

Autocatalysis Catalysis in which one of the products of the reaction is a catalyst for the reaction. Usually the catalysis starts slowly and increases as the quantity of the catalyst increases, falling off as the product is used up.

Autoclave 1. An enclosed chamber in which substances are heated under pressure to sterilize utensils, liquids, glassware, etc., using steam. The routine method uses steam pressure of $103.4X10^3$ Pa at 121°C for 15 minutes, or longer to allow large volumes to reach the critical temperature.

2. A pressure cooker used to sterilize growth medium and instruments for tissue culture work.

Autologous cells Cells that are taken from an individual, cultured (or stored), and, possibly, genetically manipulated before being infused back into the original donor.

Autolysis The process of self destruction of a cell, cell organelle, or tissue. It occurs by the action of lysosomic enzymes.

Automated engineering Engineering design done by a computer system, generating detailed designs from broad specifications with little or no human help.

Automated manufacturing As used here, nanotechnology-based manufacturing requiring little human labour.

Autonomous A term applied to any biological unit that can function on its own, i.e., without the help of another unit, such as a transposable element that encodes an enzyme for its own transposition.

Autopolyploid A polyploid that has multiple and identical or nearly identical sets of chromosomes (genomes) all derived from the same species. A polyploid species with genomes derived from the same original species.

Autoradiography A technique that captures the image formed in a photographic emulsion as a result of the emission of either light or radioactivity from a labelled component that is placed next to unexposed film. The technique is used for detecting the location of an isotope in a tissue, cell or molecule. The sample is placed in contact with a photographic emulsion, usually an X-ray film. The emission of b-particles from the sample activates the silver halide grains in the emulsion and allows them to reduce to metallic silver when the film is developed. In genetic engineering, autoradiography is most commonly used to detect the hybridization of a radioactive DNA (probe) molecule to denatured DNA in either the Southern transfer or colony hybridization procedures.

Autosomal dominant 'Autosomal' refers to a non-sex chromosome. Autosomal dominance is when one particular form of a gene, one allele, dominates over other alleles and is always expressed when present in an individual whether they are homozygous for that allele or heterozygous.

Autosome In the human being, any one of the 22 matched pairs of chromosomes. One in each pair is inherited from the mother, the other from the father.

Autotrophic Self-nourishing organisms capable of utilizing carbon dioxide or carbonates as the sole source of carbon and obtaining energy for life processes from radiant energy or from the oxidation of inorganic elements, or compounds such as iron, sulphur, hydrogen, ammonium and nitrites.

Autotrophy Autotrophy is the capacity of an organism to use light as the sole energy source in the synthesis of organic material from inorganic elements or compounds. Autotrophic organisms include green photosynthesizing plants and some photosynthetic bacteria.

Auxin A group of plant growth regulators (natural or synthetic) which stimulate cell division, enlargement, apical dominance, root initiation, and flowering. One naturally produced auxin is indole-acetic acid (IAA).

Auxin cytokinin ratio The relative proportion of auxin to cytokinin present in plant-tissue-culture media. Varying the relative amounts of these two hormone groups in tissue culture formulae affects the proportional growth of shoots and roots *in vitro*. As the ratio is increased (increased auxin or decreased cytokinin), roots are more likely to be produced, and as it is decreased root growth declines and shoot initiation and growth are promoted. This relationship was first recognized by C.O. Miller and F. Skoog in the 1950s.

Auxotroph A mutant cell or micro-organism lacking one metabolic pathway present in the parental strain, and that consequently will not multiply on a minimal medium, but requires for growth the addition of a specific compound, such as an amino acid or a vitamin.

Availability A reflection of the form and location of nutritional elements and their suitability for absorption. In tissue culture media this is related to the abundance of each nutritional element, the osmotic concentration and pH of the medium, the stability and solubility of the item in question, the presence of absorbing agents in the media, and other factors.

Avian influenza Referred to as the "bird flu", this is a highly contagious influenza virus that can infect any bird.

Avirulent Unable to cause disease.

Axenic culture Free of external contaminants and internal symbionts; generally not possible with surface sterilization alone, and incorrectly used to indicate *aseptic culture*.

Axil The angle formed between a leaf's petiole and a stem where an axillary bud develops. The angle formed on an inflorescence, pedicel and rachis (of an Orchid).

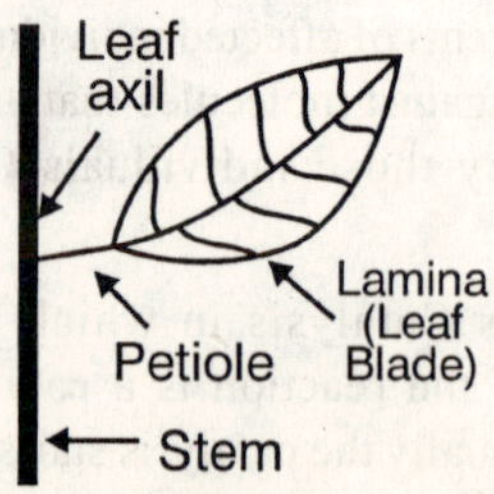

Fig. Axil

Axillary bud A bud found at the axil of a leaf (synonymous with lateral bud).

Axillary bud proliferation Propagation in culture using protocols and media which promote axillary (lateral shoot) growth. This is a technique for mass production (micro-propagation) of plantlets in culture, achieved primarily through hormonal inhibition of apical dominance and stimulation of lateral branching.

Axon The cell extension of a neuron (nerve cell) that carries an electrical signal to synapses which are secreting chemical signaling molecules called neurotransmitters to stimulate/inhibit receiving cells. Some axons in the peripheral nervous system connecting to muscle cells or connecting sensory neurons from the skin, eye, or internal organs to the central nervous system can be quite long (measure in centimeters; 1 inch = 2.54 cm) compared to the size of an average cell (measured in micrometers).

B

B carotene The most abundant of the carotenoids. b-Carotene has strong provitamin A activity. Unlike vitamin A itself, b-carotene is a strong antioxidant.

B cryptoxanthin A carotenoid pigment found in a biological system.

B lymphocytes; B cells An important class of lymphocytes that mature in bone marrow (in mammals) and the Bursa of Fabricius (in birds), that are largely responsible for the antibody-mediated or humoral immune response; they give rise to the antibody-producing plasma cells and some other cells of the immune system.

BAC (bacterial artificial chromosome) A cloning vector constructed from bacterial fertility (F) factors; like YAC vectors, they accept large inserts of size 200 to 500 kb.

Bacillus A rod-shaped bacterium.

Bacillus subtilis A bacterium commonly used as a host in recombinant DNA experiments. Important because of its ability to secrete proteins.

Bacillus thuringiensis A naturally occurring bacterium present in soil and used successfully by home gardeners and organic farmers to control certain insects. When ingested by a target insect, the protein produced by Bt. affects the insect by disturbing the digestive system, eventually killing it. Because different bt toxins affect different insects, it can be used very selectively meaning the Bt. protein is harmless to non-target insects and people or animals.

Back mutation A second mutation at the same site in a gene as the original mutation. The second mutation restores the wild-type nucleotide sequence.

Backcross Crossing an individual with one of its parents or with the genetically equivalent organism. The offspring of such a cross are referred to as the backcross generation or backcross progeny.

Bacteria Bacteria consist of only a single cell, but don't let their small size and seeming simplicity fool you. They're an amazingly complex and fascinating group of creatures. Bacteria have been found that can live in temperatures above the boiling point and in cold that would freeze your blood. They "eat" everything from sugar and starch to sunlight, sulfur and iron. There's even a species of bacteria—*Deinococcus radiodurans*—that can withstand blasts of radiation 1,000 times greater than would kill a human being.

Bacteria fall into a category of life called the Prokaryotes *(pro-carry-oats)*. Prokaryotes' genetic material, or DNA, is not enclosed in a cellular compartment called the nucleus.

Bacteria and archaea are the only prokaryotes. All other life forms are Eukaryotes *(you-carry-oats)*, creatures whose cells have nuclei.

Bacteria are among the earliest forms of life that appeared on Earth billions of years ago. Scientists think that they helped shape and change the young planet's environment, eventually creating atmospheric oxygen that enabled other, more complex life forms to develop. Many believe that more complex cells developed as once free-living bacteria took up residence in other cells, eventually becoming the organelles in modern complex cells. The mitochondria *(mite-oh-con-dree-uh)* that make energy for your body cells is one example of such an organelle. There are thousands of species of bacteria, but all of them are basically one of three different shapes. Some are rod- or stick-shaped and called bacilli *(buh-sill-eye)*. Others are shaped like little balls and called cocci *(cox-eye)*. Others still are helical or spiral in shape, like the *Borrelia* pictured at the top of this page.

Some bacterial cells exist as individuals while others cluster together to form pairs, chains, squares or other groupings.

Bacteria live on or in just about every material and environment on Earth from soil to water to air, and from your house to arctic ice to volcanic vents. Each square centimeter of your skin averages about 100,000 bacteria. A single teaspoon of topsoil contains more than a billion (1,000,000,000) bacteria.

Some bacteria move about their environment by means of long, whip-like structures called flagella. They rotate their flagella like tiny outboard motors to propel themselves through liquid environments. They may also reverse the direction in which their flagella rotate so that they tumble about in one place.

Other bacteria secrete a slime layer and ooze over surfaces like slugs. Others are fairly stationary.

Some bacteria are photosynthetic—they can make their own food from sunlight, just like plants. Also like plants, they give off oxygen. Other bacteria absorb food from the material they live on or in. Some of these bacteria can live off unusual "foods" such as iron or sulfur. The microbes that live in your gut absorb nutrients from the digested food you've eaten.

Bacterial toxin A toxin produced by a bacterium, such as Bt toxin by *Bacillus thuringiensis*.

Bacteriocide; bactericide A chemical or drug that kills bacterial cells.

Bacteriocin A protein produced by bacteria of one strain and active against those of a closely related strain.

Bacteriophage A bacteriophage is a virus that infects bacteria. Like viruses that infect eukaryotes (plants, animals and fungi), phages display a variety of different compositions. Typically, they consist of an outer protein hull and the enclosed genetic material, which consists of double-stranded DNA of 5 to 650 kbp (kilo base pairs) in the vast majority of the phages known. The dimensions of a phage are on the order of 24 to 200 nm. Several phages sport a structure called tail, which is used to inject the genetic material into the host.

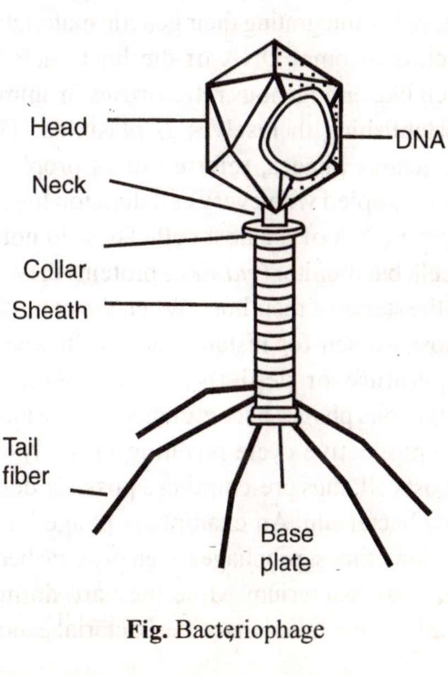

Fig. Bacteriophage

Phages are ubiquitous and can be found in many reservoirs populated by bacteria, such as soil or the intestine of animals. One of the densest natural sources for phages is seawater, where up to $2.5X10^8$ virions per cm^3 have been found.

Phages were discovered independently in 1915 by Frederick Twort and in 1917 by Fılix d'Herelle, who coined the name bacteriophage. Twort did not pursue the matter further, but d'Herelle went to Tbilisi in the Soviet Union, to continue the research on phages and phage therapy, where George Eliavi and he founded the still existing George Eliava Institute. Phages play an important role in molecular biology as cloning vectors to insert DNA into bacteria. Phage display is a test to screen for protein interactions by integrating multiple genes from a gene bank into phages.

Phages infect only specific bacteria. It is unknown whether there is a phage for every species of bacteria as only a fraction of bacteria have been studied on this detail. Some phages are virulent, meaning that upon infecting a cell they immediately begin reproducing, and within a short time lyse (destroy) the cell, releasing new phages. Some phages (so-called *temperate phages*) can instead enter a relatively harmless state, either integrating their genetic material into the chromosomal DNA of the host bacterium (much like endogenous retroviruses in animals) or establishing themselves as plasmids. These endogenous phages, referred to as prophages, are then copied with every cell division together with the DNA of the host cell. They do not kill the cell, but monitor (*via* some proteins they code for) the status of their host. When the conditions of host worsen for instance due to increase in temperature or depletion of nutrients, the endogenous phages become active. They initiate the reproductive cycle resulting in the lysis of the host cell, thus pre-empting a possible demise of the bacterium. An example is phage λ of E. coli. Sometimes, prophages even provide benefit to the host bacterium while they are dormant, by adding new functions to the bacterial genome, a phenomenon called lysogenic conversion. A famous example is the conversion of a harmless strain of *Vibrio cholerae* by a phage into a highly virulent one, which causes cholera.

Bacteriorhodopsin Pigmented protein found in the plasma membrane of a salt-loving bacterium, *Halobacterium halobium*; it pumps protons out of the cell in response to light.

Bacteriostat A substance that inhibits or slows down growth and reproduction of bacteria.

Bacterium Common name for the class Schizomycetes: minute (0.5-5mm), unicellular organisms, without a distinct nucleus. Bacteria are prokaryotes, and most of them are identified by means of Gram staining. They are classified on the basis of their oxygen requirement (aerobic vs anaerobic) and shape (spherical = coccus; rodlike = bacillus; spiral = spirillum; comma-shaped = vibrio; corkscrew-shaped = spirochaete; filamentous). Bacteria usually reproduce asexually, by simple cell division, although a few undergo a form of sexual reproduction, termed *conjugation*. A few bacteria can photosynthesize (including green-blue cyanobacteria), some are saprophytes and others are parasites and can cause diseases. They are major agents of fermentation, putrefaction and decay, and frequently a source of contamination in tissue culture. In plant pathology, strains of bacteria causing disease in specific plant cultivars are called pathovars.

Baculovirus Baculoviruses are a class of insect virus which have been used to make DNA cloning vectors for gene expression in eukaryotic cells. Baculoviruses have a gene which is expressed at very high levels late in their infection cycle, filling the nucleus of the cell with many-sided bodies full of a protein which is not needed to produce more viruses, but is necessary for the virus's spread in the wild. In a vector cloning system, this gene is replaced by one that the biotechnologist wants expressed. Production of the protein can be up to 50% of the cells' protein content, and several proteins can be made simultaneously, so that multi-sub-unit enzymes can be made by this system. Being an animal expression system, baculoviruses produce proteins that are glycosylated (addition of carbohydrates) like the proteins in animals, making it an attractive option for the production of biopharmaceuticals. In addition, baculoviruses are non-infective and non-pathogenic to vertebrates.

Bagasse The dry, fibrous residue that remains after the stalks of sugar cane have been crushed and all the juice extracted. It can be used as a source of cellulose for some paper products.

Balanced lethal system A system for maintaining a recessive lethal allele at each of two loci on the same pair of chromosomes. In a closed population with no crossing-over between the loci, only the double heterozygotes for the lethal mutations survive.

Balanced polymorphism Two or more types of individuals maintained in the same breeding population.

Balloon angioplasty A surgical procedure for re-opening a clogged artery. A balloon is attached to a tube and inserted into a blocked or narrowed artery. When it reaches the site of the blockage it is inflated to re-open the passage way.

Bar A unit used for pressure of fluid. 1 bar = 10^5 Pa. 1 bar is approximately equivalent to 1 atmosphere.

Barr body A condensed mass of chromatin found in the nuclei of female mammals. It is a late-replicating, inactive X-chromosome. Named after its discoverer, Murray Barr (1908-).

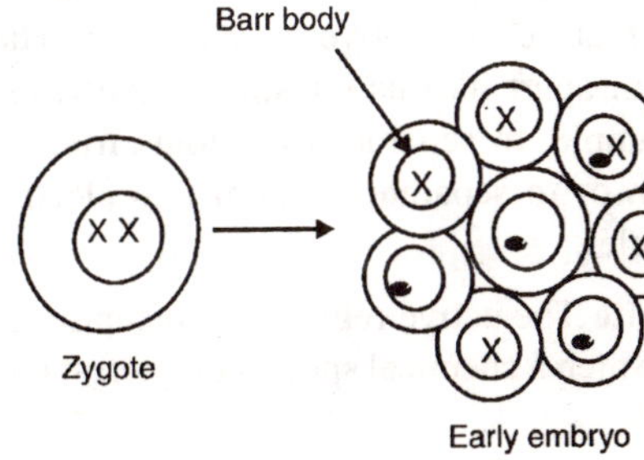

Fig. Barr body

Barrier height Roughly synonymous with activation energy.

Basal 1. Located at the base of a plant or a plant organ.

2. A fundamental formulation of a tissue culture medium.

Basal body Short cylindrical array of microtubules plus their associated proteins found at the base of a eucaryotic cell cilium or flagellum. Serves as a nucleation site for the growth of the axoneme. Closely similar in structure to a centriole.

Basal lamina (plural basal laminae) Thin mat of extracellular matrix that separates epithelial sheets, and many types of cells such as muscle cells or fat cells, from connective tissue. Sometimes called a basement membrane.

Base A cyclic, nitrogen-containing compound that is one of the essential components of nucleic acids. Exists in five main forms (adenine, A; guanine, G; thymine, T; cytosine, C; uracil, U). A and G have a similar structure and are called purines; T, C and U have a similar structure and are called pyrimidines. A base joined to a ribose sugar joined to a phosphate group is a nucleotide - the building block of nucleic acids.

Base analogues Unnatural purine or pyrimidine bases that differ slightly in structure from the normal bases, but can be incorporated into nucleic acids. They are often mutagenic.

Base collection In PGR: Defined in the International Undertaking on Plant Genetic as a collection of seed stock or vegetative propagating material (ranging from tissue cultures to whole plants) held for long-term security in order to preserve the genetic variation for scientific purposes and as a basis for plant breeding as multiplication and evaluation.

Base pair (BP) The two strands that constitute DNA are held together by specific hydrogen bonding between purines and pyrimidines (A pairs with T; and G pairs with C). The size of a nucleic acid molecule is often described in terms of the number of base pairs (symbol: bp) or thousand base pairs (kilobase pairs; symbol: kb; a more convenient unit) it contains.

Base sequence The order of the chemical units (bases) adenine, thymine, cytosine and guanine in DNA that forms the genetic code. The sequence of the bases will determine what protein is produced.

Base substitution Replacement of one base by another in a DNA molecule.

Basic Having the properties of a base.

Basophil A type of white blood cell that contributes to inflammatory reactions and symptoms of allergy.

Batch culture A suspension culture in which cells grow in a finite volume of liquid nutrient medium and follow a sigmoid pattern of growth.

Batch fermentation A process in which cells or micro-organisms are grown for a limited time. At the beginning of the fermentation, an inoculum is introduced into fresh medium, with no addition or removal of medium for the duration of the process.

Bearing A mechanical device that permits the motion of a component (ideally, with minimal resistance) in one or more degrees of freedom while resisting motion (ideally, with a stiff restoring force) in all other degrees of freedom.

Bench-scale process A small- or laboratory-scale process; commonly used in connection with fermentation.

Beta cells Insulin-producing cells in the pancreas. If these cells are destroyed, Type I diabetes develops. Also known as islet cells.

Beta DNA The normal form of DNA found in biological systems, which exists as a right-handed helix.

Beta-lactamase Ampicillin resistance gene.

Bias Systematic error resulting in over or underestimation of the strength of the association between an exposure and an outcome.

Biennial In botany, a plant which completes its life cycle within two years and then dies. For most biennial plants, the two growing seasons have to be separated by a period of cold temperature sufficient to induce flowering and fruit formation.

Bile The digestive juice released from liver (stored in gall bladder) into the digestive tract to help solubilize and absorb fat soluble nutrients. Bile contains bile acids, biochemical derivatives of cholesterol. Bile acids serve as intestinal detergents for the proper homogenization and uptake (absorption) of dietary lipids.

Binary vector system A two-plasmid system in *Agrobacterium tumefaciens* for transferring into plant cells a segment of T-DNA that carries cloned genes. One plasmid contains the virulence gene (responsible for transfer of the T-DNA), and another plasmid contains the T-DNA borders, the selectable marker and the DNA to be transferred.

Binding The ability of molecules to stick to each other because of the exact shape and chemical nature of parts of their surfaces. Many biological molecules bind extremely tightly and specifically to other molecules: enzymes to their substrates; antibodies to their antigens; DNA strands to their complementary strands; and so on. Binding can be characterized by a *binding constant* or *association constant* (Ka), or its inverse, the *dissociation constant* (Kd).

Binding energy The reduction in the free energy of a system that occurs when a ligand binds to a receptor. Generally used to describe the total energy required to remove something, or to take a system apart into its constituent particles—for example, to separate two atoms from one another, or to separate an atom into electrons and nuclei.

Binding site The active region of a receptor; any site at which a chemical species of interest tends to bind.

Binomial expansion The probability that an event will occur 0, 1, 2, ..., *n* times out of *n* is given by the successive terms of the expression $(p + q)n$, where p is the probability of the event occurring, and $q = 1 - p$.

Binomial nomenclature In biology, each species is generally identified by two terms: the first is the genus to which it belongs, and the second is the specific epithet that distinguishes it from others in that genus (e.g., *Quercus suber*, cork oak). The genus name always has an initial capital; the specific epithet is never capitalized, even though it may be derived from a proper name (e.g., keranda nut, *Elaeocarpus bancroftii*). Both terms in the binomial are italicized. Based on the system of classification developed by Carolus Linnaeus.

Bio A prefix derived from *bios* and used in scientific words to associate the concept of "living organisms." Usually written with a hyphen before vowels, for emphasis or in neologisms; otherwise usually without a hyphen.

Bio-accumulation In an organism, concentration of materials which are not components critical for that organism's survival. Usually it refers to the accumulation of metals or other compounds (e.g., DDT). Many organisms - plants, fungi, protists, bacteria, etc. - accumulate metals when grown in a solution of them, either as part of their defence mechanism against the poisonous effect of those compounds, or as a side-effect of the chemistry of their cell walls. Bio-accumulation is important as part of the microbial mining cycle, removing toxic metals from wastewater, as a purification (bioremediation) process, etc.

Bio-assay A procedure for the assessment of a substance by measuring its effect in living cells or on organisms. Animals have been used extensively in drug research in bio-assays for the pharmacological activity of drugs. However, bio-assays are now usually developed using bacteria or animal or plant cells, as these are usually much easier to handle than whole animals or plants, are cheaper to make and keep, and avoid the ethical problems associated with testing of animals. Sometimes used to detect minute amounts of substances that influence or are essential to growth.

Bioaugmentation Increasing the activity of bacteria that break down pollutants by adding more of their kind. A technique used in bioremediation.

Biocatalysis; biocatalyst Use of enzymes to catalyse chemical reactions.

Biochemical The product of a chemical reaction in a living organism.

Biochemical pathway A series of biological reactions leading up to the creation of a particular product or event.

Biochemistry The study of the chemistry of living organisms, especially focussing on metabolism.

Biochip An electronic device that uses organic molecules to form a semiconductor.

Biocide Any chemical agent that can kill a living organism. For example, pesticides that kill insects.

Biocontainment A process aimed at keeping biological organisms within a limited space or area. For example, if an outbreak of a cow disease is found on one farm, a biocontainment process would aim at stopping the disease from spreading to other farms.

Biocontrol The control of living organisms (especially pests) by biological means. Any process using deliberately introduced living organisms to restrain the growth and development of other, very often pathogenic, organisms, such as the use of spider mites to control cassava mealy bug, or the introduction of myxomatosis into Australia to control rabbits. The term also applies to use of disease-resistant crop cultivars. Biotechnology approaches biocontrol in various ways, such as using fungi, viruses or bacteria which are known to attack an insect or weed pest.

Bioconversion Conversion of one chemical into another by living organisms, as opposed to their conversion by enzymes (which is biotransformation) or by chemical processes. The usefulness of bioconversion is much the same as that of biotransformation - in particular its extreme specificity and ability to work in moderate conditions. However, bioconversion has several other properties, including the possibility of having several chemical steps. A major commercial application is in the manufacture of steroids. The "basic" steroid molecule, often isolated from plants, is itself a very complicated molecule, and not one that is easy to modify by normal chemical means to produce the very specific molecules needed for drug use. However, a particular type of bioconversion that attacks only specific bits of the molecule can be used. Bioconversion is particularly useful for introducing chemical changes at specific points in large, complex molecules.

Biodegradable A property of molecules or chemicals that refers to their usefulness as food because they can be metabolized (metabolism) by organism.

Biodegradation The breakdown by living organisms of a compound to its chemical constituents. Materials that can be easily biodegraded are colloquially termed *biodegradable.*

Biodiesel An alternative fuel for use in diesel engines that is made from natural renewable sources such animal fats or vegetable oils and does not contain petroleum. It has similar properties to petroleum but releases fewer environmental pollutants in its emissions. Biodiesel can be used in diesel engines with little or no modifications, either as a diesel fuel substitute, or added to petroleum-based fuels to reduce their polluting effect. Examples include oils such as soybeans, rapeseed, sunflowers or animal tallow.

Biodiscovery The search for plant and animal species from which useful compounds can potentially be obtained.

Biodiversity or biological diversity Refers to the variety of ecosystems and species of plants and animals that can be found in the environment. There are three important levels of biodiversity: genes, species and ecosystems. Genetic diversity is considered the total of genetic information, contained in the genes of individual plants, animals and microorganisms that live on earth. Species diversity is the variety of living organisms on earth (there are many millions, of which only a small proportion has been described). Ecosystem diversity is the variety of habitats and ecological processes in the biosphere (the regions of the surface and atmosphere of the earth).

Biodynamic agriculture An organic farming system which treats the earth as a living organism in the context of the whole cosmos. It is therefore a holistic approach which works not only with soil, weather and seasons, as external factors, but also with the movements of planetary bodies, in particular the moon against the background of the stars. The farms are treated as individualities and as far as possible as self-contained units in that external inputs are kept to an absolute minimum if not zero. This is the most rigorous definition possible for a sustainable agriculture. It includes the development of local breeds and varieties suited to the natural conditions of a locality and to biodynamic farming practices.

Bioelectricity The term bioelectricity refers to the use of charged molecules and elements (ions) in biological systems. The movement and placement of charges has a great influence on molecular interactions between molecules and thus affects structure and function of proteins, DNA, and cell membranes. The latter are able to stabilize local charge separation in form of ion gradients which are a form of energy storage but also serve as information processing device.

Bio-engineering The use of artificial tissues, organs and organ components to replace parts of the body that are damaged, lost or malfunctioning.

Bioenrichment A bioremediation strategy that involves adding nutrients or oxygen, thereby bolstering the activity of microbes as they break down pollutants.

Bio-ethics The branch of ethics that deals with the life sciences and their potential impact on society. At one extreme, it can be enormously useful in focusing attention on problems that need to be confronted; at the other extreme, it can become a name-calling argument between the "pro-biotechnology" and "anti-biotechnology" schools of thought, which, as it reduces discussion to epithets and clichıs, can make better sound bites.

Biofouling When living organisms attach to and start living on any object that is submerged in the sea. This is commonly seen as barnacles attached to the hulls of ships.

Biofuel A gaseous, liquid or solid fuel that contains energy derived from a biological source. For example, rapeseed oil or fish liver oil can be used in place of diesel fuel in modified engines. A commercial application is the use of modified rapeseed oil, which - as rapeseed methyl ester (RME) - can be used in modified diesel engines, and is sometimes named bio-diesel.

Biogas A mixture of methane and carbon dioxide resulting from the anaerobic decomposition of waste such as domestic, industrial and agricultural sewage.

Biogenesis The principle that a living organism can only arise from other living organisms similar to itself and can never originate from non-living material.

Bio-informatics The use and organization of information of biological interest. In particular, it is concerned with organizing bio-molecular databases, in getting useful information out of such databases, in utilizing powerful computers for analysing such information, and in integrating information from disparate biological sources.

Biolistics A technique to insert DNA into cells. The DNA is mixed with small metal particles - usually tungsten or gold - a fraction of a micrometre across. These are then fired into a cell at very high speed. They puncture the cell and carry the DNA into the cell. Biolistics has an advantage over transfection, transduction, etc., because it can apply to any cell, or indeed to parts of a cell. Thus use of biolistics has inserted DNA into animal, plant and fungal cells, and into mitochondria inside cells.

Biolistic device A device that shoots microscopic DNA-coated particles into target cells.

Biologic A therapeutic or prophylactic derived from a living source (human, animal or unicellular). Most biologics are complex mixtures that are not easily identified or characterized, and many are manufactured using biotechnology. Biological products often represent the cutting-edge of biomedical research and are sometimes the most effective way to prevent or treat a disease.

Biologic response modifier A substance that alters the growth or functioning of a cell. Includes hormones and compounds that affect the nervous and immune systems.

Biological containment Restricting the movement of (genetically engineered) organisms by arranging barriers to prevent them from growing outside the laboratory. Biological containment can take two forms: making the organism unable to survive in the outside environment, or making the outside environment inhospitable to the organism. The latter is rarely suitable for bacteria, which, in principle, could survive almost anywhere. Thus for bacteria and yeasts, the favoured approach is to mutate the genes in the organism so that they require a supply of a specific nutrient that is usually available only in the laboratory. If they get out, they then cannot grow. Making the environment unfriendly to the organism is partly a biological control, partly a physical one. Thus, some of the first genetically engineered rice strains were developed in England (which is too cold for rice to grow) and tried in the field in Arizona (where it is too dry). Biological containment may also involve the use of vector molecules and host organisms which have been genetically disabled such that they can survive only in the peculiar conditions provided by the experimenter and which are unavailable outside the laboratory.

Biological control The control of a population of one organism by another organism. Generally the controlling organism is a predator or disease-causing organism of the species being controlled.

Biological invasion The introduction of an organism into a new environment or geographical region followed by rapid multiplication and spread.

Biological oxygen demand (BOD) The amount of oxygen used for growth by organisms in water that contains organic matter.

Biological process A process which involves a reaction normally carried out in a living cell or organism.

Biological response modifiers Proteins that modify or enhance an immune or other biological response.

Biological value (of proteins) The biological value of a protein refers to the how much of the nitrogen content of food is retained by the body. The biological value of proteins ranges from 50 to 100 percent and is a measure of how much dietary protein source can support growth. Animal proteins have biological values of 70 percent or higher, and plant proteins have biological values of 50 to 70.

Biologically active Describes a specific substance which has an effect on the metabolic activity of living cells.

Bioluminescence The enzyme-catalyzed production of light by a number of diverse organisms (e.g. fireflies and many deep ocean marine organisms). Utilized as a reporter gene in plant transgenesis, and for the detection of food-borne pathogenic bacteria.

Biomarker An effect in a biological system. Measurements in the human body or its products. Some biomarkers, such as the levels of certain vitamins in blood serum, are used as indices of nutritional status. Others are used as indices of the risk or progression of disease.

Biomass The totality of biological matter in a given area. As commonly used in biotechnology, refers to the use of cellulose, a renewable resource, for the production of chemicals that can be used to generate energy or as alternative feedstocks for the chemical industry to reduce dependence on nonrenewable fossil fuels.

Biomass concentration The amount of biological material in a specific volume.

Biomass proteins The total weight of all protein, in a given area, organism or group of organisms.

Biomaterials Biological molecules, such as proteins and complex sugars, used to make medical devices, including structural elements used in reconstructive surgery.

Biome A major ecological community or complex of communities, extending over a large geographical area and characterized by a dominant type of vegetation.

Biometry The application of statistical methods to the analysis of biological problems.

Biomolecules Any molecule that is involved in the maintenance and metabolic processes of living organisms. Biomolecules include carbohydrate, lipid, protein, water and nucleic acid.

Biopesticide A compound that kills organisms by virtue of specific biological effects rather than as a broader chemical poison. Specific types include bio-insecticides and bio-fungicides. Bio-pesticides differ from biocontrol agents in that bio-pesticides are passive agents, whereas biocontrol agents are active, seeking out the pest to be destroyed. There are some extremely attractive anti-pest materials, such as *Bacillus thuringiensis* (Bt) toxin, which specifically interferes with the absorption of food from the guts of some insects but is harmless to mammals. The rationale behind developing bio-pesticides is that they are more likely to be biodegradable and are targeted at specific elements of the pest's metabolism.

Biopharming The production of pharmaceuticals such as edible vaccines and antibodies in plants or domestic animals.

Biopolymer Any large polymeric molecule (protein, nucleic acid, polysaccharide, lipid) produced by a living organism.

Bioprocess Any process that uses complete living cells or their components (e.g. enzymes, chloroplasts) to effect desired physical or chemical changes.

Bioreactor A tank in which cells, cell extracts or enzymes carry out a biological reaction. Often refers to a growth chamber (fermenter, fermentation vessel) for cells or micro-organisms.

Bioremediation The use of plants and micro-organisms to consume or otherwise help remove materials (such as toxic chemical wastes and metals) from contaminated sites (especially from soil and water) A natural process in which environmental problems are treated by the use of bacteria or other micro-organisms that break down a problem substance, such as oil, into less harmful molecules.

Biosafety The assessment of the impact and safety of genetically improved/modified organisms and the development of protective policies and procedures for adoption to ensure this.

Biosensor A device that uses an immobilized agent (such as an enzyme, antibiotic, organelle or whole cell) to detect or measure a chemical compound. A reaction between the immobilized agent and the molecule being analysed is transduced into an electric signal.

Biosphere The part of the earth and its atmosphere that is inhabited by living organisms.

Biosynthesis Synthesis of compounds by living cells, which is the essential feature of anabolism.

Biotechnology 1. A broad term generally used to describe the use of biology in industrial processes such as agriculture, brewing and drug development. The term also refers to the production of genetically modified organisms or the manufacture of products from genetically modified organisms.

2. The use of plants, animals and micro-organisms to create products or processes. Traditional applications include animal breeding, brewing beer with yeast, and cheese making with bacteria. Recent developments include the use of enzymes or bacteria in a wide range of applications, including waste management, industrial production, food production and remediation of contaminated land. Modern biotechnology also includes the use of gene technology, which allows us to move genetic material from one species to another.

Biotic Relating to life or specific life conditions.

Biotic factor Other living organisms that are a component of an organism's environment, and form the biotic environment, affecting the organism in many ways.

Biotic stress Stress resulting from living organisms which can harm plants, such as viruses, fungi, bacteria, parasitic weeds and harmful insects.

Biotin A vitamin of the B complex. It is a co-enzyme for various enzymes that catalyse the incorporation of carbon dioxide into various compounds. It is essential for the metabolism of fats. Biotin is attached to pyruvate carboxylase by a long, flexible chain like that of lipoamide in the pyruvate dehydrogenase complex. Adequate amounts are normally produced by the intestinal bacteria in animals.

Biotin labelling 1. The attachment of biotin to another molecule.

2. The incorporation of a biotin-containing nucleotide into a DNA molecule.

Biotinylated DNA A DNA molecule labelled with biotin by incorporation of biotinylated-dUTP into a DNA molecule. It is used as a non-radioactive probe in hybridization experiments, such as Southern transfer. The detection of the labelled DNA is achieved by complexing it with streptavidin (an antibiotic with a high affinity for biotin) to which is attached a colour-generating agent such as horseradish peroxidase that gives a fluorescent green colour upon reaction with various organic reagents.

Biotope A small habitat in a large community.

Biotoxin A naturally produced toxic compound which shows pronounced biological activity and presumably has some adaptive significance to the organism which produces it.

Biotransformation The conversion of one chemical or material into another using a biological catalyst: a near synonym is biocatalysis, and hence the catalyst used is called a biocatalyst. Usually the catalyst is an enzyme, or a whole, dead micro-organism that contains an enzyme or several enzymes.

Biotreatment The treatment of a waste or hazardous substance using organisms such as bacteria, fungi and protozoa .

Bipyridinium Polymer derived from N,N-bis[(p-trimethoxysilyl)-benzyl]-4,4'- bipyridinium bpy 2,2'-Bipyridine.

Bivalent A pair of synapsed or associated homologous chromosomes (one of maternal origin; the other of paternal origin) that have each undergone duplication. Each duplicated chromosome comprises two chromatids. Thus a bivalent comprises four chromatids.

Blastocyst After a mammalian ovum is fertilised it begins to divide. The blastocyst is an early stage of this process which consists of a sphere that is fluid-filled and surrounded by a layer of cells. surrounding a fluid-filled cavity. There is a mass of cells at one side which will become the embryo. The blastocyst is formed before implantation into the uterus.

Blastomere Any one of the cells formed from the first few cleavages in animal embryology. The embryo usually divides into two, then four, then eight blastomeres, and so on.

Blastula In animals, an early embryo form that follows the morula stage; typically, a single-layered sheet (blastoderm) or ball of cells (blastocyst).

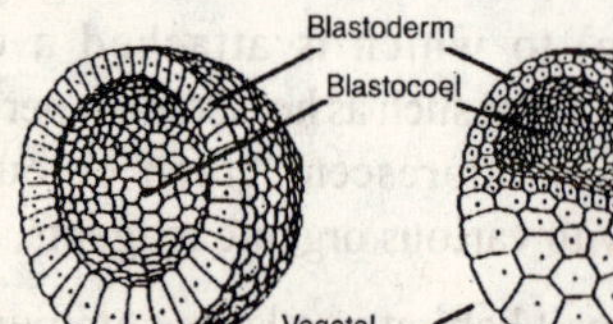

Fig. Blastula

Bleach A fluid, powder or other whitening (bleaching) or cleaning agent, usually with free chlorine ions. Commercial bleach contains calcium hypochlorite or sodium hypochlorite, and is a common disinfectant used for cleaning working surfaces, tools and plant materials in plant tissue culture and grafting.

Bleeding Used to describe the occasional purplish-black coloration of media due to phenolic products given off by (usually fresh) transfers.

Blood pressure The pressure exerted by the circulating blood on the walls of arteries and veins and in the heart chambers. When the heart contracts, the maximum pressure exerted is the systolic pressure; during relaxation, the minimum pressure is the diastolic.

Blot 1. As a verb, this means to transfer DNA, RNA or protein to an immobilizing matrix.

2. As a noun, it usually refers to the autoradiograph produced during the Southern or northern blotting procedures. The variations on this theme depend on the molecules:

– Southern blot: the molecules transferred are DNA molecules, and the probe is DNA.

– northern blot: the molecules transferred are RNA, and the probe is DNA.

– western blot: the molecules transferred are protein, and the probe is labelled antibody.

– Southwestern blot: the molecules transferred are protein, and the probe is DNA.

– dot blot: DNA, RNA or protein are dotted directly onto the membrane support, so that they form discrete spots.

– colony blot: the molecules (usually DNA) are from colonies of bacteria or yeast growing on a bacteriological plate.

Blunt end The end of a DNA duplex molecule in which neither strand extends beyond the other.

Blunt-end cut To cleave phospho-diester bonds in the backbone of duplex DNA between the corresponding nucleotide pairs on opposite strands. This cleavage process results in both strands finishing at the same residue, i.e., there are no nucleotide extensions on either strand.

Blunt-end ligation Joining (ligation) of the nucleotides that are at the ends of two blunt-ended DNA duplex molecules.

Bond Two atoms are said to be bonded when the energy required to separate them is substantially larger than the van der Waals attraction energy. Ionic bonds result from the electrostatic attraction between ions; covalent and metallic bonds result from the sharing of electrons among atoms; hydrogen bonds are weaker and result from dipole interactions and limited electron sharing. When used without modification, "bond" usually refers to a covalent bond.

Boring platform Sterile bottom half of a Petri dish used for preparing explants with a cork borer.

Bound water Water held by the cell and not released if freezing occurs in the intercellular space.

Bovine somatotrophin (BST) Bovine growth hormone, this protein is found naturally in cattle, and is the bovine counterpart of human growth hormone, one of the earliest biopharmaceutical products. It has been cloned, using recombinant DNA technology, expressed in large amounts and marketed as an agricultural product to improve the growth rate and protein:fat ratios in farm cattle, and to improve milk yield. Its use is banned in some countries.

Bract A modified leaf that subtends flowers or inflorescences and may appear to be a petal.

BRCA1 and BRCA2 (Breast Cancer genes 1 and 2) Two genes that normally help to restrain cell growth, but which can contain certain genetic mutations associated with the development of breast and ovarian cancer. Note, however, that inherited BRCA1 and BRCA2 mutations are thought to account for less than 10 percent of all breast and ovarian cancers. Recent evidence suggests that somatic cell genetic mutations (i.e., noninherited genetic mutations) in these two genes may also play a role in the development of cancer.

Breed (I) A sub-specific group of domestic livestock with definable and identifiable external characteristics that enable it to be separated by visual appraisal from other similarly defined groups within the same species or (II) A group of domestic livestock for which geographical and/or cultural separation from phenotypically similar groups has led to acceptance of its separate identity.

Breed at risk Any breed that may become extinct if the factors causing its decline in numbers are not eliminated or mitigated. Breeds may he in danger of becoming extinct for a variety of reasons. Risk of extinction may result from, *inter alia,* low population size; direct and indirect impacts of policy at the farm, country or international levels; lack of proper breed organization; or lack of adaptation to market demands. Breeds are categorized as to their risk status on the basis of, *inter alia,* the actual numbers of male and/or female breeding individuals and the percentage of pure-bred females. FAO has established categories of risk status: critical, endangered, critical-maintained, endangered-maintained, and not-at-risk.

Breeding value In quantitative genetics, the part of the deviation of an individual phenotype from the population mean that is due to the additive effects of alleles. In practical terms: if an animal is mated with a random sample of animals from a population, that animal's breeding value for a certain trait is twice the average deviation of its offspring from the population mean for that trait.

Brewer's yeast Strains of yeast, often *Saccharomyces cerevisiae,* that are used in the production of beer.

Brewing The process by which beer is made. In the first stage the barley grain is soaked in water and allowed to germinate (malting), during which the natural enzymes of the grain convert the seed starch to maltose, and then to glucose. Grain is then dried, crushed, and added to water at a specific temperature (steeping) and any remaining starch is converted to sugar. The resulting liquid (wort) is the raw material to which yeast is added to convert sugar to alcohol. Hops (female flowers of *Humulus lupulus*) are added during this process to give a characteristic flavour.

Bridge A filter paper or other substrate used as a wick and support structure for a plant tissue in culture when a liquid medium is used.

Broad host range plasmid A plasmid that can replicate in a number of different bacterial species.

Broad sense heritability In quantitative genetics, the proportion of the total phenotypic variation due to genetic variation.

Broad-spectrum Describes an agent, such as an herbicide, that has a wide range of use.

Broodstock The group of males and females from which fish are bred for aquaculture.

Brownian assembly Brownian motion in a fluid brings molecules together in various positions and orientations. If molecules have suitable complementary surfaces, they can bind, assembling to form a specific structure. Brownian assembly is a less paradoxical name for self-assembly (how can a structure assemble itself, or do anything, when it does not yet exist).

Brownian motion Motion of a particle in a fluid owing to thermal agitation, observed in 1827 by Robert Brown. (Originally thought to be caused by a vital force, Brownian motion in fact plays a vital role in the assembly and activity of the molecular structures of life.)

Browning Discoloration due to phenolic oxidation of freshly cut surfaces of explant tissue. In later stages of culture, such discoloration may indicate a nutritional or pathogenic problem, generally leading to necrosis.

Brucellosis Disease caused by infection with organisms of the genus *Brucella.*

BT Abbreviation for Bacillus thuringiensis.

Bt crops Crops that are genetically engineered to carry a gene from the soil bacterium Bacillus thuringiensis (Bt). The bacterium produces proteins that are toxic to some pests but non-toxic to humans and other mammals. Crops containing the Bt gene are able to produce this toxin, thereby providing protection for the plant. Bt corn and Bt cotton are examples of commercially available Bt crops.

BT toxin Proteins produced by Bacillus thuringiensis (a group of soil bacteria). These proteins are called "cry" proteins, and are toxic to insects. When the DNA coding for "cry" proteins is inserted into the genomes of crop plants, the crop becomes resistant to insects.

Bubble column fermenter A fermentation vessel, or bioreactor, in which the cells or micro-organisms are kept suspended in a tall cylinder by rising air, which is introduced at the base of the vessel.

Bud A region of meristematic tissue with the potential for developing into leaves, shoots, flowers or combinations; generally protected by modified scale leaves. A terminal (or apical) bud exists at the tip of a stem or branch, while axillary (or lateral) buds develop in the axils of leaves.

Bud scar A scar left on a shoot when the bud or bud scales drop.

Bud sport A somatic mutation arising in a bud and producing a genetically different shoot. Bud sports includes changes due to gene mutation, somatic reduction, chromosome deletion or polyploidy.

Budding 1. A method of asexual reproduction in which a new individual is derived from an outgrowth (bud) that becomes detached from the body of the parent.

2. Among fungi, budding is characteristic of the yeast *Saccharomyces cerevisiae.*

3. A form of grafting in which a single vegetative bud is taken from one plant and inserted into stem tissue of another plant so that the two will grow together. The inserted bud develops into a new shoot.

Buffer A solution that resists change in pH when an acid or alkali is added, or when solutions are diluted.

Bulimia An eating disorder consisting of an insatiable craving for food, often resulting in voracious eating followed by self-induced vomiting.

Bulk technology Technology in which atoms and molecules are manipulated in bulk, rather than individually.

Buoyant density The intrinsic density which a molecule, virus or sub-cellular particle has when suspended in an aqueous solution of a salt, such as CsCl, or a sugar, such as sucrose. DNA from different species has a characteristic buoyant density, which reflects the proportion of G=C base pairs. The greater the proportion of G=C, the greater the buoyant density of the DNA.

By product An incidental or secondary product made in the manufacture of a primary product.

C

C Cytosine residue in either DNA or RNA.

C kinase $Ca2^+$-dependent protein kinase that, when activated by diacylglycerol and an increase in the concentration of $Ca2^+$, phosphorylates target proteins on specific serine and threonine residues.

$Ca2^+$ release channel Ion channel in the membrane of the endoplasmic reticulum and sarcoplasmic reticulum (in muscle cells) that releases $Ca2^+$ into the cytosol when activated.

CAAT box (also CAT box) A conserved sequence found within the promoter region of the protein-encoding genes of many eukaryotic organisms. It has the consensus sequence GGCCAATCT; it occurs around 75 bases prior to the transcription initiation site; and it is one of several sites for recognition and binding of regulatory proteins called transcription factors.

CAD Computer-aided design.

Cadherin Member of a family of proteins that mediate $Ca2^+$-dependent cell-cell adhesion in animal tissues.

Caged compound Organic molecule designed to change into an active form when irradiated with light of a specific wavelength. An example is caged ATP.

Calf scours. A watery diarrhoea in calves.

Calicivirus The virus that causes Rabbit Calicivirus Disease, RCD, in rabbits. It is spread by mosquitoes and fleas.

Callus 1. A protective tissue, consisting of parenchyma cells, that develops over a cut or damaged plant surface.

2. Mass of unorganized, thin-walled parenchyma cells induced by hormone treatment.

3. Actively dividing non-organized masses of undifferentiated and differentiated cells often developing from injury (wounding) or in tissue culture in the presence of growth regulators.

Callus culture A technique of tissue culture; it is usually on solidified medium and initiated by inoculation of small explants or sections from established organ or other cultures (the inocula). Callus culture is used as the basis for organogenic (shoot, root) cultures, cell cultures or proliferation of embryoids. Callus cultures can be indefinitely maintained through regular sub-culturing.

Calmodulin Ubiquitous $Ca2^+$-binding protein whose binding to other proteins is governed by changes in intracellular $Ca2^+$ concentration. Its binding modifies the activity of many target enzymes and membrane transport proteins.

Calorie (cal) Chemical energy in foods is expressed in calories (Cal). The scientific definition of a calorie is the amount of energy needed to raise the temperature of 1 gram of water by one degree Celsius from 15° to 16° at atmospheric pressure. This is the true calorie, sometimes referred to as a "small calorie". A kilocalorie is equal to 1000 calories. In reference to food energy, the term "calorie" is actually a 1,000 calories or 1 kilocalorie. The USDA Nutrient Database for Standard Reference contains values for both kilocalories and kilojoules (1 kcal equals 4.184 kJ).

Calvin cycle Major metabolic pathway by which CO_2 is fixed during photosynthesis.

Calyx All the sepals of a flower considered collectively. The outermost whorl of flower parts.

Cam A component that translates or rotates to move a contoured surface past a follower; the contours impose a sequence of motions (potentially complex) on the follower.

Cambial zone Region in stems and roots consisting of the cambium and its recent derivatives.

Cambium A layer, usually regarded as one or two cells thick, of persistently meristematic tissue between the xylem and phloem tissues, and which gives rise to secondary tissues, thus resulting in an increase in diameter. The two most important cambia are the vascular (fascicular) cambium and the cork cambium.

Cancer Cancer is an abnormal, uncontrolled and rapid growth of cells that invade and destroy surrounding tissues. It is a broad term for more than 100 diseases characterised by this growth. Cells from the tumour can break away (metastasize) and spread through the bloodstream or lymph system to other parts of the body creating new tumours.

Candidate gene A gene whose function suggests that it may be involved in the genetic variation observed for a particular trait, e.g., the gene for growth hormone is a candidate gene for body weight.

Candidate-gene strategy An experimental approach in which knowledge of the biochemistry and/or physiology of a trait is used to draw up a list of genes whose protein products could be involved in the trait.

Canola Any of several cultivars of oilseed rape (more fully: canola oil); the vegetable oil high in mono-unsaturated fatty acid obtained from these cultivars.

CAP (catabolite gene activator protein) Gene regulatory protein in procaryotes that, when glucose is absent, activates genes responsible for the breakdown of alternative carbon sources.

Cap site The site in a gene where translation is initiated.

Capacitance An element (capacitor) in an electrical circuit capable of separating charges and storing electrical energy. In cells, membranes have capacitor properties contributing to the storage of electrochemical energy (ion gradients).

Capsid Protein coat of a virus, formed by the self-assembly of one or more protein subunits into a geometrically regular structure

Carbanion A highly reactive anionic chemical species with an even number of electrons and an unshared pair of electrons on a tetravalent carbon atom.

Carbene A highly reactive chemical species containing an electrically neutral, divalent carbon atom with two nonbonding valence electrons; a prototype is CH_2.

Carbohydrates Biochemical name for sugar containing molecules including single sugar (monosaccharides) like glucose and galactose, but also polysaccharides (complex carbohydrates) like starch (poly-glucose), cellulose (plant fiber material, also poly-glucose with a different chemical bond structure linking glucose units than those found in starch/glycogen and enzymatically indigestible by humans), chitin (hard shells of insects), and more complex carbohydrate components part of lipids and proteins such as blood serum glycoproteins (antibodies and blood group determinants A, O, B, rhesus positive or negative). All microorganisms (bacteria, viruses) contain carbohydrate surfaces (glycolipids) being the major determinants of immunogenic reactions during infections.

Carbon The element that defines the chemical properties of all life. All molecules that contain carbon are known as organic molecules and studies by organic chemistry. Carbon is the third most common element in cells, after hydrogen and oxygen, which are the most common biological elements because they are found in water. Also water makes up to 70% of a cells weight, it is not an organic molecule, since it lacks carbon.

Carbonium ion A highly reactive cationic chemical species with an even number of electrons and an unoccupied orbital on a carbon atom.

Carbonyl A chemical moiety consisting of O with a double bond to C. If the C is bonded to N, the resulting structure is termed an amide; if it is bonded to O, it is termed a carboxylic acid or an ester linkage.

Carboxylic acid A molecule that includes a C having a double bond to O and a single bond to OH.

Carboxypeptidases Two enzymes (A and B) found in pancreatic juice. Their role is to remove the C-terminal amino-acid from a peptide; the A form removes any amino acid; the B form removes only lysine or arginine. Used when sequencing peptides.

Carcinogenesis The complex, multistep process of cancer causation.Cardiovascular disease (CVD)Any one of numerous abnormal conditions characterised by dysfunction of the heart and blood vessels.Other types of CVD mentioned in this blood are congestive heart failure and sudden cardiac death.

Carcinoma A malignant tumor derived from epithelial tissue, which forms the skin and outer cell layers of internal organs.

Carotene A reddish-orange plastid pigment involved in light reactions in photosynthesis.

Carotenoid A group of chemically similar red to yellow pigments responsible for the characteristic colour of many plant organs or fruits, such as tomatoes, carrots, etc. Oxygen-containing carotenoids are calledxanthophylls. Carotenoids serve as light-harvesting molecules in photosynthetic assemblies and also play a role in protecting prokaryotes from the deleterious effects of light.

Carpel Female reproductive organ of flowering plants, consisting of stigma, style and ovary. In some plants, one or more carpels unite to form the pistil.

Fig. Carpel

Carrier In genetics, typically an individual that has one recessive mutant allele for some defective condition that is "masked" by a dominant normal allele at the same locus, i.e., an individual that is heterozygous for a recessive harmful allele and a dominant normal allele; the phenotype is normal, but the individual passes the defective (recessive) allele to half of its offspring.

Carrier DNA DNA of undefined sequence content which is added to the transforming (plasmid) DNA used in physical DNA-transfer procedures. This additional DNA increases the efficiency of transformation in electroporation and chemically mediated DNA-delivery systems. The mechanism responsible for this effect is not known.

Carrier gas The gas that carries the sample in gas chromatography.

Carrier molecule 1. A molecule that plays a role in transporting electrons through the electron transport chain. Carrier molecules are usually proteins bound to non-protein groups and able to undergoing oxidation and reduction relatively easily, thus allowing electrons to flow.

2. A lipid-soluble molecule that can bind to lipid-insoluble molecules and transport them across membranes. Carrier molecules have specific sites that interact with the molecules they transport. Efficiency of carrier molecules may be modified by modifying the interacting sites through genetic engineering.

Carrying capacity The maximum number of organisms of a given species that can be supported in a given area or habitat.

Case control study A study design in which persons with a disease (cases) are compared with those without the disease (controls) to see how their exposures to causative factors may have differed.

Casein A group of proteins found in milk.

Casein hydrolysate Mixture of amino acids and peptides produced by enzymatic or acid hydrolysis of casein.

Catabolic pathway A pathway by which an organic molecule is degraded in order to release energy for growth and other cellular processes; degradative pathway.

Catabolism The metabolic breakdown of large molecules in living organism, with accompanying release of energy.

Catabolite repression Glucose-mediated reduction in the rates of transcription of operons that encode enzymes involved in catabolic pathways (such as the *lac* operon).

Catalyst A substance that promotes a chemical reaction by lowering the activation energy of a chemical reaction, without itself undergoing any permanent chemical change. The process is catalysis.

Catalytic antibody (abzyme) An antibody selected for its ability to catalyse a chemical reaction by binding to and stabilizing the transition-state intermediate.

Catalytic RNA (ribozyme; gene shears) A natural or synthetic RNA molecule that cuts an RNA substrate.

Cataract A disorder in which the lens of the eye becomes partly or completely opaque as a result of the precipitation of proteins.

Cation A positively charged ion such as a sodium ion, Na^+.

Caulogenesis Stem organogenesis; induction of shoot development from callus.

CD molecules; cluster of differentiation molecules Any group of antigens that is associated with a specific sub-population of T cells. There are designations for surface molecules on various cells of the immune system, e.g., CD4 is present on the surface of helper T cells.

CDNA DNA synthesized from an RNA template using reverse transcriptase.

CDNA cloning A method of cloning the coding sequence of a gene, starting with its mRNA transcript. It is normally used to clone a DNA copy of a eukaryotic mRNA. The cDNA copy, being a copy of a mature messenger molecule, will not contain any intron sequences and may be readily expressed in any host organism if attached to a suitable promoter sequence within the cloning vector.

CDNA; complementary DNA The double-stranded DNA complement of an mRNA sequence; synthesized *in vitro* from a mature RNA template using reverse transcriptase (to create a single strand of DNA from the RNA template) and DNA polymerase (to create the double-stranded DNA). Preparation of cDNAs is often the first step in cloning DNA sequences of interest. Used as specific and sensitive probes in hybridization studies, because cDNAs usually do not include regulatory or other controlling sequences, and so they can be used to identify (probe) and isolate genes and their associated sequences from genomic DNA.

CDNA library A collection of cDNA clones that were generated *in vitro* from the mRNA sequences isolated from an organism or a specific tissue or cell type or population of an organism.

CDR (complementarity-determining regions) These are regions of the variable (V) regions of light and heavy antibody chains that make contact with the antigen. The primary amino acid sequences of these regions are highly variable among antibodies of the same class.

CEC Process in which a chemical reaction precedes and follows the electron-transfer process.

Cell Smallest unit of life (single cell organism or bacteria) or unit of higher organisms, i.e., multicellular organisms. Cells are surrounded by a cell membrane (and cell wall in bacteria and plants = a membrane plus some chemically more stable structures, often mixtures of proteins and polysaccharides) and contain all necessary elements to sustain life; proteins, nucleic acids, lipids, minerals, and a diverse class of metabolites. Cells of higher organisms (known as eukaryotes) are subdivided into subcellular compartments called organelles such as the mitochondrion, the cell nucleus, the endoplasmatic reticulum, the Golgi apparatus and many smaller organelles with highly specialized functions. While all these organelles are found in animal cells, plant cells in addition contain a central vacuole that controls pressure to stabilize the cell and chloroplasts, the site of photosynthesis or light depended biosynthesis of sugars (carbohydrates).

Cell-based therapies Therapies involving the transplantation of stem cells into damaged tissues to regenerate the various cell types of that tissue. For example, bone marrow transplants are a form of cell-based therapy that has been used for over 30 years to treat leukaemia. New stem cell research may lead to cell-based therapies to treat a range of conditions, including heart disease, spinal injuries, diabetes and Parkinson disease.

Cell culture The *in vitro* growth of cells derived from multi-cellular organisms. The cells are usually of one type.

Cell cycle Sequence of stages that a cell passes through between one division and the next. The cell cycle oscillates between mitosis and the interphase, which is divided into G, S, and G2. In the G phase there is a high rate of biosynthesis and growth; in the S phase there is the doubling of the DNA content as a consequence of chromosome replication; in the G2 phase the final preparations for cell division (cytokinesis) are made.

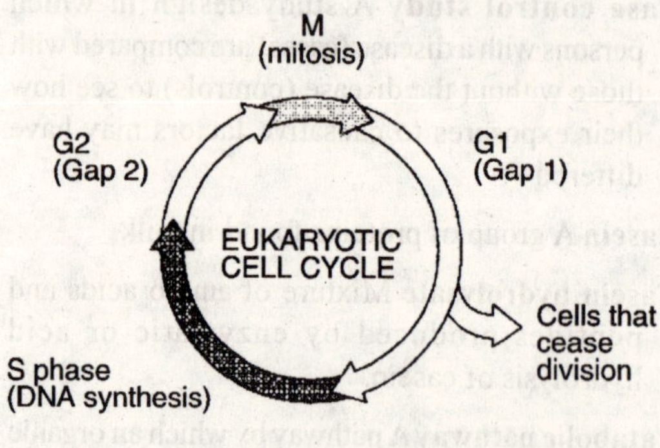

Fig. Cell cycle

Cell differentiation Continuous loss of physiological and cytological characters of young cells, resulting in getting the characters of adult cells. The unspecialized cells become modified and specialized for the performance of specific functions. Differentiation results from the controlled activation and de-activation of genes.

Cell division Formation of two or more daughter cells from a single parent cell. The nucleus divides first, followed by the formation of a cell membrane between the daughter nuclei. Division of somatic cells is termed mitosis; egg and sperm precursors are formed following meiosis.

Cell fusion Formation of a single hybrid cell from two cells of different species, cultured *in vitro*. The cells fuse and coalesce, but their nuclei may remain separated. During subsequent cell division, a single spindle is formed so that each daughter cell has a single nucleus containing sets of chromosomes from each parental line.

Subsequent divisions often result in the loss of chromosomes and therefore of genes. The cell fusion technique can be used to determine the control of specific genes and their assignment to chromosomes.

Cell generation time The interval between the beginning of consecutive divisions of a cell. The time that it takes for a population of single-celled organisms to double its cell number. Successive generations of cells or organisms within a population are separated by a time interval called generation time. The cell regeneration time can be determined with the aid of time-lapse microcinematography.

Cell hybridization The fusion of two or more dissimilar cells leading to the formation of a somatic hybrid.

Cell line A cell lineage that can be maintained in culture. A cell line arises from a primary culture. It implies that cultures from it consist of several lineages of the cells originally present in the primary culture.

Cell-mediated immunity Acquired immunity in which T lymphocytes play a predominant role. Development of the thymus in early life is critical to the proper development and functioning of cell-mediated immunity.

Cell mediated immune response The activation of T cells of the immune system in response to the presence of a foreign antigen.

Cell membrane The membrane that separates the cell wall and the cytoplasm, and regulates the flow of material into and out of the cell.

Cell number The number of cells per unit volume of a culture.

Cell pharmacology Delivery of drugs by medical nanomachines to exact locations in the body.

Cell plate The precursor of the cell wall, formed as cytokinesis starts during cell division. The cell plate develops in the region of the equatorial plate and arises from membranes in the cytoplasm.

Cell sap Water and dissolved substances, sugar, amino acids, waste substances, etc., in the plant cell vacuole.

Cell selection The process of selecting cells within a group of genetically different cells. Select cells or cell lines are sub-cultured onto fresh medium for continued selection and often are exposed to an increased level of the selection agent. The final objective is to regenerate plants exhibiting the traits selected for at the cellular level.

Cell strain A strain of cells having specific properties or markers derived from a primary culture of a cell line by selection or cloning. The selected properties must persist during subsequent cultivation.

Cell surgery Modifying cellular structures using medical nanomachines.

Cell suspension Cells in culture in moving or shaking liquid medium, often used to describe suspension cultures of single cells and cell aggregates.

Cell wall A rigid external coat which surrounds plant cells. It is formed outside the plasmalemma and consists primarily of cellulose.

Cellular automaton A program the applies a simple rule of what to do repeatedly. Depending on the rule of what to do next, the pattern or behavior generated can look (i) repetitive, simple, and symmetric, (ii) nested (fractal), (iii) random and without any symmetry or repetition whatsoever, or (iv) complex with local patterns but overall broken symmetry.

Cellular oncogene (proto-oncogene) A normal mammalian or avian gene that when mutated or improperly expressed contributes to the development of cancer.

Cellularity Characterizes the physical and chemical properties of cells found within a specific tissue.

Cellulose A complex carbohydrate composed of long, unbranched chains of beta-glucose ((1.4)-linked-b-d-glucose) molecules, which contribute to the structural framework of plant cell walls. It comprises 40% to 55% by weight of the plant cell wall.

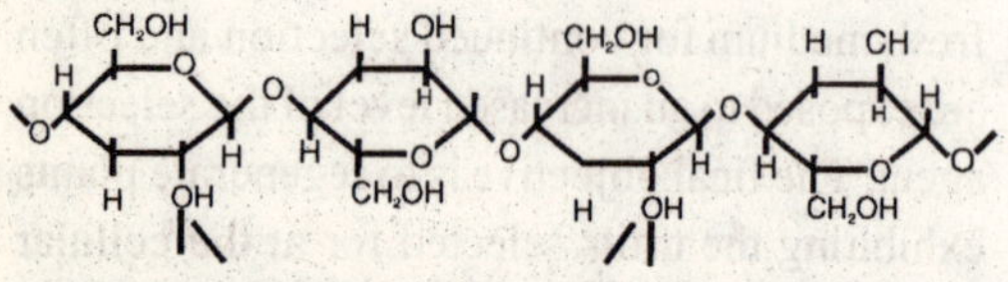

Fig. Cellulose

Cellulolytic The capacity to digest the components of plant tissues and fibers used in making paper and textiles.

Cellulosome A multi-protein aggregate that is present in some cellulolytic micro-organisms and contains multiple copies of all the enzymes required to completely break down cellulose. This complex is often found on the outer surface of cellulolytic micro-organisms.

Centers of origin Usually the location in the world where the oldest cultivation of a particular crop has been identified.

Centimorgan (CM) One percent recombination between two loci.

Central dogma The basic concept that, in nature, genetic information generally can flow only from DNA to RNA to protein. It is now known, however, that information contained in RNA molecules of certain viruses (called retroviruses) can also flow back to DNA.

Central mother cell A subsurface cell located in a plant apical meristem and characterized by a large vacuole.

Central processor unit The central processing unit of a computer, responsible for executing instructions to process information.

Centres of origin The locations in the world where particular domesticated plants originated. These areas show the highest variation, and are rich in wild alleles.

Centrifuge A device in which solid or liquid particles of different densities are separated by rotating them in a tube in a horizontal circle. The denser particles tend to move along the length of the tube to a greater radius of rotation, displacing the lighter particles to the other end.

Centriole An organelle in many animal cells that appears to be involved in the formation of the spindle during mitosis. During cell division, the two centrioles move to opposite sides of the nucleus to form the ends of the spindle.

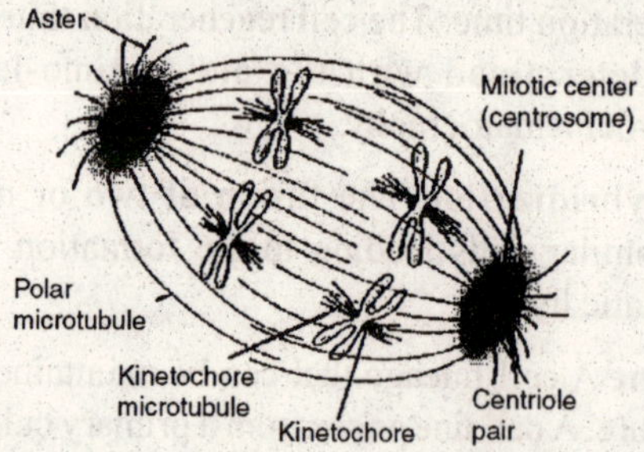

Fig. Centriole

Centromere The portion of the chromosome to which the spindle fibres attach during mitotic and meiotic division. It appears as a constriction when chromosomes contract during cell division. After chromosomal duplication, which occurs at the beginning of every mitotic and meiotic division, the two resultant chromatids are joined at the centromere.

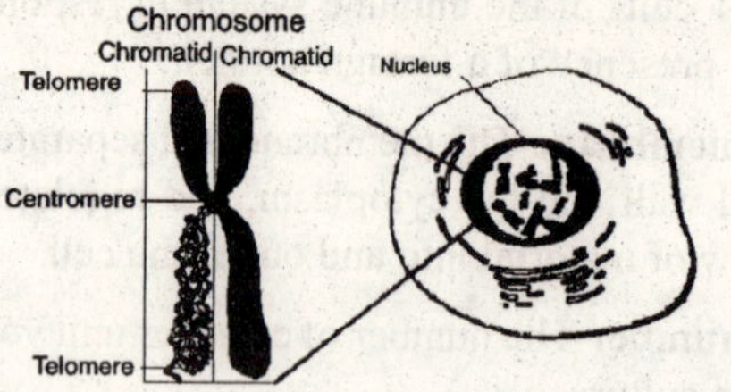

Fig. Centromere

Centrosome A specialized region of a living cell, situated next to the nucleus, where micro-tubules are assembled and broken down during cell division. The centrosome of most animal cells contains a pair of centrioles. During metaphase the centrosome separates into two regions, each containing one of the centrioles.

Cephem type antibiotic An antibiotic that shares the basic chemical structure of cephalosporin.

Chain terminator 1. Codons which do not code for an amino acid. They signal ribosomes to terminate protein synthesis. The codons are UAA, UAG and UGA, and have been termed ochre, amber and opal, respectively. Also known as stop codons or termination codons. Often two of these codons are found together at the end of a coding sequence of RNA.

2. In the Sanger method of DNA sequencing, di-deoxynucleoside triphosphates are added as chain terminators in the synthesis of a complementary DNA strand.

Chakrabarty decision A landmark legal case in the U.S.A., in which it was held that the inventor of a new micro-organism whose invention otherwise met the legal requirements for obtaining a patent, could not be denied a patent solely because the invention was alive. This has set the precedent for the patenting of life forms.

Challenge test Reintroduction of a food previously eliminated from the diet on suspicion that it caused an adverse reaction.

Channel (ion channel) A membrane protein that allows the passive flow of ions across a cell membrane. Ion channels are usually selective for a specific ion type (e.g. Na, or K, or Cl, or Ca) and or either open or closed. These are two structural states of the protein and the change from the open to the closed state (gating) is regulated by the cell. Several regulatory mechanisms have been described including voltage-gating, ligand-gating, heat, and mechano-sensation (touch sensitive). Channels are a category of transporters.

Character A distinctive feature of an organism.

Characterization 1. All activities associated with the description of AnGR aimed at better knowledge of these resources and their state. Characterization by a country of its AnGR will incorporate development of necessary descriptors for use; identification of the country's sovereign AnGR; baseline and advanced surveying of these populations, including their enumeration and visual description, their comparative genetic description in one or more production environments, their valuation, and ongoing monitoring of those AnGR at risk.

2. Systematic recording of descriptors that are independent of environmental factors.

Charcoal The black porous residue of partly burnt wood, bones, etc; a form of carbon. When treated to purify it and increase its adsorptive power, it is called activated charcoal in which form it is added to nutrient medium in order to prevent or decrease the effect of browning.

Chelate Complex organic molecule that can combine with cations and does not ionize. Chelates can supply micronutrients to plants at slow, steady rates. Usually used to supply iron to plant cells.

Chemical genomics Using structural and functional genomic information about biological molecules, especially proteins, to identify useful small molecules and alter their structure to improve their efficacy.

Chemical metabolism genes (CYP 2D6) Genes which code for enzymes governing the metabolism of chemical compounds in the body.

Chemical mutagen A chemical or product capable of causing genetic mutation in living organisms exposed to it.

Chemically-defined medium When all of the chemical components of a plant tissue culture medium are fully known and defined.

Chemiluminescence The emission of light from a chemical reaction.

Chemostat A continuous and open culture in which growth rate and cell density are maintained constant by a fixed rate of input of a growth-limiting nutrient.

Chemotaxis The movement of a cell, or the whole or part of an organism, towards or away from an increasing concentration of a particular substance.

Chemotherapy The treatment of disease, especially infections or cancer, by means of chemicals. In treating cancers, it involves administering chemicals toxic to malignant cells.

Chiasma A visible point of junction between two non-sister chromatids of homologous chromosomes during the first meiotic prophase. In the diplotene stage of prophase I of meiosis, the four chromatids of a bivalent are associated in pairs, but in such a way that one part of two chromatids is exchanged.

Chimera (or chimaera) From chimera, a mythological creature with the head of a lion, the body of a goat and the tail of a serpent. An organism whose cells are not all derived from the same zygote. 1. Animal. An individual exhibiting two or more genotypes in patches derived from two or more embryos. An individual derived from two embryos by experimental intervention.

2. Plant. Part of a plant with a genetically different constitution as compared with other parts of the same plant. It may result from different zygotes that grow together, or from artificial fusion (grafting); it may either be periclinal chimera, in which one tissue lies over another as a glove fits a hand; mericlinal chimera, where the outer tissue does not completely cover the inner tissue; and sectoral chimera, in which the tissues lie side by side.

3. A recombinant DNA molecule that contains sequences from different organisms.

Chimeric DNA A recombinant DNA molecule containing unrelated genes.

Chimeric gene A semi-synthetic gene, consisting of the coding sequence from one organism, fused to promoter and other sequences derived from a different gene. Most genes used in transformation are chimeric.

Chimeric selectable marker gene A gene that is constructed from parts of two or more different genes and allows the host cell to survive under conditions where it would otherwise die.

Chi-squared test (C^2 test) A significance test used to statistically assess the goodness of fit of observed data to a prediction.

Chitin A nitrogenous polysaccharide occurring as skeletal material in many invertebrates and fungi.

Chloramphenicol An antibiotic that interferes with protein synthesis.

Chlorenchyma Tissue containing chloroplasts, including leaf mesophyll and other parenchyma cells.

Chlorophyll One of the two pigments responsible for the green colour of most plants. It is essential in the absorption of light energy for photosynthesis.

Chloroplast Specialized cytoplasmic organelle that contains chlorophyll. Lens-shaped and bounded by a double membrane, chloroplasts contain membranous structures (thylakoids) piled up into stacks, surrounded by a gel-like matrix (stroma). They are the site of solar energy transfer and important reactions of starch or sugar synthesis. Chloroplasts have their own DNA and are inherited cytoplasmically, independent of nuclear genes.

Chlorosis Failure of chlorophyll development, and appearance of yellow colour in plants, because of a nutritional disturbance or because of an infection by a virus, bacteria or fungus.

Cholesterol Important lipid found only in animals. Cholesterol is important as cell membrane component, but also serves as a biosynthetic precursor for steroid hormones (e.g. sex hormones) and the active gall bladder ingredients bile acids (detergents). The human liver can synthesize all the necessary levels of cholesterol and will reduce its own synthesis if cholesterol is taken in during a meal (only from animal sources). 'Bad' and 'Good' cholesterol refers to special transport particles of lipids in our blood serum called lipoprotein particles. The low density form or LDL is high in cholesterol and chronically high concentration of LDL in blood results in insoluble deposits that can clog arteries and restrict blood flow contributing to heart problems.

Chorionic villus sampling (CVS) A small piece of tissue from part of the placenta which shared the genetic make up of the fetus is removed for prenatal diagnosis.

Chromatid Each of the two daughter strands comprising a duplicated chromosome. The term remains in use while the two chromatids are still joined at the centromere. As soon as the centromere divides, setting the two chromatids adrift (during anaphase of mitosis; and during anaphase II of meiosis), they are called chromosomes.

Chromatin Substance of which eukaryotic chromosomes are composed. It consists of primarily DNAm with some proteins (mainly histones), and small amounts of RNA. Originally named because of the readiness with which it stains with certain dyes (chromaticity).

Chromatin fibres A basic organizational unit of eukaryotic chromosomes, consisting of DNA and associated proteins assembled into strands of 30 nm average diameter.

Chromatography 1. A method for separating and identifying the components of mixtures of molecules having similar chemical and physical properties.

2. The term originally used by Mikhail Tswett (1906) to describe the separation of a mixture of leaf pigments on a calcium carbonate column.

Chromocentre Body produced by fusion of the heterochromatic regions of the chromosomes in the polytene tissues (e.g., the salivary glands) of certain *Diptera*.

Chromogenic substrate A compound or substance that contains a colour-forming group.

Chromomeres Small bodies, described by J. Belling, that are identified by their characteristic size and linear arrangement along a chromosome.

Chromonema (pl: chromonemata) An optically single thread forming an axial structure within each chromosome.

Chromoplast Plastid containing pigments, such a chloroplast, or one in which carotenoids predominate.

Chromosomal aberration Any change in chromosome structure or number. Although it can be a mechanism for enhancing genetic diversity, such alterations are usually fatal or ill-adaptive, especially in animals.

Chromosomal integration site A chromosomal location where foreign DNA can be integrated, often without impairing any essential function in the host organism.

Chromosomal polymorphism The occurrence of one to several chromosomes in two or more alternative structural forms within a population; the structurally changed chromosomes are the result of chromosome mutations (i.e., any structural change involving the gain, loss or re-location of chromosome segments).

Chromosome 1. A single DNA molecule, a tightly coiled strand of DNA, condensed into a compact structure *in vivo* by complexing with accessory histones or histone-like proteins.

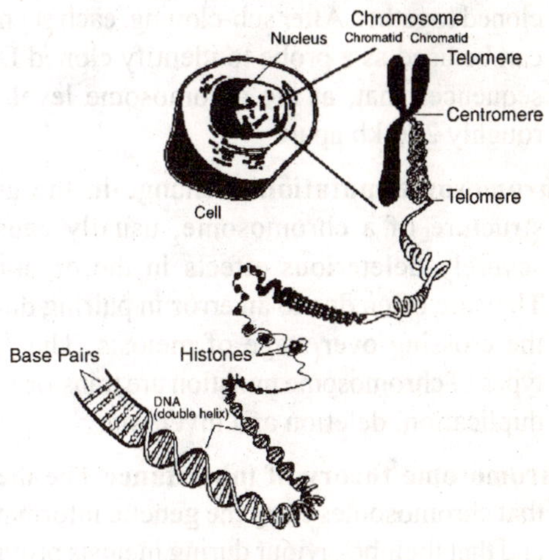

Fig. Chromosomes

2. A group of nuclear bodies containing genes which are largely responsible for the differentiation and activity of a eukaryotic cell; one of the bodies into which the nucleus resolves itself at the beginning of mitosis and from which

it is derived at the end of mitosis. Chromosomes contain most of the cell's DNA. Chromosomes exist in pairs in eukaryotes - one paternal (from the male parent) and one maternal (from the female parent). Each eukaryotic species has a characteristic number of chromosomes. Bacterial and viral cells contain only a single chromosome, consisting of a single or double strand of DNA or, in some viruses, RNA, without histones.

Chromosome aberration Abnormal structure or number of chromosomes; includes deficiency, duplication, inversion, translocation, aneuploidy, polyploidy, or any other change from the normal pattern.

Chromosome banding Staining of chromosomes in such a way that light and dark areas occur along the length of the chromosomes in repeatable patterns. Lateral comparisons identify pairs. Each chromosome can be identified by its banding pattern.

Chromosome jumping A technique that allows two segments of duplex DNA that are separated by thousands of base pairs (about 200 kb) to be cloned together. After sub-cloning, each segment can be used as a probe to identify cloned DNA sequences that, at the chromosome level, are roughly 200 kb apart.

Chromosome mutation A change in the gross structure of a chromosome, usually causing severely deleterious effects in the organism. They are often due to an error in pairing during the crossing-over stage of meiosis. The main types of chromosome mutation are translocation, duplication, deletion and inversion.

Chromosome theory of inheritance The theory that chromosomes carry the genetic information and that their behaviour during meiosis provides the physical basis for segregation and independent assortment.

Chromosome walking A technique that identifies overlapping cloned DNA fragments that form one continuous segment of a chromosome. These fragments can be generated either by random shearing or by partial digestion with a four-base-pair cutter such as *Sau*3A. A series of colony hybridizations is then carried out, starting with some cloned fragment which has already been identified and which is known to be in the region encompassed by the overlapping clones. This identified fragment is used as a probe to pick out clones containing adjacent sequences. These are then used as probes themselves to identify clones carrying sequences adjacent to them and so on. At each round of hybridization one "walks" further along the chromosome from the initial fragment.

Chronic obstructive pulmonary disease General term that refers to two diseases of the lung - chronic bronchitis and emphysema. Also known as chronic obstructive lung disease.

Chylomicrons Large particles composed mainly of triglycerides stabilised with a protein and phospholipid surface, which carry fats recently absorbed from a meal in the bloodstream.

Chyme Digested content of the stomach released for further digestion in the small intestine.

Chymosin One of the enzymes which causes milk to clot and turn into curds and whey during cheese making.

Cilium Hairlike locomotor structure on certain cells; a locomotor structure on a ciliate protozoan.

Circadian Of physiological activity, etc.: occurring or recurring about once a day.

Circularization A DNA fragment generated by digestion with a single restriction endonuclease will have complementary 5′ and 3′ extensions (sticky ends). If these ends are annealed and ligated, the DNA fragment will have been converted to a covalently-closed circle, or circularized.

Cirrhosis A chronic degenerative disease of the liver. Fat infiltration occurs, the cells degenerate, the liver architecture is destroyed by fibrous tissue, blood supply is reduced and liver function deteriorates.

Cis acting sequence A nucleotide sequence that only affects the expression of genes located on the same chromosome.

Cis fatty acid The form of most naturally occurring unsaturated fatty acids, where the hydrogen bonds are on adjacent sides of double bonds, resulting in a bend in the hydrocarbon chain at that point.

Cis heterozygote A heterozygote that contains two mutations arranged in a cis configuration (e.g., *a+ b+ / a b)*.

Cistron A DNA sequence that codes for a specific polypeptide; a gene.

Claims The section of a patent that states, in detail, the uses and possible applications of the invention described in the patent.

Class switching The process during which a plasma cell stops producing antibodies of one class and begins producing antibodies of another class.

Classical mechanics Classical mechanics describes a mechanical system as a set of particles (which in a limiting case can form continuous media) having a well-defined geometry at any given time, and undergoing motions determined by applied forces and by the initial positions and velocities of the particles. The forces themselves may have electromagnetic or quantum mechanical origins. Classical statistical mechanics uses the same physical model, but treats the geometry and velocities as uncertain, statistical quantities subject to random thermally-induced fluctuations. Classical mechanics and classical statistical mechanics give a good account of many mechanical properties and behaviors of molecules; but for describing the electronic properties and behaviors of molecules, they are often useless.

Cleave To break phospho-diester bonds of double-stranded DNA, usually with a type II restriction endonuclease.

Clinical studies Human studies that are designed to measure the efficacy of a new drug or biologic. Clinical studies routinely involve the use of a control group of patients that is given an inactive substance (placebo) that looks like the test product.

Clinical trials Study involving human participants that is intended to discover or verify the effects of a drug for use in humans, to identify any adverse events in respect of a drug or to study the absorption, distribution, metabolism and excretion of a drug, with the objective of ascertaining its safety or efficacy.

Clonal propagation Asexual propagation of many new plants (ramets) from an individual (ortet); all have the same genotype.

Clonal selection The production of a population of plasma cells all producing the same antibody in response to the interaction between a B lymphocyte producing that specific antibody and the antigen bound by that antibody.

Clone 1. A group of cells or organisms that are genetically identical as a result of asexual reproduction, breeding of completely inbred organisms, or forming genetically identical organisms by nuclear transplantation.

2. Group of plants genetically identical in which all are derived from one selected individual by vegetative propagation, without the sexual process.

3. A population of cells that all carry a cloning vehicle with the same insert DNA molecule.

4 To clone. To insert a DNA segment into a vector or host chromosome.

5. genetic replica of another organism obtained through a non-sexual (no fertilization) reproduction process.

Cloning by nucleus transfer involves the transfer of a donor cell (from (cultured) cells of embryonic, foetal or adult origin) into the recipient cytoplasm of an enucleated oocyte or zygote, and the subsequent development of embryos and animals.

Cloned strain or line A strain or line descended directly from a clone.

Cloning vector A small, self-replicating DNA molecule - usually a plasmid or viral DNA chromosome - into which foreign DNA is inserted in the process of cloning genes or other DNA sequences of interest. It can carry inserted DNA and be perpetuated in a host cell. Also called a cloning vehicle, vector, or vehicle.

Closed continuous culture A continuous culture in which inflow of fresh medium is balanced by outflow of corresponding volumes of spent medium. Cells are separated mechanically from outflowing medium and added back to the culture.

CMOS An acronym for *complementary metal-oxide-semiconductor*, as in CMOS transistor and CMOS logic.

Co carcinogen A chemical substance that enhances the action of a carcinogen but does not itself initiate cancer.

Co culture The joint culture of two or more types of cells, such as a plant cell and a micro-organism, or two types of plant cells. Used in various dual-culture systems or in nurse-culture.

Co dominance The situation in which both alleles in a heterozygous individual are expressed, so that the phenotype of heterozygotes incorporates the phenotypic effect of each allele. For example, roan coat colour in cattle results from a mixture of red hairs and white hairs, caused by heterozygosity for the red allele and the white allele. Also, protein polymorphisms and microsatellites show co-dominance: heterozygotes have two bands, whereas homozygotes have only one band.

Co dominant alleles Alleles that produce independent effects when in the heterozygous condition.

Co enzymes or co factors Non-protein substances necessary for the function of some enzymes. Essential coenzymes are also known as vitamins.

Co evolution The evolution of complementary adaptations in two species caused by the selection pressure that each exerts on the other. It is common in symbiotic associations, in insect-pollinated plants, etc.

Co fermentation The simultaneous growth of two micro-organisms in one bioreaction.

Co generation Production of both electricity and process heat (steam) in an industrial plant.

Co integrate A DNA molecule formed by the fusion of two different DNA molecules, usually mediated by a transposable element.

Co integrate vector system A two-plasmid system for transferring cloned genes to plant cells. The cloning vector has a T-DNA segment that contains cloned genes. After introduction into *Agrobacterium tumiefasciens*, the cloning vector DNA undergoes homologous recombination with a resident disarmed Ti plasmid to form a single plasmid carrying the genetic information for transferring the genetically engineered T-DNA region to plant cells.

Co linearity A relationship in which the units in one molecule occur in the same sequence as the units in another molecule which they specify; e.g., the nucleotides in a gene are co-linear with the amino acids in the polypeptide encoded by that gene.

Co metabolism A microbe oxidizing not only its main energy source but also another organic compound.

Co polymers Mixtures consisting of more than one monomer; for example, polymers of two kinds of organic bases, such as uracil and cytosine (poly-UC) have been combined for studies of the genetic code.

Co repressor An effector molecule that forms a complex with a repressor and turns off the expression of a gene or set of genes.

Co suppression A naturalgene silencing phenomenon, which probably evolved as part of plants' defence against viral attack, but which has become important in the context of plant transgenesis. Operates by inhibiting the expression of transgenes with homology to native DNA through the interaction of native and transgenic mRNA.

Co transformation In genetic engineering experiments, it is often necessary to transform with a plasmid for which there is no selectable phenotype and then screen for the presence of that plasmid within the host cell. Co-transformation is a technique in which host cells are incubated with two types of plasmid, one of which is selectable and the other not. Cells which have been transformed with the first plasmid are then selected. If transformation has been carried out at high DNA concentration, then it is probable that these cells will also have been transformed with the second (non-selectable) plasmid. The technique is frequently used in experiments with mammalian cells.

Co variance A measure of the statistical association between variables; the extent to which two variables vary together.

Coat protein (capsid) The coating protein that encloses the nucleic acid core of a virus.

Coccus (pl: cocci) A spherical bacterium. Cocci may occur singly, in pairs, in groups of four or more, and in cubical packets.

Coconut milk Liquid endosperm of the coconut, often used to supply organic nutrients to cultured cells and tissues.

Cocoon A protective coverage for eggs and/or larvae produced by many invertebrates, such as the silkworm moth.

Coding The specification of a peptide sequence, by the code contained in DNA molecules.

Coding sequence That portion of a gene which directly specifies the amino acid sequence of its protein product. Non-coding sequences of genes include control regions, such as promoters, operators and terminators, as well as the intron sequences of certain eukaryotic genes.

Coding strand The strand of aDNA double helix that contains the same base sequence (after substituting U for T) found in the mRNA molecule resulting from transcription of that segment of DNA. Sometimes called the sense strand. The mRNA molecule is transcribed from the other strand, known as the template or antisense strand.

Codon A set of three nucleotides in mRNA, functioning as a unit of genetic coding by specifying a particular amino acid during the synthesis of polypeptides in a cell. A codon specifies a transfer RNA carrying a specific amino acid, which is incorporated into a polypeptide chain during protein synthesis. The specificity for translating genetic information from DNA into mRNA, then to protein, is provided by codon-anticodon pairing.

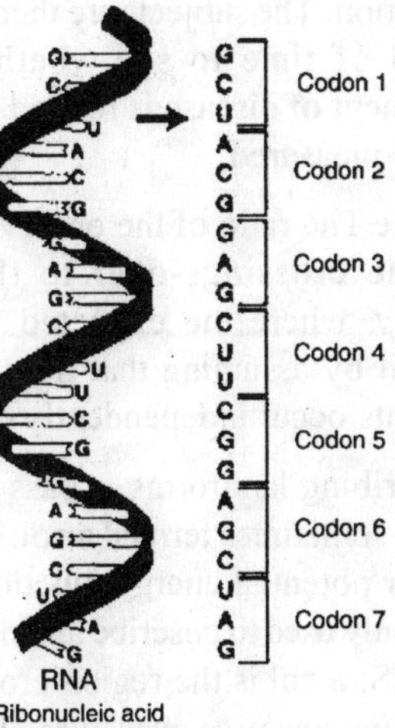

Fig. Codon

Codon optimization An experimental strategy in which codons within a cloned gene - ones not generally used by the host cell translation system - are changed by *in vitro* mutagenesis to the preferred codons, without changing the amino acids of the synthesized protein.

Coefficient A number expressing the amount of some change or effect under certain conditions (e.g., the coefficient of inbreeding).

Coenzyme (cofactor) An organic molecule, such as a vitamin, that binds to an enzyme and is required for its catalytic activity.

Cognitive ability Knowledge, perception.

Cognitive, cognition Cognition refers to the physical recognition of external stimuli that lead to conscious experiences.

Cohesion Holding together; a force holding a solid to a solid or a solid to a liquid, owing to attraction between like molecules.

Cohesive ends Double-stranded DNA molecules with single-stranded ends which are complementary to each other, enabling the different molecules to join each other.

Cohort or follow up study A study design in which data on exposures to possible risk factors for disease are collected from a group of people (cohort) who do not have the disease under investigation. The subjects are then followed for a period of time to see whether the later development of disease is related to the factors that were measured.

Coincidence The ratio of the observed frequency of double crossings-over to the expected frequency, where the expected frequency is calculated by assuming that the two crossing-over events occur independently of each other.

Col In describing landforms, a pass between two valleys is sometimes termed a col. In describing molecular potential energy functions, this term is commonly used to describe analogous features of the PES; a col is the region around a saddle point having negative curvature along one axis and positive curvature along all orthogonal axes.

Colchicine An alkaloid obtained from *Colchicum autumnale*, autumn crocus or meadow saffron, which inhibits spindle formation in cells during mitosis, so that chromosomes cannot separate during anaphase, thus inducing multiple sets of chromosomes. Also used to halt mitosis at metaphase - the stage when chromosomes are most visible.

Cold hardy Ability to grow in cold temperatures or resistant to cold weather damage.

Coleoptile Protective sheath covering the shoot apex of the embryo in monocotyledenous plants. It protects the plumule as it emerges through the soil.

Coleorhiza A protective sheath surrounding the radicle of monocotyledenous plants.

Collagen A fibrous protein found in skin, tendons, ligaments, cartilage, and bones.

Collenchyma A tissue of living cells, the walls being unevenly thickened with cellulose and hemicellulose, but never lignified; it functions in mechanical support in young, short-lived or non-woody organs and is thus found in midribs and leaf petioles.

Fig. Collenchyma

Colony 1. An aggregate of identical cells (clones) derived from a single progenitor cell.

2. A group of interdependent cells or organisms.

Colony hybridization A technique that uses a nucleic acid probe to identify a bacterial colony with a vector carrying a specific cloned gene or genes.

Colony stimulating factors (CSFs) A group of lymphokines that induce the maturation and proliferation of white blood cells from the primitive cell types present in bone marrow.

Combinatorial chemistry A product discovery technique that uses robotics and parallel synthesis to generate and screen quickly as many as several million molecules with similar structure in order to find chemical molecules with desired properties.

Combinatorial library During the ligation reaction with cDNAs of light and heavy antibody chains into a bacteriophage lambda (1) vector, many novel combinations consisting of one heavy and one light chain coding region are formed. The library comprises these combinations, each in a separate vector.

Commensalism The close association of two or more dissimilar organisms where the association is advantageous to one and doesn't affect the other(s).

Companion cell Living cell associated with the sieve cell of phloem tissue in vascular plants.

Fig. Companion cell

Comparative gene mapping The comparison of map locations of genes between species. The results of these comparisons indicate substantial conservation of blocks of genes and even large segments of chromosomes between species. Great use can be made of this conservation of map position. For example, in the case of mammals, it means that if a gene has been mapped in one or both of the intensely-mapped species (humans and mice), then the likely location of that gene in other mammals can be predicted with considerable confidence. Conversely, if a mapped anonymous DNA marker has an effect on a quantitative trait (this being indicative of the marker being linked to a quantitative trait locus (QTL)) in, say, cattle, then knowledge of the comparative map between cattle and humans can identify genes in the homologous region of the human genome that could correspond to the QTL. Such genes are called comparative positional candidate genes.

Comparative positional candidate gene A gene that is likely to be located in the same region as a DNA marker that has been shown to be linked to a single-locus trait or to a quantitative trait locus (QTL), where the gene's likely location in the genome of the species in question is based on its known location in the map of another species, i.e., is based on the comparative map between the two species.

Competency An ephemeral state, induced by treatment with cold cations, during which bacterial cells are capable of uptaking foreign DNA.

Competent A competent cell is capable of developing into a fully functional embryo. The opposite is non-competent.

Complement Blood proteins that amplify the action of antibodies.

Complement proteins Proteins that bind to antibody-antigen complexes and help degrade the complexes by proteolysis.

Complementarity The relationship of the nucleotide bases on two different strands of DNA or RNA. When the bases are paired properly (adenine with thymine [DNA] or uracil [RNA]; guanine with cytosine), the strands are complementary.

Complementary entity 1. One of a pair of nucleotide bases that form hydrogen bonds with each other. Adenine (A) pairs with thymine (T) [or with Uracil (U) in RNA], and guanine (G) pairs with cytosine (C).

2. One of a pair of segments or strands of nucleic acid that will hydridize (join by hydrogen bonding) with each other.

Complementary genes Two or more interdependent genes, such that (in the case of dominant complementarity) the dominant allele from either gene can only produce an effect on the phenotype of an organism if the dominant allele from the other gene is also present; or (in the case of recessive complementarity) only double homozygous recessive show the effect.

Complementary homopolymeric tailing The process of adding complementary nucleotide extensions to different DNA molecules, e.g., dG (deoxyguanosine) to the 3′-hydroxyl ends of one DNA molecule and dC (deoxycytidine) to the 3′-hydroxyl ends of another DNA molecule to facilitate, after mixing, the joining of the two DNA molecules by base pairing between the complementary extensions. Also called dG - dC tailing, dA - dT tailing.

Complementary metal oxide semiconductor As used in CMOS transistors and CMOS logic.

Complementary nucleotides Members of the pairs adenine-thymine, adenine-uracil, and guaninecytosine that have the ability to hydrogen bond to one another.

Complementary strands of RNA/DNA Complementary RNA (called nuclear RNA) is made when cells are making proteins from genes. It is made from the single-stranded DNA code and eventually changed into messenger RNA (mRNA). The mRNA carries its coded message to another part of the cell where it is translated into proteins. Complementary DNA (cDNA) is an artificial DNA made in the lab from mRNA strands, and is often used as probes.

Complementation test; trans test Introduction of two mutant chromosomes into the same cell to determine whether the mutants are alleles of the same gene. If the mutations are non-allelic, the genotype will be *m1* +/+ *M2* , and the phenotype will be wild-type (normal), because each chromosome "covers" for the other. In contrast, if they are allelic, the mutant phenotype will result.

Complete digest The treatment of a DNA preparation with an endonuclease for sufficient time for all of the potential target sites within that DNA to have been cleaved.

Complexity Complexity is the measure of the number and strength of interactions of its components. The components are organized not in a linear chain, but a network with specific connectivity, branches and loops. Network components affect each other through their interactions (molecular interactions). Complexity in biology is the result of dynamic interactions that follow each other in time and with both forward and feedback loops. Because of loops, the output of a network will affect a future input, thus continuously adjusting the physical output value of the system. Biological networks have evolved as stable systems. Stability means that biological systems are in homeostatic equilibrium, with information constantly flowing through the system and the output kept within a narrow range.

Compliance The reciprocal of stiffness; in a linear elastic system, displacement equals force times compliance.

Composite transposon A transposable element formed when two identical or nearly identical transposons insert on either side of a non-transposable segment of DNA, such as the bacterial transposon Tn5.

Compound chromosome A chromosome formed by the union of two separate chromosomes, as in attached-X chromosomes or attached-X-Y chromosomes.

Computational biology A subdiscipline within bioinformatics concerned with computation-based research devoted to understanding basic biological processes.

Concatemer A DNA segment composed of repeated sequences linked end to end.

Concordance Identity of matched pairs or groups for a given trait, such as sibs expressing the same trait.

Conditional lethal mutation A mutation that is lethal under one set of environmental conditions - the restrictive conditions - but is viable under another set of environmental conditions - the permissive conditions, e.g., temperature-sensitive mutations.

Conditioning 1. The effects on phenotypic characters of external agents during critical developmental stages.

2. The undefined interaction between tissues and culture medium resulting in the growth of single cells or small aggregates. Conditioning may be accomplished by immersing cells or callus contained within a porous material (such as dialysis tubing) into fresh medium for a period dependent on cell density and a volume related to the amount of fresh medium.

Conductance Conductance refers to the flow of ions (charges molecules and elements and measured in 'Siemens') such as sodium, potassium, and chloride and is the proportionality factor relating current to a voltage difference. In electricity theory conductance is the inverse of the resistance as defined by Ohm's law V = RI, where V is the voltage, R the resistance, and I the current).

Confidence interval The range of values within which a variable is likely to lie.

Configuration space A mathematical space describing the three-dimensional configuration of a system of particles (e.g., atoms in a nanomechanical structure) as a single point; the configuration space for an *N* particle system has 3*N* dimensions.

Confined field trials The release of novel organisms under terms and conditions of confinement such as reproductive isolation, site-monitoring and restrictions on use of the post experimental environment.

Conformation A molecular geometry that differs from other geometries chiefly by rotation about single or triple bonds; distinct conformations (termed con formers) are associated with distinct potential wells. Typical biomolecules and products of organic synthesis can interconvert among many conformations; typical diamondoid structures are locked into a single potential well, and thus lack conformational flexibility.

Congeneric Belonging to the same generation.

Congenital Present at birth.

Congenital hypothyroidism An inherited trait that results in reduced activity of the thyroid gland, generally due to reduced production of thyroid stimulating hormone. The trait results in a reduced base rate of the body's chemical reactions, tissue swelling and weight gain. It can cause neurological and development problems in affected children.

Conidium (pl: conidia) An asexual spore produced by a specialized hypha in certain fungi.

Conjugation 1. Union of gametes or unicellular organisms during fertilization.

2. The unidirectional transfer of plasmid DNA from one bacterium cell to another, involving cell-to-cell contact. The plasmid usually encodes the majority of the functions necessary for its own transfer.

3. Attachment of sugar and other polar molecules to less polar compounds, thus making them more water soluble.

Conjugative functions Plasmid-based genes and their products that facilitate the transfer of a plasmid from one bacterium to another.

Consanguinity Related by descent from a common ancestor.

Consensus sequence The nucleotide sequence that is present in the majority of genetic signals or elements that perform a specific function.

Conservation of farm animal genetic resources Refers to all human activities, including strategies, plans, policies and actions, undertaken to ensure that the diversity of farm animal genetic resources is being maintained to contribute to food and agricultural production and productivity; now and in the future.

Conservative In design and analysis, a conservative model or a conservative assumption is one that departs from accuracy in such a way that it reduces the chances of a false-positive assessment of the feasibility of the system in question. Conservative assumptions overestimate problems and underestimate capabilities.

Conspecific Belonging to the same species.

Constant domains Regions of antibody chains that have the same amino acid sequence in different members of a particular class of antibody molecules.

Constitutive enzyme An enzyme that is synthesized continually regardless of growth conditions.

Constitutive gene A gene that is continually expressed in all cells of an organism.

Constitutive promoter An unregulated promoter that allows for continual transcription of its associated gene.

Constitutive synthesis Continual production of RNA or protein by an organism.

Contaminant Bacterial, fungal or algal micro-organism accidentally introduced into a culture or culture medium. Contaminant may overgrow the plant cells and consequently inhibit their growth. Working under aseptic conditions with a rigorous exclusion of potential contaminants must be practised in plant tissue culture.

Contig A set of overlapping clones that provide a physical map of a portion of a chromosome.

Contiguous map; contig map The alignment of sequence data from large, adjacent regions of the genome to produce a continuous nucleotide sequence across a chromosomal region.

Continuous cropping The growing of two or more crops, one after the other, on the same land in one growing season. Also known as succession cropping.

Continuous culture A suspension culture continuously supplied with nutrients by the inflow of fresh medium. The culture volume is normally constant.

Continuous fermentation A process in which cells or micro-organisms are maintained in culture in the exponential growth phase by the continuous addition of fresh medium that is exactly balanced by the removal of cell suspension from the bioreactor.

Continuous variation Variation not represented by distinct classes. Phenotypes grade into each other, and measurement data are required for analysis. Multiple genes are usually responsible for this type of variation.

Control 1. Unchanged (standard) protocol or treatment for comparison with the experimental treatment. The term is commonly used for untreated organisms.

2. To direct or regulate cultures with addition of plant growth regulators.

Controlled environment A closed environment in which parameters, such as light, temperature, relative humidity and sometimes the partial gas pressure (and possibly its composition), are fully controlled.

Controlling element In eukaryotes, transposable elements which control the activity of standard genes. A controlling element may, in the simplest case, inhibit the activity of a gene through becoming integrated in, or close to, that gene. Occasionally, either in germinal or somatic tissue, it may be excised from this site, and due to excision the activity of the gene is more or less restored, while the element may become reintegrated elsewhere in the genome where it may affect the activity of another gene. For example, in maize, a controlling element such as *Ac* or *Ds* is capable of influencing the expression of a nearby gene.

Convention on biological diversity An international agreement (under the united nations environment program (UNEP). Its objectives are "the conservation of biological diversity, the sustainable use of its components and the fair and equitable sharing of the benefits arising out of the utilisation of genetic resources". Under one part of the Convention, the Biosafety Protocol, countries are currently discussing procedures to deal with the concerns about the potential risks to biodiversity and human health posed by living modified organisms.

Conversion The development of a somatic embryo into a plant.

Coordinate repression Correlated regulation of the structural genes in an operon by a molecule that interacts with the operator sequence.

Copy gene Genetic material incorporating the genetic code for a desirable trait which has been copied from DNA of the donor to the host organism. (It is not technically possible to take a gene from a donor organism and insert it directly into the host organism.)

Copy number The average number of molecules of a plasmid or gene per genome contained in a cell.

Coronary heart disease (CHD) A condition in which the main arteries supplying the heart are blocked and no longer able to supply sufficient blood, and therefore oxygen, to the heart muscle (myocardium), which may then quickly die. The main cause of reduced blood flow is the accumulation of plaques, a disease known as atherosclerosis.

Coronavirus A single-stranded RNA virus that resembles a crown when viewed under an electron microscope because of its petal-shaped projections. Of the more than 30 isolated strains of coronavirus, three or four infect humans and may cause respiratory diseases such as SARS and gastroenteritis. They are believed to cause a large percentage of all common cold cases in humans.

Corpus The corpus is found below the tunica and is a part of the apical meristem. In the corpus, cells divide in all directions, giving them an increase in volume.

Correlation A statistical association between variables.

Cortex Primary tissue of a stem or root, bounded externally by the epidermis and internally in the stem by the phloem and in the root by the pericycle.

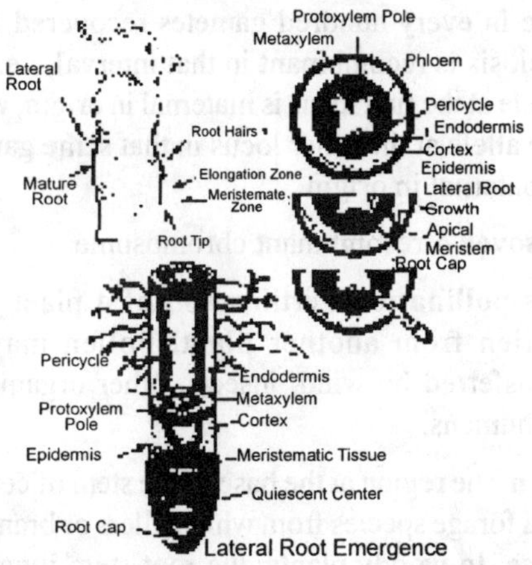

Fig. Cortex

Cos ends The 12-base, single-strand, complementary extensions of bacteriophage lambda (l) DNA.

Cosmid A plasmid vector which contains the two cos (cohesive) ends of phage lambda *(l)* and one or more selectable markers such as an antibiotic resistance gene. Cosmids exploit certain properties of phage lambda *(l)* to enable large, 40-50 kb, DNA fragments to be cloned at high efficiency. Cosmids and cosmid recombinants replicate as plasmids.

Cot curve When duplex DNA is heated, it dissociates into single strands. When the temperature is lowered, complementary strands tend to anneal or re-nature. The extent of re-naturation depends on the product of DNA concentration in moles of nucleotides per litre, and time in seconds. A graph showing the proportion of re-natured DNA against cot is known as a cot curve. The cot at which half the DNA has re-natured is the half-cot, a parameter indicating the degree of complexity of the DNA.

Cotyledons Leaflike structures at the first node of the seedling stem. In some dicotyledons, they contain stored food for the young plant not yet able to photosynthesize its own food. Often referred to as seed leaves.

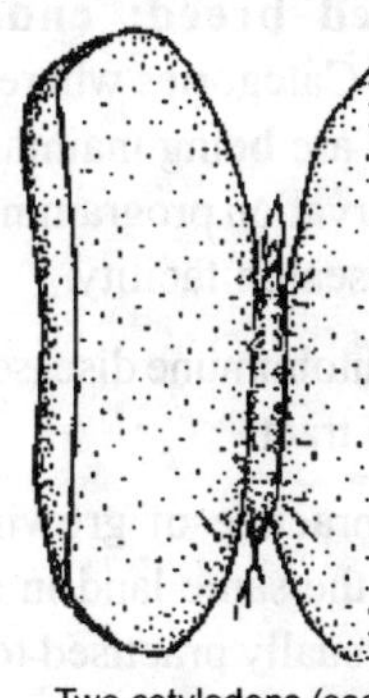

Two cotyledons (seed leaves) of a dicotyledon

Fig. Cotyledon

Coupling The phase state in which either two dominant or two recessive alleles of two different genes occur on the same chromosome.

Covalent The chemical bond between atoms.

Covalent bond A bond in which an electron pair is equally shared by protons in two adjacent atoms.

Covalent radius Given a set of *N* elements that can form covalent single bonds in molecules, with $N(N-1)$ possible elemental pairings, it has proved possible to define a covalent radius for each element such that the actual bond length between any two elements that form a covalent single bond is roughly equal to the sums of their covalent radii.

Covalently closed circle (CCC) A double-stranded DNA molecule with no free ends. The two strands are interlinked and will remain together even after denaturation. In its native form, a CCC will adopt a supercoiled configuration.

CPDNA The DNA of plant plastids, including chloroplasts.

Critical breed A breed where the total number of breeding females is less than 100 or the total number of breeding males is less than or equal to five; or the overall population size is close to, but slightly above 100 and decreasing, and the percentage of pure-bred females is below 80%.

Critical-maintained breed; endangered-maintained breed Categories where critical or endangered breeds are being maintained by an active public conservation programme or within a commercial or research facility.

Crohn's disease An autoimmune disease involving the gastrointestinal tract.

Crop rotation The practice of growing several different crops on the same land in successive years or seasons. Usually practised to replenish soil, and curb pests and diseases.

Cross In genetic studies, the mating of two individuals or populations. Also called mating.

Cross-breeding Mating between members of different populations (lines, breeds, races or species).

Cross-compatible The ability of two related organisms to exchange genes through sexual reproduction.

Cross hybridization The hydrogen bonding of a single- stranded DNA sequence that is partially but not entirely complementary to a singlestranded substrate. Often, this involves hybridizing a DNA probe for a specific DNA sequence to the homologous sequences of different species.

Cross pollination Application of pollen from one plant to another to effect the latter's fertilization.

Cross pollination efficiency Efficiency of pollen from one plant reaching the stigma of another plant.

Crossing over A process in which homologous chromosomes exchange material through the breakage and reunion of two chromatids. A single crossover represents one reciprocal breakage and reunion event. A double crossover requires two simultaneous reciprocal breakage and reunion events.

Crossing-over unit A measure of distance between two loci on genetic maps that is based on the average number of crossing-over events that take place in the interval between those two loci during meiosis. A map interval that is one crossing-over unit in length (a centiMorgan) describes an interval between two loci such that one in every hundred gametes recovered from meiosis is recombinant in that interval, i.e., the allele at the first locus is maternal in origin, while the allele at the other locus in that same gamete is paternal in origin.

Crossover A recombinant chromosome.

Cross-pollination Fertilization of a plant with pollen from another plant. Pollen may be transferred by wind, insects, other organisms, or humans.

Crown The region at the base of the stem of cereals and forage species from which tillers or branches arise. In woody plants, the root-stem junction. In forestry, the top portions of the tree.

Crown gall (A.S. *gealla*, gall) A bulbous growth that occurs at the base of certain plants as result of infection, especially by *Agrobacterium tumefaciens*; a bacterial gene carried by the Ti plasmid is transferred by the bacteria into a higher plant cell, where it causes a tumour-like growth.

Cry Designation of a gene encoding insecticidal crystal proteins in the soil bacterium Bacillus thuringiensis.

Cry proteins A class of crystalline proteins produced by strains of Bacillus thuringiensis, and engineered into crop plants to give resistance against insect pests. These proteins are toxic to certain categories of insects (e.g. corn borers, corn rootworms, mosquitoes, black flies, armyworms, tobacco hornworms, some types of beetles, etc.), but are harmless to mammals and most beneficial insects.

Cryobiological preservation; cryopreservation; freeze preservation. The preservation of germplasm resources in a dormant state by cryogenic techniques, as currently applied to storage of plant seeds and pollen, micro-organisms, animal sperm, and tissue culture cell lines.

Cryogenic At very low temperature.

Cryoprotectant Compound preventing cell damage during freezing and thawing processes. Cryoprotectants are agents with high water solubility and low toxicity. Two types of cryoprotectant agent are commonly used: permeating (glycerol and DMSO and non-permeating (sugars, dextran, ethylene glycol, polyvinyl pyrolidone and hydroxyethyl starch).

Cryptic 1. Structurally heterozygous individuals not identifiable on the basis of abnormal meiotic-chromosome pairing configurations ('cryptic structural hybrids').

2. A form of polymorphism controlled by recessive genes ('cryptic polymorphism').

3. Any mutation which is exposed by a sensitizing mutation and otherwise poorly detected (such mutations probably escape detection because of the plasticity of composition of the corresponding polypeptide).

4. Phenotypically very similar species (cryptic species) which do not hybridize under normal conditions.

5. Cryptic genetic variation refers to the existence of, for example, alleles conferring high performance for a trait, in a breed that has low performance for that trait.

Crystal In biology refers to large, regular assembly of macromolecules such as proteins and nucleic acids. It is possible to isolate and purify proteins or DNA in large quantity and let them crystallize in saturated solution. These protein or DNA crystals contain millions of regularly aligned units that allow the determination of the electron distribution from X-ray diffraction studies. Since atoms are distinguished by their specific numbers of electrons, their distribution allows an analysis of the atomic composition of proteins or DNA. This electron distribution is then used to calculate the so called high resolution structure of molecules.

CSIRO Acronym for the commonwealth scientific and industrial research organisation. This is a government-funded organisation that carries out research in science, for the benefit of the community and industry.

Cultivar A category of plants that are, firstly, below the level of a sub-species taxonomically, and, secondly, found only in cultivation. It is an international term denoting certain cultivated plants that are clearly distinguishable from others by stated characteristics and that retain their distinguishing characters when reproduced under specific conditions.

Culture A population of plant or animal cells or micro-organisms that is grown under controlled conditions.

Culture alteration A term used to indicate a persistent change in the properties of a culture's behaviour (e.g., altered morphology, chromosome constitution, virus susceptibility, nutritional requirements, proliferative capacity, etc.). The term should always be qualified by a precise description of the change which has occurred in the culture.

Culture medium Any nutrient system for the cultivation of cells of plants, bacteria or other organisms; usually a complex mixture of organic and inorganic nutrients.

Culture room Room for maintaining cultures and often in a controlled environment of light, temperature and humidity.

Curing The elimination of a plasmid from its host cell. Many agents which interfere with DNA replication, e.g., ethidium bromide, can cure plasmids from either bacterial or eukaryotic cells.

Cut Slang: to make a double-stranded break in DNA, usually with a type II restriction endonuclease. E.g., "The DNA was cut with *Eco*RI and run out on a 1% agarose gel."

Cuticle Layer of cutin or wax, formed on the outer surface of leaves and fruits, thought to have evolved to reduce evaporative water loss.

Cybrid A cytoplasmic hybrid, originating from the fusion of a cytoplast (the cytoplasm without nucleus) with a whole cell, as in nuclear transfer (although the term is not used in that context). Note that the nucleus and cytoplasm of the fused cell products are from different genetic sources.

Cyclic A structure is termed cyclic if its covalent bonds form one or more rings.

Cyclic adenosine monophosphate (cAMP) A "messenger" molecule that regulates many intracellular reactions by transducing signals from extracellular growth factors to cellular metabolic pathways.

Cycloaddition A reaction in which two unsaturated molecules (or moieties within a molecule) join, forming a ring.

Cyclodextrin Cyclic polymer of dextrose.

Cystein An amino acid.

Cystic fibrosis In the UK the most common single genetic disability affecting about 1 in 2000 children. The genetic defect results in the production of abnormally thick mucus. The mucus obstructs the airways of the lung and leads to recurrent lung damage, and it obstructs glands of the intestine and leads to impaired digestion and absorption of food. The disorder requires lifelong treatment to combat infection and to enhance nutrition. Its genetic inheritance is recessive.

Cyto Referring to cell.

Cytochrome A class of pigments in plant and animal cells, usually in the mitochondria. They function as electron carriers in respiration.

Cytochrome p450 A highly diversified set (more than 1500 known sequences) of heme-containingproteins. Frequently called hydroxylases, although P450 proteins can perform a wide spectrum of other reactions. In bacteria they are soluble and approximately 400 amino acids long; eukaryotic P450s are larger - about 500 amino acids. In mammals they are critical for drug metabolism, haemostasis, cholesterol biosynthesis and steroidogenesis; in plants they are involved in plant hormone synthesis, phytoalexin synthesis, flower petal pigment biosynthesis and many unknown functions. In fungi they make ergosterol and they are involved in pathogenesis. Bacterial P450s are key elements in antibiotic synthesis.

Cytogenetic Area of biology concerned with chromosomes and their implications in genetics, cellular activity and variability.

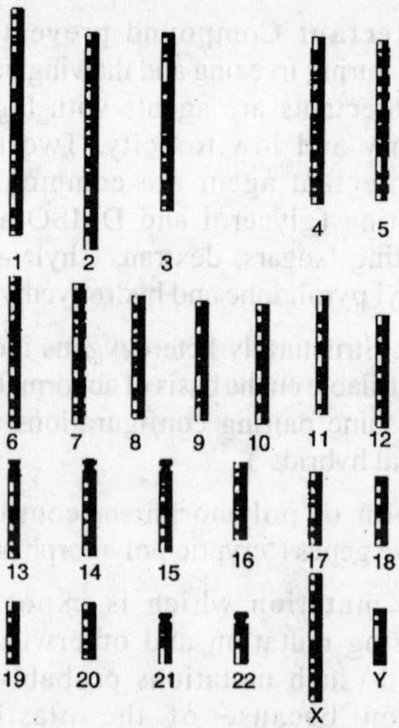

Fig. Cytogenetic map

Cytokine A generic name for a diverse group of soluble proteins and peptides which act as humoral regulators at extremely small concentrations and which, either under normal or pathological conditions, modulate the functional activities of individual cells and tissues.

Cytokinin Plant growth regulators (hormones) characterized as substances that induce cell division and cell differentiation (e.g., BA, kinetin, and 2-iP). In tissue culture, these substances are associated with enhanced callus and shoot development. The compounds are derivatives of adenine.

Cytology The study of the structure and function of cells.

Cytolysis Cell disintegration.

Cytoplasm The living material of the cell, exclusive of the nucleus, consisting of a complex protein matrix or gel. The part of the cell in which essential membranes and cellular organelles (mitochondria, plastids, etc.) reside.

Cytoplasmic genes DNA-containing bodies in the cell but external to the nucleus.

Cytoplasmic inheritance Hereditary transmission dependent on the cytoplasm or structures in the cytoplasm rather than the nuclear genes; extrachromosomal inheritance. Thus, plastid characteristics in plants are inherited by a mechanism independent of nuclear genes.

Cytoplasmic male sterility Genetic defect due to defective functions of mitochondria in the pollen. Fertilization will not occur. Exploited in certain plant breeding strategies, such as F1-hybrid maize cultivars.

Cytoplasmic organelles Discrete sub-cellular structures located in the cytoplasm of cells; these allow division of labour within the cell.

Cytosine (C) A chemical that is a basic structural unit of a living thing's DNA. It is one of 4 key bases or nucleotides (adenine (A), guanine (G) and thymine (T)) that repeat throughout the length of a DNA chain. The order of their sequence defines the arrangement of genes.

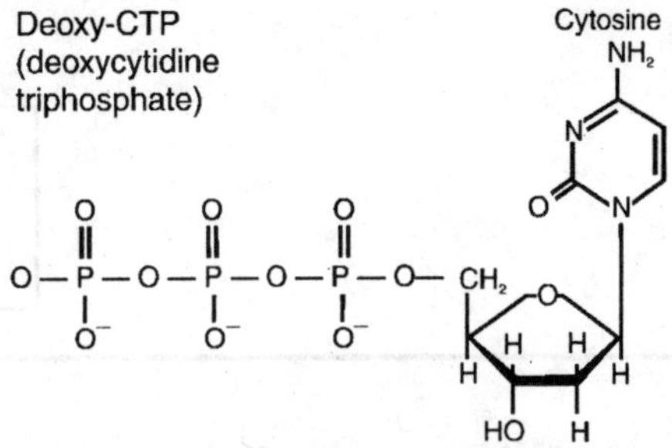

Fig. Cytosine (C)

Cytoskeleton A fibrous network made of proteins that contributes to the structure and internal organization of eukaryotic cells. The cytoskeleton is found in the cytoplasm of cells and has three major fiber types: microfilaments made of actin protein, intermediate filaments made of various kinds of proteins (e.g. keratin), and microtubules made of tubulin. The three fiber types serve different functions; actin filaments are often associated to changes in cell size and structure, contractility such as in muscle cells, and cell division, growth and motility. Microtubules are major filaments for internal transport and movement of chromosomes and organelles during cell division. Intermediate filaments contribute to flexibility, elasticity, and stiffness of cells and tissues.

Cytosol The fluid portion of the cytoplasm, i.e., the cytoplasm minus its organelles.

Cytotoxic Able to cause cell death.

Cytotype A maternally inherited cellular condition in *Drosophila* that regulates the activity of transposable P elements.

D

Dabs (single-domain antibodies) Antibodies with only one (instead of two) protein chain derived from only one of the domains of the antibody structure. Dabs exploit the finding that, for some antibodies, half of the antibody molecule binds to its target antigen almost as well as the whole molecule. The potential advantages of dabs are that they can be made easily by bacteria or yeasts, and offer a way to clone antibody-like molecules into bacteria, and hence to be able to easily screen millions of antibodies. Related ideas are single-chain antigen binding technology (SCA), biosynthetic antibody binding sites (BABS), minimum recognition units (MRUs), and complementary determining regions (CDRs).

Dalton (da) A unit of atomic mass roughly equivalent to the mass of a hydrogen atom. 1.67×10^{-24} g. Named after the famous nineteenth-century chemist, John Dalton (1766-1844)). Used in shorthand expressions of molecular weight, especially as kilo- (kDa) or megadaltons (MDa), which are equal to respectively to 1×10^3 and 1×10^6 daltons.

Darwinian cloning Selection of a clone from a large number of essentially random starting points, rather than isolating a natural gene or making a carefully designed artificial one. Molecules which are more similar to those needed are selected, mutated to generate new variants, and re-selected. The cycle proceeds until the required molecule is found. The advantage of the system is that the selection is from a vast number of possibilities.

Darwinian fitness The true measure of evolutionary change of an organism. Darwinian fitness refers to the numerical advantage of having offspring. The individual with the most offspring has the higher fitness. The reasons can be chance or natural selection and are not important to measure fitness. It is often equated with survival of the fittest, which is often meant to be the strongest or best adapted individual. However, this interpretation is wrong, if it does not explain why a certain individual has the most offspring. Overall, the genetic variation of the individuals with the most offspring will dominate the genepool of a population. The change in genetic variability in a population from generation to generation is the true measure of (micro-)evolution.

Daughterless carp Carp which only produce male fish. This slows the growth of the population with the aim of reducing overall carp numbers. Since all fish embryos start life as males, the technology works by silencing or switching off the gene responsible for stimulating the development of female embryos.

De differentiation The process by which cells lose their specialization and proliferate by cell division to form a mass of cells which, in response to appropriate stimuli, may later differentiate again to form either the same cell type or a different one. De-differentiation occurs in response to wounding and in tissue cultures.

De ionized water Water which is free of most inorganic (not completely free, since Na is present in ample quantities) and most organic compounds.

De mineralize To remove the mineral content (salts, ions) from a substance, especially water. Removal methods include distillation and electrodialysis. The process is de-mineralization.

De novo Arising, anew, afresh, once more. Also *ex novo*.

De repression The process of "turning on" the expression of a gene or set of genes whose expression has been repressed (turned off). Displacement of a repressor protein from a promoter region of DNA. When attached to the DNA, the repressor protein prevents RNA polymerase from initiating transcription. The "turning on" of a gene.

Death phase The final growth phase of cell culture, during which nutrients have been depleted and cell number decreases.

Deceleration phase The phase of declining growth rate, following the linear phase and preceding the stationary phase in most batch-suspension cultures.

Decode To spell-out the sequences of DNA and proteins into their basic units.

Defined 1. Fixed conditions of medium, environment and protocol for growth.

2. Precisely known and stated elements of a tissue culture medium.

Definite diagnosis The diagnosis of a condition based on the unique characteristics of that condition. Definite diagnosis is difficult for conditions that share symptoms with other conditions.

Degeneracy (of the genetic code) The specification of one amino acid by more than one codon. It arises from the inevitable redundancy resulting from 64 triplets in a triplet code (4 x 4 x 4 = 64) encoding only 20 amino acids.

Degenerate Not a moral judgment but an adjective that describes multiple states that amount to the same thing: different triplet combinations of nucleotide bases (codons) that code for the same amino acid, for example.

Degeneration 1. Changes in cells, tissues or organs due to disease.

2. The reduction in size or complete loss of organs during evolution.

Dehalogenation The removal of halogen atoms (chlorine, iodine, bromine, fluorine) from molecules, usually during biodegradation.

Dehiscence The spontaneous and often violent opening of a fruit, seed pod or anther to release and disperse the seeds or pollen.

Dehydrogenase An enzyme that catalyses the removal of hydrogen atoms in biological reactions.

Deletion A mutation involving the removal of one or more base pairs in a DNA sequence. Large deletions are sometimes microscopically visible in karyotype analyses.

Deliberate release Putting something into the outside world; in biotechnology it means putting a genetically modified organism (GMO) into field trials.

Delta-endotoxins Bt insecticidal proteins.

Deme A group of organisms in the same taxon.

Denatured DNA Double-stranded DNA that has been converted to single strands by breaking the hydrogen bonds linking complementary nucleotide pairs. Often reversible. Usually achieved by heating.

Denaturation Loss of native configuration of a macro-molecule (protein or nucleic acid) by physical or chemical means, usually accompanied by loss of biological activity. Denatured proteins often unfold their polypeptide chains and express changed properties of solubility. The separation of duplex nucleic acid molecules into single strands. Most commonly used by genetic engineers to describe the destruction of hydrogen bonds maintaining the double-stranded nature of all or part of a DNA molecule.

Denatured protein Altering the in vivo conformation of a protein by heat or salt treatment, thereby destroying its biological activity. Unlike denatured DNA, denatured proteins are seldom able to be renatured.

Dendrite Extension of a nerve cell, typically branched and relatively short, that receives stimuli from other nerve cells

Denitrification A chemical process in which nitrates in the soil are reduced to molecular nitrogen, which is released to the atmosphere.

Density gradient centrifugation High-speed centrifugation in which molecules "float" at a point where their density equals that in a gradient of caesium chloride or sucrose. The density gradient may either be formed before centrifugation by mixing two solutions of different density (as in sucrose density gradients) or it can be formed by the process of centrifugation itself (as in CsCl and Cs2SO4 density gradients).

Dental caries A disease affecting the hard tissues of the teeth resulting in progressive decay. Bacteria that accumulate in a dense mass known as plaque on the surface of the teeth ferment dietary carbohydrates to form acids that de-mineralise the hard tissues underneath. Hence, cariogenicity refers to the capacity of a food or drink to lead to caries in those who consume it. Periodontal disease is a related bacterial infection that affects the softer supporting tissues of the teeth.

Depolarization A process of changing the membrane potential from negative to more positive values. The sign of the potential refers to the inside of the cell. Depolarization causes action potentials in neurons and muscle cells. Depolarization is the result of inward currents carried by Na^+ and Ca^{++} ions.

Derivative 1. Resulting from or derived from.

2. Term used to identify a variant during meristematiccell division.

Descriptive epidemiology The study of variations in the occurrence of disease in terms of person, place and time, without the purpose of establishing causal inference.

Desiccant Any compound used to remove moisture or water.

Desiccate To dry, exhaust or deprive of water or moisture. Any chemical used for this purpose is called a desiccant. An apparatus for drying and preventing hygroscopic samples from rehydrating is a desiccator. The process is desiccation.

Designer foods Foods that are enriched with nutraceuticals, antioxidants, and secondary metabolites to improve the physical performance of the body.

Desmosome Specialized cell-cell junction, usually formed between two epithelial cells, characterized by dense plaques of protein into which intermediate filaments in the two adjoining cells insert.

Dessicator Apparatus for drying or depriving of moisture.

Desulphurization (USA: desulfurization) Technology for removing sulphur from oil and coal by use of bacteria. Sulphur residues in fuels end up as sulphur dioxide when the fuel is burned, resulting in acid rain. Bacteria may oxidize sulphites (insoluble) into sulphates (soluble), which can be washed away with the bacteria.

Detergent Type of small amphipathic molecule that tends to coalesce in water, with its hydrophobic tails buried and its hydrophilic heads exposed; widely used to solubilize membrane proteins.

Determinate growth Growth determined and limited in time, as in most floral meristems and leaves. The differentiation process is irreversibly established. Determinate growth contrasts with the usual culture growth, which is infinite and indeterminate.

Determination Commitment by an embryonic cell to a particular specialized path of development; it reflects a change in the internal character of the cell.

Development The sum total of events that contribute to the progressive elaboration of an organism. The two major aspects of development are growth and differentiation.

Developmental asynchrony A pattern of development within sub-populations that allows different sub-populations to reach sexual maturity at different times.

Deviation 1. In statistics: the difference between an actual observation and the mean of all observations.

2. An alteration from the typical form, function or behaviour. Mutation or stress are the common reasons behind deviation.

Dextrin An intermediate polysaccharide compound resulting from the hydrolysis of starch to maltose by amylase enzymes.

DI deoxynucleotide (ddNTP) A deoxynucleotide that lacks a 3′- hydroxyl group, and is thus unable to form the 3′-5′ phospho-diester bond necessary for chain elongation. Di-deoxynucleotides are used in DNA sequencing and the treatment of viral diseases. Also sometimes referred to as didN.

DIsulphide bond A chemical bond that stabilizes the three-dimensional structure of proteins, and hence the protein's normal function. They form between cysteine residues in the same or different peptide molecules.

Diabetes A gradual wearing down or away. Also with regard to soil: a lowering of the nutrient content and associated ability to support continuing crop growth.

Diabetes mellitus A metabolic disorder in which the hormone insulin is ineffective either because of a failure of the pancreas to secrete it (type I, insulin-dependent, (IDDM), or juvenile onset) or because target tissues are relatively insensitive to its action (type II, non-insulin-dependent (NIDDM), or maturity-onset, diabetes). In IDDM, patients require regular administration of insulin. In contrast, patients with NIDDM may actually have a high blood concentration of insulin (hyperinsulinaemia) and the condition is frequently associated with obesity and hyperlipidaemia.

Diagnostic A product used for the diagnosis of disease or medical condition. Both monoclonal antibodies and DNA probes are useful diagnostic products.

Diagnostic procedure A test or assay used to determine the presence of an organism, substance or nucleic acid sequence alteration.

Diakinesis A stage of meiosis just before metaphase I, in which the separation of homologous chromosomes is almost completed.

Diamondoid As used in this volume, this term describes structures that resemble diamond in a broad sense: strong, stiff structures containing dense, three-dimensional networks of covalent bonds, formed chiefly from first and second row atoms with a valence of three or more. Many of the most useful diamondoid structures will in fact be rich in tetrahedrally coordinated carbon. *Diamondoid* is used more narrowly elsewhere in the literature.

Diazotroph An organism that can fix atmospheric nitrogen.

Dicentric chromosome A chromosome having two centromeres.

Dichogamy The condition in which the male and the female reproductive organs of a flower mature at different times, thereby making self-fertilization improbable or impossible.

Dicotyledon A plant with two cotyledons, or seed leaves. One of the two classes of plants in the

Angiosperms (the other class is the monocotyledons). Colloquially called a dicot. Examples include many crop plants (potato, pea, beans), ornamentals (rose, ivy) and timber trees (oak, beech, lime).

Dideoxynucleotide (DIDN) A deoxynucleotide that lacks a 3' hydroxyl group, and is thus unable to form a 3'-5' phosphodiester bond necessary for chain elongation. Dideoxynucleotides are used in DNA sequencing and the treatment of viral diseases.

Diet history A method of dietary assessment in which subjects are asked open-ended questions about their usual dietary intakes.

Dietary recall A method of dietary assessment in which subjects are asked to recall their food consumption over a specific period of time.

Differential centrifugation A method of separating sub-cellular particles according to their sedimentation coefficients, which are roughly proportional to their size. Cell extracts are subjected tc a succession of centrifuge runs at progressively faster rotation speeds. Large particles, such as nuclei or mitochondria, will be precipitated at relatively slow speeds; higher G forces will be required to sediment small particles, such as ribosomes.

Differentially permeable Referring to a membrane, through which different substances diffuse at different rates; some substances may be unable to diffuse through such a membrane.

Differentiation The process by which cells and tissues undergo a series of changes resulting in their specialisation to a specific form or function. A differentiated cell such as a muscle cell or a skin cell contains a full set of genes of that organism, but only expresses the genes necessary for its specific function. In animals, stem cells (both embryonic and adult) are the only cells capable of undergoing differentiation to form more specialised cell types found in the body.

Diffusion coefficient D The diffusion coefficient D describes the relationship between a concentration gradient DC/Dx and the flow of matter per unit area (flux rate J)

Digest To cut DNA molecules with one or more restriction endonucleases.

Dihybrid An individual that is heterozygous for two pairs of alleles; the progeny of a cross between homozygous parents differing at two loci.

Dimer Association of two molecules.

Dimethyl sulphoxide; dimethyl sulfoxide (C_2H_6OS) A highly hygroscopic liquid and powerful solvent with little odour or colour. It is an organic co-solvent used in small quantities to dissolve neutral organic substances in tissue culture media preparation. DMSO also has uses as a cryoprotectant.

Dimorphism The existence of two distinctly different types of individuals within a species. An obvious example is the sexual dimorphism in certain animals.

Diploid 1. The status of having two complete sets of chromosomes, most commonly one set of paternal origin and the other of maternal origin.

2. An organism or cell with a double set (2n) of chromosomes (most commonly one of paternal origin, and the other of maternal origin), or referring to an individual containing a double set of chromosomes per cell. Somatic tissues of higher plants and animals are ordinarily diploid in chromosome constitution, in contrast with the haploid gametes.

Diploid cell A cell which contains two sets of chromosomes.

Diplonema Stage in prophase of meiosis I following the pachytene stage, but preceding diakinesis, in which one pair of sister chromatids begin to separate from the other pair, i.e., the centromeres begin to disjoin.

Diplophase Phase with 2n chromosomes.

Dipolar bond A covalent bond in which one atom supplies both bonding electrons, and the other atom supplies an empty orbital in which to share them. Also termed a dative bond.

Direct organogenesis Formation of organs directly on the surface of cultured intact explants. The process does not involve callus formation.

Direct repeat Two or more stretches of DNA within a single molecule which have the same nucleotide sequence in the same orientation. Direct repeats may be either adjacent to one another or far apart on the same molecule. For example

TATTA...TATTA ATAAT...ATAAT.

Directed mutagenesis The process of generation of nucleotide changes in cloned genes by any one of several procedures, including site-specific and random mutagenesis. Also called *in vitro* mutagenesis.

Directional cloning The technique by which DNA insert and vector molecules are digested with two different restriction enzymes to create non-complementary sticky ends at either end of each restriction fragment, so allowing the insert to be ligated to the vector in a specific orientation and preventing the vector from re-circularizing.

Disaccharide A carbohydrate consisting of two linked sugar units.

Disarm To delete from a plasmid or virus genes that are cytotoxic or tumour inducing.

Disassembler An instrument able to take apart structures a few atoms at a time, recording structural information at each step.

Discontinuous variation Phenotypic variation involving distinct classes, such as red versus white, tall versus dwarf.

Discordant Members of a pair showing different, rather than similar, characteristics.

Disease The opposite of ease. Any alteration from the state of metabolism necessary for the normal development and functioning of an organism, usually associated with infection by a pathogen or the malfunction or absence of one or more genes.

Disease resistance The genetically determined ability to prevent the reproduction of a pathogen, thereby remaining healthy. Some resistances operate by pathogen exclusion, some by preventing pathogen spread, and some by tolerating pathogen toxin.

Disease-free A plant or animal certified through specific tests as being free of specified pathogens. Disease-free should be interpreted to mean "free from any known diseases" as "new" diseases may yet be discovered to be present.

Disease-indexing Disease-indexed organisms have been assayed for the presence of known diseases according to standard testing procedures.

Disinfectation Full elimination of internal micro-organisms from a culture; disinfectation is rarely obtained.

Disjunction Separation of homologous chromosomes during anaphase I of meiosis; separation of sister chromatids during anaphase of mitosis and anaphase II of meiosis. As soon as the sister chromatids have separated, they are each called a chromosome.

Disomy The presence of a pair of specific chromosome. This is the normal condition, and abnormal occurrences are monosomy, trisomy and nullisomy (with respectively one chromosome of a pair, three or none). There are also abnormal disomic conditions, such as when both chromosomes of the pair were inherited from the same parent.

Dispense Portion out a nutrient medium into containers, such as test tubes, jars, Erlenmeyer flasks, Petri dishes, etc.

Dissecting microscope A microscope with a low magnifying power of about 50X, used to examine or excise small plant or animal parts.

Dissection Separation of a tissue by cutting for analysis or observation.

Dissociation constant For systems in which ligands of a particular kind bind to a receptor in a solvent there will be a characteristic frequency with which existing ligand-receptor complexes dissociate as a result of thermal excitation, and a characteristic frequency with which empty receptors bind ligands as a result of Brownian encounters, forming new complexes. The frequency of binding is proportional to the concentration of the ligand in solution. The dissociation constant is the magnitude of the ligand concentration at which the probability that the receptor will be found occupied is 1/2.

Dissolve Pass chemicals into solution.

Distillation The process of heating a mixture to separate the more volatile from the less volatile parts, and then cooling and condensing the resulting vapour so as to produce a more nearly pure or refined substance.

Ditype In fungi, a tetrad that contains two kinds of meiotic products (spores), e.g., 2AB and 2ab.

Diurnal Term describing the occurrence of an event at least once every 24 hours.

Diversity (within a species) The genetic differences that can be found within a population in a species.

Dizygotic twins Two-egg twins, i.e. a pair of individuals that shared the same uterus at the same time, but which arose from separate and independent fertilization of two ova.

DNA (deoxyribonucleic acid; formerly spelt desoxyribonucleic acid) The long chain of molecules in most cells that carries the genetic message and controls all cellular functions in most forms of life. The information-carrying genetic material that comprises the genes. DNA is a macro-molecule composed of a long chain of deoxyribonucleotides joined by phosphodiester linkages. Each deoxyribonucleotide contains a phosphate group, the five-carbon sugar 2-deoxribose, and a nitrogen-containing base. The genetic material of most organisms and organelles so far examined is double-stranded DNA; a number of viral genomes consist of single-stranded DNA or single-or double-stranded RNA. In double-stranded DNA, the two strands run in opposite (anti-parallel) directions and are coiled round one another in a double helix. Purine bases on one strand specifically hydrogen bond with pyrimidine bases on the other strand, according to the Watson-Crick rules (A pairs with T; G pairs with C). Hence a constant width for the double helix of 20 E (2.0 nm) is maintained. In the B-form, DNA adopts a right-handed helical conformation, with each chain making a complete turn every 34 E (3.4 nm), or once every ten bases.

DNA amplification Multiplication of a piece of DNA in a test-tube into many thousands of millions of copies. The most commonly used process is the polymerase chain reaction (PCR) system, but other systems are being developed, including ligase chain reaction (LCR), nucleic acids sequence-dependent amplification, and the Q-b system.

DNA bank Storage of DNA, which may or may not be the complete genome, but should always be accompanied by inventory information.

DNA carriers 1. Substances or particles that can transfer genes into a cell. These include viruses, liposomes (fat globules) and artificial chromosomes (sequences of DNA created in a laboratory) that can transport large amounts of DNA.

2. An individual who carriers one copy of a recessive gene for a hereditary condition.

DNA chip A small piece of glass or silicon that has small pieces of DNA arrayed on its surface.

DNA cloning The process of creating exact copies of a DNA strand using various technologies.

DNA construct A DNA molecule inserted into a cloning vector, usually a plasmid.

DNA delivery system A generic term for any procedure that transports DNA into a recipient cell.

DNA diagnosis The use of DNA polymorphisms to detect the presence of a specific allele (often associated with a disease or syndrome) or DNA sequence.

DNA fingerprinting The process of identifying the specific sequence of DNA molecules. Because all DNA is unique, this allows matching of biological samples with its host - ie is has been used to determine if a crop disease developed locally or arrived from a different location where it already existed.

DNA helicase (gyrase) An enzyme that catalyses the unwinding of the complementary strands of a DNA double helix.

DNA hybridization The pairing of two DNA molecules, often from different sources, by hydrogen bonding between complementary nucleotides. This technique is frequently used to detect the presence of a specific nucleotide sequence in a DNA sample.

DNA library A collection of cloned DNA fragments that collectively represent the genome of an organism.

DNA ligase An enzyme that catalyses a reaction that links two DNA molecules via the formation of a phospho-diester bond between the 3′ hydroxyl and 5′ phosphate of adjacent nucleotides. It plays an important role in DNA repair and replication. DNA ligase is one of the essential tools of recombinant DNA technology, enabling (among other things) the incorporation of foreign DNA into vectors. The ligase enzyme encoded by phage T4 is commonly used in gene-cloning experiments. It requires ATP as a co-factor. T4 is used *in vitro* to join the vector and insert DNAs.

DNA marker sequence A DNA sequence that can almost always be seen each time a certain characteristic associated with a specified trait is seen. This marker sequence is "linked" to the sequence that causes the trait.

DNA micro-array A small glass surface to which has been fixed an array of DNA fragments, each with a defined location. A typical DNA chip would contain 10 000 discrete spots (each containing a different DNA fragment) in an area of just a few square centimetres. When a solution of fluorescently labelled DNA fragments is hybridized to the chip, spots to which hybridization occurs are visible as fluorescence. If the spots on the chip are genes (expressed sequence tags), hybridization with cDNA from a particular tissue shows which genes are expressed in that tissue. If the spots are short, synthesized oligonucleotides (approximately 25 bases) corresponding to that part of a gene containing a single nucleotide polymorphisms (SNP) with a separate spot for each of the 4 possible bases at that site, hybridization with genomic DNA from an individual plant or animal enables that individual to be genotyped at as many SNP loci as are represented on the chip. The big advantage of DNA chips is the extent to which the process of genotyping can be automated, thereby enabling huge numbers of plants or animals to be genotyped for a huge number of loci.

DNA polymerase An enzyme that catalyses the synthesis of double-stranded DNA, using single-stranded DNA as a template.

DNA primase An enzyme that catalyses the synthesis of short strands of RNA that initiate the synthesis of DNA strands.

DNA probe A labelled (tagged) segment of DNA that is able, after a DNA hybridization reaction, to detect a specific DNA sequence in a mixture of sequences. If the tagged sequence is complementary to any one in the mixture, the two sequences will form a double helix. This will be identified thanks to its label (either by radioactivity or fluorescence).

DNA repair A variety of mechanisms that repair errors (e.g. the incorporation of a non-complementary nucleotide) that occur naturally during DNA replication.

DNA repair enzymes Proteins that recognize and repair certain abnormalities in DNA.

DNA replication The process whereby DNA copies itself, under the action of and control of DNA polymerase.

DNA sequencing Procedures for determining the nucleotide sequence of a DNA fragment. Two common methods available: 1. The Maxam Gilbert technique, which uses chemicals to cleave DNA into fragments at specific bases; or, most commonly,

2. the Sanger technique (also called the di-deoxy or chain-terminating method) which uses DNA polymerase to make new DNA chains, in the presence of di-deoxynucleotides (chain terminators) to stop the chain randomly as it grows. In both cases, the DNA fragments are separated according to length by polyacrylamide gel electrophoresis, enabling the sequence to be read directly from the gel. The procedure has become increasingly automated and large-scale in recent years.

DNA topo-isomerase An enzyme that catalyses the introduction or removal of supercoils in DNA.

DNA vaccines Pieces of foreign DNA that are injected into an organism to trigger an immune response.

Dolly The [name of the] first mammal to be created by cloning a cell from an adult animal. In this particular case, the cell came from the mammary tissue of an adult ewe. The creation of Dolly showed that the process of differentiation into adult tissue is not, as previously thought, irreversible. The result was achieved by nuclear transfer. Dolly's birth was announced in 1997. Since then, cattle and mice have also been cloned from adult cells.

Domain A segment of a protein that has a discrete function or conformation. At the protein level, a domain can be as small as a few amino acid residues or as large as half of the entire protein.

Domestic animal diversity (DAD) The spectrum of genetic differences within each breed, and across all breeds within each domestic animal species, together with the species differences; all of which are available for the sustainable intensification of food and agriculture production.

Dominance The type of gene action exhibited by a dominant allele.

Dominant 1. Of alleles, one whose effect with respect to a particular trait is the same in heterozygotes as in homozygotes. The opposite is recessive.

2. Of an individual animal, one that is allowed priority in access to food, mates, etc., by others of its species because of its success in previous aggressive encounters.

3. Of an animal or plant species, the most conspicuously abundant and characteristic in a particular location or environment.

Dominant gene or allele Most genes in the human body come in pairs. The pair's members can have codes which match each other or be different. If they are different, one is usually dominant, the other is described as 'recessive'. This means that it is this gene which is expressed in the phenotype (eg the gene implicated in Huntington's disease).

Dominant marker selection Selection of cells via a gene encoding a product that enables only the cells that carry the gene to grow under particular conditions. For example, plant and animal cells that express the introduced Neor gene are resistant to the compound G418, while cells that do not carry the Neor gene are killed by G418.

Dominant oncogene A gene that stimulates cell proliferation and contributes to oncogenesis when present in a single copy.

Dominant selectable marker gene A gene that allows the host cell to survive under conditions where it would otherwise die.

Donor plant (mother plant) An explant, graft or cutting used as a source of plant material for micro-propagation purposes.

Dormancy (F. *dormir*, from L. *dormire*, to sleep) An inactive period in the life of an animal or plant during which growth slows or completely ceases. Physiological changes associated with dormancy help the organism survive adverse environmental conditions. Annual plants survive the winter as dormant seeds, while many perennial plants survive as dormant tubers, rhizomes, or bulbs. Hibernation and aestivation in animals help them survive extremes of cold and heat, respectively.

Dosage compensation A phenomenon whereby inactivation of all but one of the X chromosomes in female mammals results in males and females producing the same quantity of peptide from X-linked genes.

Dose-response assessment The determination of the relationship between the magnitude of exposure (dose) to a biological, chemical or physical agent and the severity and\or frequency of associated adverse health effects (response).

Double bond Two atoms sharing electrons as in a single bond (that is, a sigma bond) may also share electrons in an orbital with a node passing through the two atoms. This adds a second, weaker bonding interaction (a pi bond); the combination is termed a double bond. A twisting motion that forces the nodal plane at one atom to become perpendicular to the nodal plane on the other atom eliminates the (signed) overlap between the atomic orbitals, destroying the pi bond. The energy required to do this creates a large barrier to rotation about the bond.

Double crossing over The formation of two chiasmata within a chromosome arm, leading to the generation of a double recombinant gamete with respect to genes located within the segment defined by the two genes concerned.

Double fertilization A process, unique to flowering plants, in which two male nuclei, which have travelled down the pollen tube, separately fuse with different female nuclei in the embryo sac. The first male nucleus fuses with the egg cell to form the zygote; the second male nucleus fuses with the two polar nuclei to form a triploid nucleus that develops into the endosperm.

Double helix Describes the coiling of the two strands of the double-stranded DNA molecule, resembling a spiral staircase in which the base pairs form the steps and the sugar-phosphate backbones form the rails on each side. One strand runs 3' to 5', while the complementary one runs 5' to 3'.

Double recessive An organism homozygous for a recessive allele at each of two loci.

Double stranded complementary DNA (dscDNA) A double-strand DNA molecule created from a cDNA template.

Double-blind food challenge Food challenge test where neither the patient nor the clinical investigator knows the identity of the administered substance.

Doubled haploid Once parent plants with desirable traits are identified using molecular markers, the parent plants are bred using a special plant breeding process that produces identical chromosome sets in the offspring. This process locks in desirable features in the offspring plants in one season, rather than selecting plants and manually crossing until the same outcome is achieved, a process that can take over four years.

Double-helix The physical structure of DNA, comprising two parallel strands of DNA coiled helically.

Doublet The electronic state of a molecule having one unpaired spin is termed a doublet. This term is derived from spectroscopy: an unpaired spin can be either up or down with respect to a magnetic field, and these states have different energy, resulting in field-dependent pairs, or doublets, of spectral lines.

Down's syndrome A genetic disability in which characteristic physical features are accompanied by learning difficulty. The disorder is the consequence of a chromosome abnormality and is often called trisomy-21 because most commonly there is an additional copy of chromosome 21 present.

Downstream 1. In molecular biology, the stretch of nucleotides of DNA that lie in the 3′ direction from the site of initiation of transcription, which is designated as +1 (remembering the convention that the sequence of a DNA molecule is written from the 5′ end to the 3′ end). Downstream nucleotides are marked with plus signs, e.g., +2, +10. Also, to the 3′ side of a particular gene or sequence of nucleotides.

2. In chemical engineering, those phases of a manufacturing process that follow the

biotransformation stage. Usually refers to the recovery and purification of the product of a fermentation process.

Downstream processing A general term for all the things which happen in a biotechnological process after the biology, be it fermentation of a micro-organism or growth of a plant. It is particularly relevant to fermentation processes, which produce a large quantity of a dilute mixture of substances, products and micro-organisms. These must be separated, and the product must be concentrated and purified, and converted into a form which is useful.

Driving force A terminology used in thermodynamics expressing the availability of energy to 'drive' a process such as mechanical work or chemical synthesis. Driving forces exist where a potential gradient exist. A potential gradient can be in form of a temperature gradient causing heat to flow, an electrical gradient causing electrons or ions to flow, or a concentration gradient causing diffusion.

Drosophila melanogaster The fruit fly, used for many years as a model for eukaryotic genetics. Of the nearly 300 disease-causing genes in the human genome, more than half have an analogous gene in the Drosophila genome.

Drug delivery Method by which a therapeutic agent is delivered to its site of action. For traditional therapeutic agents this is another name for *formulation*. However, biotechnology has allowed the development of a range of new therapeutic-agent delivery systems, such as liposomes and other encapsulation techniques, and a range of mechanisms that target a therapeutic agent to a particular cell or tissue.

Drug metabolism The process by which enzymes break down drugs into different chemical components. The breakdown allows the drug to carry out its therapeutic role and eventually be eliminated from the body.

Drug resistance A condition in which disease-causing microorganisms develop an ability to tolerate drugs, which were once effective against them. Results in the loss of effectiveness in a drug.

Dry weight The weight of tissue obtained following sufficiently prolonged oven-drying at high temperature to remove all water. Freeze-drying may also be employed but generates a slightly different result because bound water is not removed.

Dual culture A culture made of a plant tissue and one organism (such as a nematode) or an obligate parasite/micro-organism (such as a fungus). Dual culture techniques are used for a variety of purposes, including assessing host-parasite interactions and the production of axenic cultures.

Duchenne muscular dystrophy A single gene disorder which affects 1 in 5000 boys. The condition produces progressive weakness of proximal muscles, beginning in early childhood. Most affected children become chairbound before adulthood. The inheritance is X-linked i.e. it is transmitted from heterozygous females who are unaffected and are described as carriers, to sons who are affected, and to daughters who, like their mothers, are unaffected carriers.

Duplex DNA Double-stranded DNA.

Duplication Multiple occurrence of: 1. A DNA sequence within a defined length of DNA; or

2. A specific segment in the same chromosome or genome.

E

Ecdysone A steroid hormone in insects that stimulates moulting and metamorphosis. It acts on specific genes, stimulating the synthesis of proteins involved in these bodily changes.

Eclosion 1. Emergence of an adult insect from the pupal stage.

2. Beginning of germination of fungal spores.

Eco centrism A view that considers the whole environment or ecosphere as being important and deserving of consideration, not just that of organisms such as animals and humans. It states that all elements of the environment have worth and should be valued and cared for.

Ecology The study of the interactions between organisms and their natural environment, both living and non-living.

Ecological amplitude The range of environmental conditions in which an organism can survive and reproduce.

Ecological study A study that compares the rates of exposures and diseases in different populations using aggregate data on exposure and disease, not individual data.

Economic trait locus (ETL) A locus influencing a trait that contributes to income. The plural form (economic trait loci) is also abbreviated as ETL.

Ecosystem A specific characteristic biological system in a location or area with a unique mix of living organisms and physical consistency such as minerals, soil and air.

Ecosystem protector A nanomachine for mechanically removing selected imported species from an ecosystem in order to protect native species.

Ecotype A population or a strain of an organism that is adapted to a particular habitat.

Ecozones Large, generalized ecologically distinctive areas that are distinguished based on the interplay of landform, water, soil, climate, plants, and human factors.

Edible vaccines Transgenic plants that contain agents that are able to trigger an animals immune system.

Effective mass In a vibrating system, a particular vibrational mode can be described as a harmonic oscillator with some mass and stiffness. Given some measure of vibrational amplitude, there exists a unique choice of mass and stiffness that yields the correct values for both frequency and energy; these are the effective mass and effective stiffness.

Effector cells Cells of the immune system that are responsible for cell-mediated cytotoxicity.

Effector molecule A molecule that influences the behaviour of a regulatory molecule, such as a repressor protein, thereby influencing gene expression.

Egg 1. The fertilized ovum (zygote) in egg-laying animals after it emerges from the body.

2. The mature female reproductive cell in animals and plants.

Elastic An object behaves elastically if it returns to its original shape after a force is applied and then removed. (If an applied force causes a permanent deformation, the behavior is termed *plastic*.) In an elastic system, the internal potential energy is a function of shape alone, independent of past forces and deformations.

Elasticity Refers to the case where one economic variable causes a change in another (e.g. price and quantity demanded). In general, elasticity measures the responsiveness of one economic variable to a change in another. It is unit free and always expressed in percentage terms.

Elastin A fibrous protein that is the major constituent of the yellow elastic fibres of connective tissue.

Electro-blotting The electrophoretic transfer of macromolecules (DNA, RNA or protein) from a gel, in which they have been separated, to a support matrix, such as a nitrocellulose sheet. A transfer usually used in techniques such as Southern and northern blotting.

Electrochemical gradient Driving force that causes an ion to move across a membrane due to the combined influence of a difference in its concentration on the two sides of the membrane and the electrical charge difference across the membrane.

Electrochemical sensor Type of biosensor in which a biological process is harnessed to an electrical sensor system, such as an enzyme electrode. Other types couple a biological event to an electrical one via a range of mechanisms, such as those based on oxygen and pH.

Electrolytes Positively or negatively charged ions dissolved in the extracellular or intracellular water.

Electromagnetic radiation Electromagnetic waves, including ultraviolet (UV), X-rays, and gamma radiation (g rays). Electromagnetic radiation is used to produce mutant cells or organisms, or, in the case of UV, disinfestation and sterilization, in tissue culture.

Electromagnetic spectrum The range of wavelengths or frequencies over which electromagnetic radiation extends.

Electron Subatomic particle carrying a negative electric charge in atoms or molecules.

Electron acceptor Atom or molecule that takes up electrons readily, thereby gaining an electron and becoming reduced.

Electron carrier Molecule such as cytochrome c that transfers an electron from a donor molecule to an acceptor molecule.

Electron density The location of an electron is not fixed, but is instead described by a probability density function. The sum of the probability densities of all the electrons in a region is the electron density in that region.

Electron donor Molecule that easily gives up an electron, becoming oxidized in the process.

Electron microscope A microscope that uses an electron beam focused by magnetic "lenses."

Electron transport Movement of electrons from a higher to a lower enery level along a series of electron carrier molecules, as in oxidative phosphorylation and photosynthesis.

Electronegativity A measure of the tendency of an atom (or moiety) to withdraw electrons from structures to which it is bonded. In most circumstances, for example, sodium tends to donate electron density (it has a low electronegativity) and fluorine tends to withdraw electron density (it has a high electronegativity).

Electronic Pertaining to the energies, distributions, and behaviors of electrons.

Electrophoresis A ubiquitous molecular biology technique, with many variants, used to resolve complex mixtures of macromolecules into their components. Its principle is to subject samples to an electric field applied across a porous matrix. Molecules will migrate under these conditions at a rate dependent on their net electric charge and/or their molecular weight.

Electrophysiology The technique of recording and stimulating currents and voltages across cell membranes using microelectrodes. Electrodes can be used to measure membrane potentials and inject currents. The latter charges the membrane and changes the membrane potentials. If the recorded voltage change is fed back to the stimulating electrode, the current can be adjusted such that the measured membrane potential stays constant. This is called the 'voltage-clamp' technique and has largely been responsible to elucidate the mechanism underlying the electrical phenomena of neurons and muscle tissue.

Electroporation 1. An electrical treatment of cells that induces transient pores, through which DNA can enter the cell.

2. The introduction of DNA or RNA into protoplasts or other cells by the momentary disruption of the cell membrane through exposure to an intense electric field.

Note: Although the precise mechanism of electroporation is poorly understood, pores are thought to form by the local polarization of the cell membrane when it is exposed to a high electric potential. These openings persist for a variable amount of time, depending upon the temperature at which the cell is treated. Macro-molecules, such as DNA or RNA, enter through these openings either through diffusion or through electrophoretic movement. The membrane openings then re-seal, capturing introduced DNA and preventing escape of the cell contents.

Element An atom with a unique number of protons (atomic number). There are 102 different elements and some additional synthetic elements that are not found in nature. Elements have different physical and chemical properties and can be combined to molecules (two or more atoms linked through covalent bonds). The elements are listed according to atomic number and chemical properties in the periodic table.

ELISA (enzyme linked immunosorbent assay) A sensitive technique for accurately determining specific molecules in a mixed sample. The amount of protein or other antigen in a given sample is determined by means of an enzyme-catalysed colour change, avoiding both the hazards and expense of radioactive techniques.

It takes various forms. In the most common form, two antibody preparations are used in ELISA. An antibody (primary) specific to the test protein is adsorbed onto a solid substrate, and a known amount of the sample is added; all molecules of the test protein in the sample are bound by the antibody. A second antibody specific for a second site on the test protein is added; this is conjugated with an enzyme, which catalyses a colour change in the fourth reagent, added finally. The colour change is measured photometrically and compared against a standard curve to give the concentration of protein in the sample. ELISA is widely used for diagnostic and other purposes.

Elongation factor Protein required for the addition of amino acids to growing polypeptide chains on ribosomes.

Embryo An immature organism in the early stages of development. In mammals, this occurs in the first months in the uterus. In plants, it is the structure that develops in the megagametophyte, as result of the fertilization of an egg cell, or without fertilization. In aseptic cultures, adventitious embryos show polarization, followed by the growth of a shoot from one end and a root from the other end.

Embryo cloning The creation of identical copies of an embryo by embryo splitting or by nuclear transfer from undifferentiated embryonic cells.

Embryo culture The culture of embryos on nutrient media.

Embryo multiplication and transfer (EMT) The cloning of animal embryos and their subsequent transfer to recipients (via artificial inembryonation.The cloned embryos can be clones of an embryo or of an adult.

Embryo sac A large thin-walled space within the ovule of the seed plant in which the egg and, after fertilization, the embryo develop; the mature female gametophyte in angiosperms. Generally a seven-celled structure. The seven cells are two synergids, one egg cell, three antipodal cells (each with a single haploid nucleus) and one endosperm mother cell with two haploid nuclei.

Embryo sexing The determination of the sex of an embryo, typically by means of PCR involving amplification from a small sample of embryonic tissue, using primers specific for a locus on the Y chromosome.

Embryo splitting The splitting of young embryos into several sections, each of which develops into an animal. A form of animal cloning, i.e., of producing animals that are genetically identical. In practice, the number of identicals (identical organisms) that can be produced from a single embryo is less than 10.

Embryo storage Cryogenic preservation of animal embryos, allowing inembryonation or other manipulations long after embryo formation.

Embryo technology Generic name for any modification of mammalian embryos. It encompasses embryo cloning, embryo splitting, *in vitro* fertilization, and embryo storage.

Embryogenesis 1. (General) Development of an embryo.

2. (In plants) *In vitro* formation of plants from plant tissues, through a pathway closely resembling normal embryogeny from the zygote; if this development in culture involves somatic cells and not the zygote, it can be indicated by using the term *adventitious embryogenesis* or *somatic embryogenesis*. The generation of embryos has two stages: initiation and maturation. Initiation needs a high level of the group of plant hormones called auxins; maturation needs a lower level. Other chemicals have to be at suitable levels. The procedure involves the explanting of a piece of plant tissue and putting it on a high-auxin medium, where the cells grow into a mass of callus. This is then transferred to a maturation medium, where the callus starts to initialize organs, ultimately growing a root and a shoot.

Embryoid An embryo-like body developing *in vitro*. It forms a complete, self-contained platelet with no vascular connection with the callus. The term embryoid is no longer commonly used.

Embryonic stem cells Cells that can give rise to any type of differentiated cell. They can be derived from two sources: the inner cell mass from a blastocyst or the primordial germ cells (eggs and sperm) of an older embryo.

Empirical Relating to or based upon practical experience, trial and error, direct observation or observation alone, without benefit of scientific method, knowledge or theory.

Enabling science and technologies Areas of research relevant to a particular goal, such as nanotechnology.

Encapsidation The process by which a virus' nucleic acid is enclosed in a capsid.

Encapsulating agents Anything which forms a shell around an enzyme or bacterium, although the agents used are usually polysaccharides such as alginate or agar. The agents are inert and allow nutrients and oxygen to diffuse into and out of the sphere readily, and are easy to convert from gel (solid) to sol (liquid) or solution form by altering the temperature or the concentration of ions.

Encapsulation Any method of getting something, usually an enzyme or bacterium, into a small package or capsule while it is still working or alive. It is a method for immobilizing cells for use in a bioreactor.

Encode To specify, after decoding by transcription and translation, the sequence of amino acids in a protein.

End product inhibition The inhibition of an enzyme by a metabolite. The enzyme is sometimes the first enzyme in a biosynthetic pathway, and the metabolite is generally the product of the last step in the pathway.

Endangered A species with such a low population number, that the population is in danger of extinction.

Endangered breed A breed where the total number of breeding females is between 100 and 1 000 or the total number of breeding males is less than or equal to 20 and greater than five; or the overall population size is close to, but slightly above 100 and increasing, and the percentage of pure-bred females is above 80%; or the overall population size is close to, but slightly above, 1 000 and decreasing, and the percentage of pure-bred females is below 80%.

Endangered species A plant or animal species in immediate danger of extinction because its population numbers have reached a critical level or its habitats have been drastically reduced.

Endemic 1. Describing a plant or animal species whose distribution is restricted to one or a few localities.

2. Describing a disease or a pest that is always present in an area.

End-labelling The introduction of a radioactive atom at the end of a DNA or RNA molecule. A commonly used method is to use T4 polynucleotide kinase to introduce a 32P atom onto the end of a DNA molecule.

Endocrine cell Specialized animal cell that secretes a hormone into the blood; usually part of a gland, such as the thyroid or pituitary gland.

Endocrine glands Any gland in the body that secretes hormones into the blood. There are five major endocrine glands in the human body: the pituitary, the gonads, the adrenals, the thyroid, and the parathyroids.

Endocrine interference Interference with the normal balance hormones.

Endocytosis The process by which materials enter a cell without passing through the cell membrane. The membrane folds around material outside the cell, resulting in the formation of a saclike vesicle into which the material is incorporated. This vesicle is then pinched off from the cell surface so that it lies within the cell.

Endoderm The internal layer of cells of the gastrula, which will develop into the alimentary canal (gut) and digestive glands of the adult.

Endodermis The layer of living cells, with various characteristically thickened walls and no intercellular spaces, which surrounds the vascular tissue of certain plants and occurs in nearly all roots and certain stems and leaves. The endodermis separates the cortical cells from cells of the pericycle.

Endoergic A transformation is termed endoergic if it absorbs energy; such a reaction increases molecular potential energy. (Sometimes wrongly equated to the narrower term endothermic.)

Endogamy The fusion of reproductive cells from closely related parents, i.e., inbreeding.

Endogenote The part of the bacterial chromosome that is homologous to a genome fragment (exogenote) transferred from the donor to the recipient cell in the formation of a merozygote.

Endogenous Developed or added from within the cell or organism.

Endomitosis Duplication of chromosomes without division of the nucleus, resulting in increased chromosome number within a cell. Chromosome strands separate, but the cell does not divide.

Endonuclease An enzyme that breaks strands of DNA at internal positions; these enzymes are important tools in recombinant DNA technology.

Endophyte An organism that lives inside another.

Endoplasmic reticulum A cytoplasmic net of membranes, adjacent to the nucleus, made visible by the electron microscope. Any system of paired membranes within the cytoplasm. Frequently abbreviated to ER. They are sites of protein synthesis.

Endopolyploidy The result of nuclear divisions without subsequent cytoplasmic division (cytokinesis); the polyploids so obtained are called endopolyploids.

Endoprotease An enzyme that cleaves the peptide bonds between amino acids within a protein. Cleavage is usually at one or more specific sites.

Endoreduplication Chromosome reproduction during interphase. 4-chromatid chromosomes (diplochromosomes) are seen during this phase.

Endosome Membrane-bounded organelle in animal cells that carries materials newly ingested by endocytosis and passes many of them on to lysosomes for degradation.

Endosperm Nutritive tissue that develops in the embryo sac of most angiosperms. It usually forms after the fertilization of the two fused primary endosperm nuclei of the embryo sac with one of the two male gamete nuclei. In most diploid plants, e.g., cereals, the endosperm is triploid *(3n)*, but in some (e.g., lily) it is often consumed as the seed matures.

Endosperm mother cell One of the seven cells of the mature embryo sac, containing the two polar nuclei and, after reception of a sperm cell, gives rise to the primary endosperm cell from which the endosperm develops.

Endostatin An endogenous protein that blocks the proliferation of blood vessels.

Endothelium Single sheet of highly flattened cells (endothelial cells) that forms the lining of all blood vessels. Regulates exchanges between the bloodstream and surrounding tissues and is usually surrounded by a basal lamina.

Endothermic A transformation is termed endothermic if it absorbs energy in the form of heat. A typical endothermic reaction increases both entropy and molecular potential energy (and is thus analogous to a gas expanding while absorbing heat and compressing a spring).

Endotoxin A component of the cell wall of Gram-negative bacteria that elicits, in mammals, an inflammatory response and fever.

Energy A conserved quantity that can be interconverted among many forms, including kinetic energy, potential energy, and electromagnetic energy. Sometimes defined as "the capacity to do work," but in an environment at a uniform nonzero temperature, thermal energy does not provide this capacity. (Note, however, that all energy has mass, and thus can be used to do work by virtue of its gravitational potential energy; this caveat, however, is of no practical significance unless a *really* deep gravity well is available.)

Energy density Energy content per gram of food.

Energy expenditure Utilisation by the body of chemical energy from food components or body stores during the process of metabolism which is eventually dissipated as heat plus the heat generated by muscular activity, either in shivering or in physical activity. Usually used to mean the day's total energy (calorie) loss as heat.

Energy intake The chemical energy in foods that can be metabolised to produce energy available to the body; usually used to mean the day's total energy (calories) supplied by all the food and drink consumed.

Engineered antibodies Antibodies that are produced through cells that have been genetically engineering to produce antibodies.

Enhancer element; enhancer sequence 1. A sequence found in eukaryotes and certain eukaryotic viruses which can increase transcription of a gene when located (in either orientation) up to several kilobases from the gene concerned. These sequences usually act as enhancers when on the 5′ side (upstream) of the gene in question. However, some enhancers are active when placed on the 3′ side (downstream) of the gene. In some cases enhancer elements can activate transcription of a gene with no (known) promoter.

2. A substance or object that increases a chemical activity or a physiological process.

3. A major or modifier gene that increases a physiological process.

Entelechy That which makes something, e.g. an organism, what it is. The essence of a thing, organism.

Enterocytes Cells which line the small intestine. They have three major functions of secretion, digestion and absorption.

Enterotoxin A bacterial protein that, following release into the intestine, causes cramps, diarrhoea and nausea.

Enthalpy The enthalpy of a system is its actual energy (termed the *internal energy*) plus the product of its volume and the external pressure. Though sometimes termed "heat content," the enthalpy in fact includes energy *not* contained in the system. Enthalpy proves convenient for describing processes in gases and liquids in laboratory environments, if one does not wish to account explicitly for energy stored in the atmosphere by work done when a system expands. It is of little use, however, in describing processes in nanomechanical systems, where work can take many forms: internal energy is then more convenient. Enthalpy is to energy what the Gibbs free energy is to the Helmholtz free energy.

Entropy A measure of uncertainty regarding the state of a system: for example, a gas molecule at an unknown location in a large volume has a higher entropy than one known to be confined to a smaller volume. Free energy can be extracted in converting a low-entropy state to a high-entropy state: the (time-average) pressure exerted by a gas molecule can do useful work as a small volume is expanded to a larger volume. In the classical configuration space picture, any molecular system can be viewed as a single-particle gas in a high-dimensional space. In the quantum mechanical picture, entropy is described as a function of the probabilities of occupancy of different members of a set of alternative quantum states. Increased information regarding the state of a system reduces its entropy and thereby increases its free energy, as shown by the resulting ability to extract more work from it.

An illustrative contradiction in the simple textbook view of entropy as a local property of a material (defining an entropy per mole, and so forth) can be shown as follows: The third law of thermodynamics states that a perfect crystal at absolute zero has zero entropy*; this is true regardless of its size. A piece of disordered material, such as a glass, has some finite entropy *G*0 > 0 at absolute zero. In the local-property view, *N* pieces of glass, even (or especially) if all are atomically identical, must have an entropy of *N*G0. If these *N* pieces of glass are arranged in a regular three-dimensional lattice, however, the resulting structure constitutes a perfect crystal (with a large unit cell); at absolute zero, the third law states that this crystal has zero entropy, not *N*G0. To understand the informational perspective on entropy, it is a useful exercise to consider. 1. what the actual entropy of such crystal is as a function of *N*, with and without information describing the structure of the unit cell,

2. how the third law can be phrased more precisely, and

3. what this more precise statement implies for the entropy of well-defined aperiodic structures. Note that any one unit cell in the crystal can be regarded as a description of all the rest.

Enucleated ovum Egg cell from which the nucleus has been removed.

Environment The aggregate of all the external conditions and influences affecting the life and development of an organism.

Environmental protection agency (EPA) The U.S. regulatory agency for biotechnology of microbes. The major laws under which the agency has regulatory powers are the Federal Insecticide, Fungicide, and Rodenticide Act (FIFRA); and the Toxic Substances Control Act (TSCA).

Environmental stewardship This is a view that humans have a duty to manage and care for the whole natural environment. That we are responsible for the continued health of the whole ecosystem, not just the parts that benefit the human race. It involves integrating and applying environmental values into a process.

Environmental toxicology assessment A measurement of a toxin in a given area. For example, testing drinking water for arsenic levels.

Enzyme A protein produced in living cells, which, even in very low concentration, catalyses specific chemical reactions but is not used up in the reaction. Enzymes are classified into six major groups, according to the type of reaction they catalyse: 1. Oxidoreductase;

2. Transferases;

3. Hydrolases;

4. Lyases;

5. Isomerases;

6. Ligases. The names of most individual enzymes are usually derived from the substrate on which they act, with the suffix -ase. Thus lactase is the enzyme that acts to breakdown lactose; it is classified as a hydrolase.

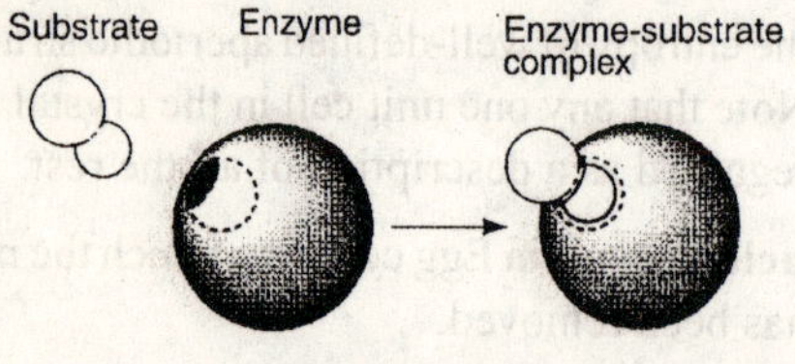

Fig. Enzyme

Enzyme bioreactor A reactor in which a chemical conversion reaction is catalysed by an enzyme.

Enzyme commission (EC) number Systematic name and number which identify an enzyme in technical literature. Assigned by the Enzyme Commission, the EC Number consists of four numbers separated by dots: the first classifies the enzyme into one of the six broad groups: 1. Oxidoreductase;

2. Transferases;

3. Hydrolases;

4. Lyases;

5. Isomerases;

6. Ligases. Each group is subdivided into sub-groups, each sub-group into sub-sub-groups, and the last number is specific for the enzyme, e.g., EC 3.1.21.1 is deoxyribonuclease I.

Enzyme electrode A type of biosensor, in which an enzyme is immobilized onto the surface of an electrode. When the enzyme catalyses its reaction, electrons are transferred from the reactant to the electrode, and so a current is generated. There are two types of enzyme electrodes:

ampometric, where the electrode is kept as near zero voltage as possible. When the enzyme catalyses its reaction, electrons flow into the electrode, and so a current flows; and potentiometric, when the electrode is held at a voltage which counteracts the voltage determined by the enzyme's tendency to push electrons into its.

Usually enzymes transfer their electrons inefficiently to the electrode, so a mediator compound is coated onto the electrode to help the transfer.

Enzyme linked immunoabsorbent assay (ELISA) A lab test that can give a simple positive or negative result for a protein using the complementary relationship of antigens and antibodies.

Enzyme stabilization using antibodies A method of stabilizing enzymes by binding antibodies to them. The antibodies should not block the active site of the enzyme, as otherwise the protein is stabilized but is inactive as a catalyst. Monoclonal antibodies are usually used as they bind to specific bits of the protein surface. If the enzyme tries to unfold into an inactivate structure, it must not only overcome its own binding energy but also throw off all the bound antibodies; this requires more energy, and so is a correspondingly slower process.

Epicotyl The upper portion of the axis of a plant embryo or seedling, above the cotyledons.

Epidemic A widespread incident of a disease in a specific area at a particular time.

Epidemiology The study of associations between patterns of disease in populations and environmental, lifestyle or genetic factors.

Epidermis 1. The outmost layer of cells of the body of an animal. In invertebrates the epidermis is normally only one cell thick and is covered by an impermeable cuticle. In vertebrates the epidermis is the thinner of the two layers of skin.

2. The outermost layer of cells covering a plant. It is overlaid by a cuticle and its functions are principally to protect the plant from injury and to reduce water loss. Some epidermal cells are modified to form guard cells or hairs of various types. In woody plants the functions of the shoot epidermis are taken over by the periderm tissues and in mature roots the epidermis is sloughed off and replaced by the hypodermis.

Epigenetic A term referring to the non-genetic causes of a phenotype.

Epigenetic factors Factors having an influence on the form, function or behaviour of an organism but originating outside the genome.

Epigenetic variation Non-hereditary and reversible variation; often the result of a change in gene expression.

Epinasty A process by which the growth of branches or petioles is abnormally pointing downward. This phenomenon is caused by the more rapid growth of the upper side. Epinasty may result from either nutritional deficiencies or irregularities at the plant growth regulator level. Not to be confused with wilting, as epinastic tissues are turgid.

Epiphyte A plant that grows upon another plant, but is neither parasitic on it nor rooted in the ground.

Episome A genetic extrachromosomal element (e.g., the fertility factor (F) in *Escherichia coli*) which replicates within a cell independently of the chromosome and is able to integrate into the host chromosome. The step of integration may be governed by a variety of factors and so the term episome has lost favour and been superseded by the wider term *plasmid.* Plasmids and F factors are episomes.

Epistasis Interaction between genes at different loci, e.g., one gene suppresses the effect of another gene that is situated at a different locus. Suppressed genes are said to be hypostatic. Dominance is associated with members of allelic pairs, whereas epistasis is interaction among products of non-alleles.

Epistatic A dependent relationship between genes.

Epithelium The surface layer of the skin or the lining layer of the intestinal mucosa or air passages of the lungs.

Epitope A specific chemical domain on an antigen that stimulates the production of, and is recognized by, an antibody. Each epitope on a molecule such as protein elicits the synthesis of a different antibody.

Epizootic A disease affecting a large number of animals simultaneously.

Epsp synthase An enzyme in plant metabolic pathways that leads to aromatic amino acid production necessary for the development of proteins essential to plants growth.

Equational division Mitotic-type division that is usually the second division in the meiotic sequence; somatic mitosis and the non-reductional division of meiosis. A chromosome division in which the two chromatids of each duplicated chromosome separate longitudinally, prior to being incorporated into two daughter nuclei.

Equatorial plate The figure formed by the chromosomes in the centre (equatorial plane) of the spindle in mitosis.

Equilibrium (chemical) At equilibrium, the state of a system does not change over time, although internal fluctuation may exist. Thermodynamically, the equilibrium is the physical state of a system that has the lowest total energy content. To lower the energy content, a system may give up energy in form of heat or work or entropy, and usually it is a combination of all three. The laws of thermodynamic dictate that the total amount of energy cannot be changed and whatever energy in a system is lost, is taken up by the surroundings (the first law). The second law dictates that at equilibrium the entropy portion of the energy content of a system must be at its maximum. Interestingly, life is characterized by avoiding equilibrium states and minimizing entropy. This is used as an explanation of how life can maintain highly organized structures (minimal entropy) at the expense of the environment, whose entropy or disorder increases. Importantly, the energy content of a system and its surroundings are interdependent and the equilibrium state of a system can be changed by changing the internal composition or external conditions (e.g. rising temperature). This shift from one equilibrium (state 1) to a second equilibrium (state 2) is often characterized as chemical equilibrium describing the relationship between two states of a system that exist under slightly different conditions. Mathematically, the equilibrium can be described with the equilibrium constant where both states are equally probable. This constant allows the quantification of system properties such as the binding of a drug to its receptor protein.

Equilibrium density gradient centrifugation A procedure used to separate macro-molecules based on their density (mass per unit volume).

Erlenmeyer flask A conical flat-bottomed laboratory flask with a narrow neck, designed by E. Erlenmeyer. Widely used for culturing micro-organisms.

Erucic acid A fatty acid having 22 carbons and one double bond that is common to traditional rapeseed oil. Canola oil contains less than 2 percent erucic acid.

Erythropoietin A protein that boosts production of red blood cells. It is clinically useful in treating certain types of anemia.

Escherichia coli A commensal bacterium inhabiting the colon of many species. *E. coli* is widely used in biology, both as a simple model of cell biochemical function and as a host for molecular cloning experiments. In environmental studies, it is a key indicator of water pollution due to human sewage effluent.

Essential amino acid or fatty acid An amino acid or fatty acid that cannot be synthesized by our own cells and need to be part of our diet to stay healthy.

Essential component Chemical compounds that are required for life, growth, and tissue repair.

Essential element Any of a number of elements required by living organisms to ensure normal growth, development and maintenance.

Essential nutrients Chemical compounds that can't be made by a specific organism but are important for the organism to survive.

Essential requirement A nutrient is essential when it is mandatory for growth, development and reproduction. In tissue culture, it comprises inorganic salts, including all of the elements necessary for plant metabolism; organic factors (amino acids, vitamins); usually also endogenous plant growth regulators (auxins, cytokinins and often gibberellins); as well as a carbon source (sucrose or glucose).

Established culture 1. An aseptic viable explant 2. A suspension culture subjected to several passages with a constant cell number per unit time.

Ester A molecule containing an ester linkage, a carbonyl group bonded to an O that is in turn bonded to a C.

Estimated breeding value (EBV) Twice the expected progeny difference. The difference is doubled because breeding value is a reflection of all the genes of an animal, in contrast to progeny difference, which is a reflection of a sample half of an animal's genes. The predicted performance of the offspring of the mating between any two animals is the average of their EBVs (averaged because each parent makes an equal contribution to each offspring).

ESTs (expressed sequence tags) Markers for parts of the genome which are known to be expressed in RNA or protein and which are used in mapping the genome.

Ethanol (ethyl alcohol) (C_2H_6O) Commonly used to disinfest plant tissues, glassware utensils and working surfaces in tissue culture manipulations. The concentration used is 70% (v/v) for disinfecting and 95% (v/v) when flaming tools. Ethanol is also used to dissolve water-insoluble additions (addendums) to culture media.

Ethephon (2 chloroethyl) Phosphonic acid ($C_1C_2PO_3H_6$) Through a spontaneous degradation of ethephon, ethylene is produced. Ethephon is a synthetic compound commonly used to treat cultured cells or unripened fruit with ethylene.

Ether A molecule containing a C-O-H double bond to another O, making this part of an ester linkage, or some other exception holds).

Ethics Ethics is a branch of philosophy that deals with morality. It is concerned with distinguishing between right and wrong human actions, both at an individual and societal level. Ethics may also apply to the rules or standards that specify how particular members of an organisation should conduct themselves.

Ethidium bromide A fluorescent dye used to stain DNA and RNA. The dye fluoresces when exposed to UV light.

Ethylene (C_2H_4) A gaseous plant growth regulator regulating various aspects of vegetative growth, fruit ripening and abscission of plant parts.

Ethylenediamine tetraacetic acid (EDTA) A chelating compound. In tissue culture it is used to keep nutrients, such as iron, bound in a form that leaves them still available to the plant but which prevents them from precipitating out.

Etiolation An abnormal increase in stem elongation, accompanied by poor or absent leaf development. Physiological etiolation is caused by a lack of chlorophyll, and is typical of plants growing under low light intensity or in complete darkness. It can also be caused by disease.

Euchromatin Genetic material that is stained less intensely by certain dyes during interphase, and that comprises many different kinds of genes.

Eugenics Deliberate manipulation of the genetic makeup of human populations, traditionally by selective birth control, infanticide, mass murder, genocide. Genetic engineering offers now opportunities for the eugenicist including genetic screening, in vitro fertilisation plus preimplantation screening, germ line genetic modification etc. Eugenic ideas can be traced to the ancient Greek philosopher Plato, but are more recently associated with Nazi Germany and Hitler.

Eukaryote Any organism characterized by having the nucleus enclosed by a membrane. Eukaryotic organisms include animals, plants, fungi and some algae. They also possess membrane-bound functional organelles, such as mitochondria and chloroplasts, in the cytoplasm of their cells.

Euploid An organism or cell having a chromosome number that is an exact multiple of the monoploid *(n)* number. Terms used to identify different levels in an euploid series are diploid, triploid, tetraploid, and so on.

Eutactic Characterized by precise molecular order, like that of a perfect crystal, the interior of a protein molecule, or a machine-phase system; contrasted to the disorder of bulk materials, solution environments, or biological structures on a cellular scale. Borderline cases can be identified, but perfection is not necessary. As a crystal with sparse defects is best described as a crystal (rather than as amorphous), so a eutactic structure with sparse defects is best described as (imperfectly) eutactic, rather than as disordered.

Eutrophication The dying off of organisms in a lake or pond due to an overabundance of algae which consume all of the dissolved oxygen in the water. This usually happens when the water becomes rich in mineral and organic nutrients, often due to run off of fertilisers from farms.

Evaluation Measurement of the characteristics that are important for production and adaptation, either of individual animals or of populations, most commonly in the context of comparative evaluation of the traits of animals or of populations.

Evolution, theory of The theory of evolution as initially formulated by Charles Darwin in 1859 is the central theory of biology. All processes that enable life are the result of the process of evolution over a period estimated to be more than 3 billion years. The mechanism of evolution are mutation and natural selection. These two processes result in changes at the genetic (mutation) and physiological level (selection of function).

Evapotranspiration The process of water loss in vapour form from a unit surface of land both directly and through leaf surfaces during a specific period of time.

Ex *situ* conservation of farm animal genetic diversity All conservation of genetic material *in vivo,* but out of the environment in which it developed, and *in vitro* including, *inter alia,* the cryoconservation of semen, oocytes, embryos, cells or tissues. Note that *ex situ* conservation and *ex situ* preservation are considered here to be synonymous.

Ex vitro Organisms removed from tissue culture and transplanted; generally plants to soil or potting mixture.

Ex vivo gene therapy The delivery of a gene or genes to the isolated cells of an individual. After culturing, the transformed cells are introduced back into the individual by transfusion, infusion or injection, to alleviate a genetic disorder.

Excinuclease The endonuclease-containing protein complex that excises a segment of damaged DNA during excision repair.

Excision 1. The natural or *in vitro* enzymatic release (removal) of a DNA segment from a chromosome or cloning vector.

2. Cutting out and preparing a tissue, organ, etc., for culture.

3. Removing adventitious shoots from callus tissue.

Excision repair DNA repair processes that involve the removal of a damaged or incorrect segment of DNA and its replacement by the synthesis of a new strand using the complementary strand of DNA as template.

Excluded volume The presence of one molecule (or moiety) reduces the volume available for other molecules (or moieties); resulting reductions in their entropy are termed *excluded volume effects*.

Excrete To transport a compound out of a cell.

Exit site (E site) The ribosome binding site that contains the free tRNA prior to its release.

Exocrine gland In animals, a gland that secretes through a duct.

Exoergic The opposite of endoergic; describes a transformation that releases energy.

Exogamy The fusion of reproductive cells from distantly related or unrelated organisms, i.e., outbreeding.

Exogenote Chromosomal fragment homologous to an endogenote and donated to a merozygote.

Exogenous DNA DNA that has been derived from a source organism and has been cloned into a vector and introduced into a host cell. Also referred to as foreign DNA or heterologous DNA.

Exon A segment of a eukaryotic gene that is transcribed as part of the primary transcript and is retained, after processing, with other exons to form a functional mRNA molecule.

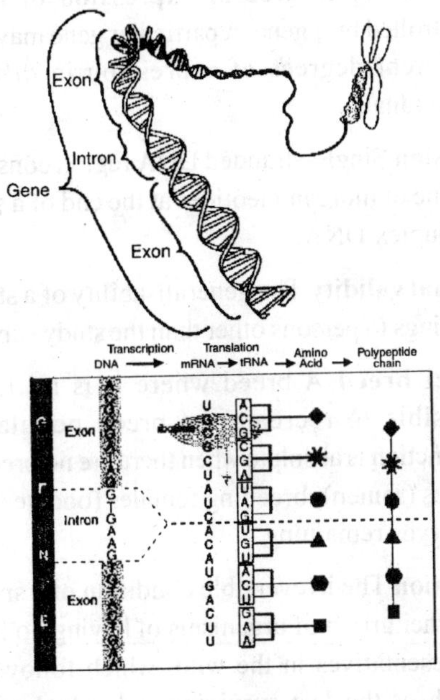

Fig. Excon

Exon amplification A procedure that is used to amplify exons.

Exonuclease An enzyme that digests DNA or RNA, beginning at the ends of strands. It requires a free end in order to degrade a DNA or RNA molecule. 5′ exonucleases require a free 5′ end and degrade the molecule in the 5′®3′ direction. 3′ exonucleases require a free 3′ end and degrade the molecule in the opposite direction..

Exopolysaccharide A high-molecular-weight polymer that is composed of sugar residues and is secreted by a micro-organism into the surrounding environment.

Exothermic The opposite of endothermic; describes an exoergic transformation in which energy is released as heat. Exoergic reactions in solution are commonly exothermic.

Exotic Non-native; refers to an organism that has been introduced into an area.

Exotoxin A toxin released by a bacterium into the medium in which it grows.

Expected progeny difference (EPD) The predicted performance of the future offspring of an animal for a particular trait, calculated from measurement(s) of the animal's own performance and/or the performance of one or more of its relatives, for the trait in question and/ or for one or more correlated traits. Typically, the prediction is expressed as a deviation from a well-defined base population, assuming the animal in question is mated to a sample of animals whose average genetic merit equals that of the base population. The predicted performance of the offspring of the mating between any two animals is the sum of their EPDs.

Experimental design Nutrition science advances by observation (for example, nutritional epidemiologists observe statistical associations between what people eat and their patterns of health and disease) and by experiment, in which subjects are given specifically formulated diets and certain effects are noted. For example, to test the metabolic effects of components of diets, experiments are performed in which a diet containing the test substance (treatment) is compared against a reference diet (control) which, if possible, is the same in all respects other than the test substance and in which the test substance is substituted by a suitable placebo. Subjects should be allocated randomly to diets so that factors other than those tested, and which may otherwise confound the results, are equalised between treatments. Ideally, neither the subjects nor the experimenters should be aware of the identity of treatment and placebo groups (double-blind).

Explant A plant part aseptically excised and prepared for culture in a nutrient medium.

Explant donor The source plant or mother plant from which is taken the explant used to initiate a culture.

Explantation The removal of cells, tissues or organs of animals and plants for observation of their growth and development in appropriate culture media.

Exploratory engineering Design and analysis of systems that are theoretically possible but cannot be built yet, owing to limitations in available tools.

Exponential phase The growth stage where cells undergo their maximum rate of cell division. The exponential phase follows the lag phase, and precedes the linear growth phase.

Export To transport a compound out of a cell.

Exposure assessment The qualitative and\or quantitative evaluation of the likely intake of biological, chemical and physical agents via food as well as exposures from other sources, if relevant.

Express To transcribe and translate a gene's message into a peptide product.

Expression In genetics, manifestation of a characteristic that is specified by a gene. With hereditary disease, for example, a person can carry the gene for the disease but not actually have the disease. In this case, the gene is present but not expressed. In industrial biotechnology, the term is often used to mean the production of a protein by a gene that has been inserted into a new host organism.

Expression library A population of different DNA molecules encoding peptides, that has been cloned into one kind of expression vector.

Expression system Combination of host and vector which provides a genetic context for making a cloned gene function, i.e., produce peptide, in the host cell.

Expression vector A cloning vector that has been constructed in such a way that, after insertion of a DNA molecule, its coding sequence is properly transcribed and the RNA is translated. The cloned gene is put under the control of a promoter sequence for the initiation of transcription, and often also has a transcription termination sequence at its end. Such promoters are termed *high level*; examples include P1 (the leftward promoter of phage l) and the promoter of the yeast PGK (phosphoglycerate kinase) gene.

Expressivity Degree of expression of a trait controlled by a gene. A particular gene may show different degrees of expression in different individuals.

Extension Single-stranded DNA region consisting of one or more nucleotides at the end of a strand of duplex DNA.

External validity The generalisability of a study's findings to persons other than the study subjects.

Extinct breed A breed where it is no longer possible to recreate the breed population. Extinction is absolute when there are no breeding males (semen), breeding females (oocytes), nor embryos remaining.

Extinction The irreversible condition of a species or other group of organisms of having no living representatives in the wild, which follows the death of the last surviving individual of that species or group. Extinction may occur on a local or global level; it can result from various human activities, including the destruction of habitats or the overexploitation of species that are hunted or harvested as a resource.

Extracellular Inside the body but not within cells; hence, extracellular compartment refers to the space, mainly water, outside the cells.EczemaAn itching red rash, often on the face, hands or skins folds, which oozes and crusts if scratched.

Extrachromosomes Self-replicative genetic elements separate from main chromosome(s) of a cell. This definition usually excludes viruses, but the division is somewhat arbitrary. In bacteria, plasmids are the principal extrachromosomes; they encode functions which are not essential to the growth and division of the host cell. In eukaryotes, extrachromosomes may be either essential or dispensable. They may inhabit (i) the nucleus, e.g., extrachromosomal rDNA molecules, yeast 2mm plasmid;

(ii) the cytosol, e.g., dsRNA molecules in fungi; or

(iii) the cytoplasmic organelles, e.g., mitochondrial DNA, chloroplast DNA. Eukaryotic extrachromosomal elements may be recognized genetically by their failure to show segregation at meiosis.

Extremophiles Microorganisms that live at extreme levels of pH, temperature, pressure and salinity.

Exude Slowly discharge; leak liquid material (exudate such as tannins or oxidized polyphenols) through pores or cuts, or by diffusion into the medium. In some woody plant species, exudation is associated with a lethal browning of explants.

F

F factor A bacterial episome that confers the ability to function as a genetic donor in conjugation; the fertility factor in bacteria.

F1 Abbreviation for filial generation 1. The initial hybrid generation resulting from a cross between two parents.

F1 hybrid First generation after a hybrid has been formed.

F2 The second filial generation, produced by crossing two members of the F1, or by self-pollinating the F1. The "grandchildren" of a given mating.

Factor VII and IX Soluble blood proteins that forms part of the cascade of the 12 reactions of blood clotting. Factor VII deficiency is associated with haemophilia A while factor IX deficiency is associated with haemophilia B.

Factor VIII A large, complex protein that aids in blood clotting and is used to treat hemophilia.

Factorial mating A mating scheme in which each male parent is mated with each female parent. Made possible in animals by means of *in vitro* embryo production. Such a mating scheme substantially reduces the rate of inbreeding in a selection programme.

Facultative anaerobe An organism that will grow under either aerobic or anaerobic conditions.

FAD (flavin adenine dinucleotide) A co-enzyme important in various biochemical reactions. It comprises a phosphorylated vitamin B2 (riboflavin) molecule linked to the nucleotide adenine monophosphate (AMP). It functions as a hydrogen acceptor in dehydrogenation reactions, being reduced to $FADH_2$. This in turn is oxidized to FAD by the electron transport chain, thereby generating ATP (two molecules of ATP per molecule of $FADH_2$).

Fail-stop Describes a component or subsystem that, in the event of a failure, produces no output (e.g., of material or data) rather than producing a damaged or incorrect output.

False-negative A negative assay result that should have been positive.

False-positive A positive assay result that should have been negative.

Familiar(ity) The knowledge of the characteristics of a given organism and experience with the use of that organism in Canada.

Farm animal genetic resources (AnGR) Those animal species that are used, or may be used, for the production of food and agriculture, and the populations within each of them. These populations within each species can be classified as wild and feral populations, landraces and primary populations, standardized breeds, selected lines, and any conserved genetic material.

Fat Triglycerides (triacylglycerol) that are either solid (e.g. in margarine, shortening, lard, etc.) or liquid (e.g. vegetable or fish oil) at room temperature.

Fat mimetics Compounds that carry out only some of the functions of fat in foods.

Fat soluble vitamins Vitamins - organic substances that cannot be produced in the body but are essential for cellular functions and must be obtained from the diet - carried in fat and usually acquired from fatty foods. Vitamins A, D, E and K are the fat-soluble vitamins.

Fat substitutes Compounds that can replace conventional fat in foods and carry out all of the same functions.

Fatty acid Compound such as palmitic acid that has a carboxylic acid attached to a long hydrocarbon chain. Used as a major source of energy during metabolism and as a starting point for the synthesis of phospholipids.

Fatty liver An abnormal accumulation of fat in the liver associated, in extreme cases, with abdominal discomfort. It often precedes other liver diseases such as cirrhosis.

Fatty streaks Slightly raised fatty deposits in arteries that do not cause obstruction; possibly but not definitely precursors of atherosclerotic plaque.

Fault-tolerant Describes a system that can suffer failure in a component or subsystem, yet continue to function correctly.

Fed-batch fermentation Culture of cells or micro-organisms where nutrients are added periodically to the bioreactor.

Feed conversion A productivity measure of an animal. Defined as the ratio of feed consumed to weight gained.

Feedback inhibition The process by which the accumulated end product of a biochemical pathway stops synthesis of that product. A late metabolite of a synthetic pathway regulates synthesis at an earlier step of the pathway.

Feedstock The raw material used for chemical or biological processes.

Feral A term used to describe domestic or introduced animals living in wild conditions or plants that have become wild.

Fermentation Metabolism to extract energy from substrates. In the context of dietary fibre, fermentation involved anaerobic (without oxygen) degradation of indigestible carbohydrates by the microflora (mainly bacteria) of the large intestine. Fermentation may also involve the degradation of sugars or starches (following enzymatic digestion by saliva) in the mouth by cariogenic bacteria.

Fermentation substrates Materials used as food for growing micro-organism. The fermentation substrates and the trace materials needed, together with chemicals added to make the fermentation easier, form the culture medium.

Fermented foods Foods that have been subjected to processing involving the action of yeasts or bacteria.

Fertile Of an organism: Capable of breeding and reproduction.

Fertilization The union of two gametes from opposite sexes to form a zygote; it involves the fusion of nuclei of gametes (karyogamy) and the fusion of cytoplasm (plasmogamy). Typically, each gamete contains a haploid set of chromosomes. Hence, after fusion of the nuclei, the resulting nucleus contains a diploid set of chromosomes.

Several categories are distinguished: 1. Self-fertilization (selfing) - fusion of male and female gametes from the same euploid organism.

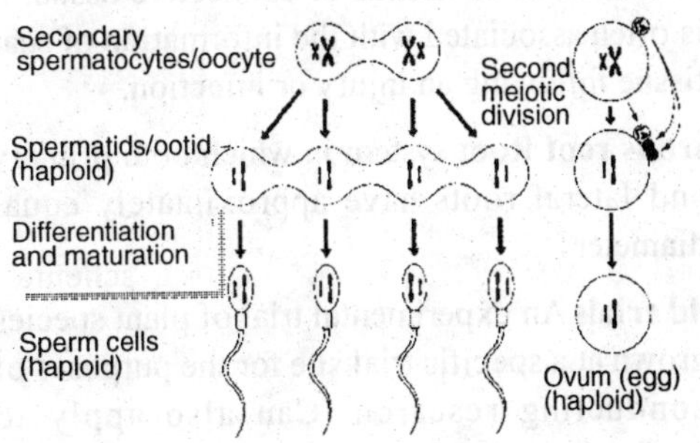

Fig. Fertilization

2. Cross-fertilization (crossing) - fusion of male and female gametes from different euploid individuals.

3. Double fertilization - in angiosperms, the fusion of one male gamete with the ovum at about the same time as the second male gamete fuses with the female polar nuclei (or secondary nucleus) to form the endosperm.

Feulgen's test A histochemical test in which the distribution of DNA in the chromosomes of dividing cell nuclei can be observed. A tissue section is first treated with dilute hydrochloric acid to remove the purine bases of the DNA, thus exposing the aldehyde groups of the sugar deoxyribose. The section is then immersed in Schiff's reagent, which combines with the aldehyde groups to form a magenta-coloured compound.

Fibres Elongated cells with tapering, pointed ends; the cells interlock to form a strong, rigid tissue; pits in the walls are usually very narrow and not very numerous.

Fibril Microscopic to sub-microscopic cellulose thread that is part of the cellulose matrix of plant cell walls.

Fibroblast Common cell type found in connective tissue. Secretes an extracellular matrix rich in collagen and other extracellular matrix macromolecules. Migrates and proliferates readily in wounded tissue and in tissue culture.

Fibrosis A condition in which there is a proliferation of fibrous or connective tissue. It is often associated with the information of scar tissue following an injury or infection.

Fibrous root Root system in which both primary and lateral roots have approximately equal diameters.

Field trials An experimental trial of plant species grown at a specific trial site for the purposes of conducting research. Can also apply to environmental products, such as pesticides. The trial can be unconfined or confined.

Field selection trait Observable characteristics which make it possible to select certain varieties of an organism from a group of them grown in their natural habitat.

Filter bioreactor; mesh bioreactor Cells are grown on an open mesh of an inert material, which allows the culture medium to flow past it but retains the cells. This is similar in idea to membrane and hollow fibre reactors, but can be much easier to set up, being similar to conventional tower bioreactors, but with the meshwork replacing the central reactor space.

Filter sterilization Process of sterilizing a liquid by passage through a filter with pores so small that they prevent the passage of micro-organisms and microbial spores.

Filtration 1. Separation of solids from liquids by using a porous material that only allows passage of the liquid or of solids of a smaller than the pore size of the filter. The material passing the filter forms the filtrate.

2. Removal of cell aggregates to obtain a filtrate of single cells that can be utilized as plating inocula.

Fission Asexual reproduction involving the division of a single-celled individual into two new single-celled individuals of equal size.

Fitness The number of offspring left by an individual, often compared with the average of the population or with some other standard, such as the number of offspring left by a particular genotype.

Fixation The situation in which only one allele for a given gene/locus is present in a population. This can occur as a result of direct selection where the allele delivers a greater level of fitness; because of indirect selection, where the locus is linked to a gene that is subject to direct selection; or because of genetic drift.

Fixative Chemical reagent such as formaldehyde or osmium tetroxide used to preserve cells for microscropy. Samples treated with these reagents are said to be "fixed," and the process is called fixation.

Flagellum Long, whiplike protrusion whose undulations drive a cell through a fluid medium. Eucaryotic flagella are longer versions of cilia; bacterial flagella are completely different, being smaller and simpler in construction.

Flaming A technique for sterilizing instruments, to remove live micro-organism contaminants. The instrument is dipped in alcohol, and the alcohol remaining on the instrument is ignited, thereby heat-sterilizing the surface.

Flanking region The DNA sequences extending on either side of a specific locus or gene.

Flavr savr (trademark) A tomato produced by Calgene, a biotechnology company now owned by Monsanto. It is genetically engineered so the enzyme that causes softening and rotting is blocked. This means the tomato can be picked ripe yet stay firm during transport.

Floccule A micro-organism aggregate or colloidal particle floating in or on a liquid. The cloudy appearance of micro-organism contaminated liquid media illustrates the flocculation phenomenon.

Flow cytometry A technique used to sort cells or other biological materials by means of flow through apertures of defined size or by laser sorting.

Flower The structure in angiosperms (flowering plants) that bears the organs for sexual reproduction.

Fluid mosaic membrane The fluid mosaic model of cell membranes describes the structural and dynamical organization of biological membranes. It is composed of phospholipids that form large planar bilayers. In-between phospholipids exist membrane proteins and the alternating composition of phospholipids and proteins found in membranes has been compared to a mosaic structure. In addition, both components are not fixed in space but can freely move within the plane of the membrane. This 'fluidity' similar to the fluid or liquid state of water if it is nor frozen is essential for the proper function of proteins in membranes.

Fluorescein Fluorescent dye that fluoresces green when illuminated with blue light or ultraviolet light.

Fluorescence activated cell sorting (FACS) The use of laser beams to detect differences in fluorescence between different types of cells in a mixture, and the subsequent deflection of cells into separate bins corresponding to each type of cell in the mixture. One of the popular uses of this technology is in sperm sexing.

Fluorescence immunoassay (FIA) An immunoassay based on the use of fluorescence-labelled antibody.

Fluorescence in situ hybridization (FISH) Hybridization of cloned, fluorescently labelled DNA or RNA, to intact biological materials, notably chromosome spreads and thin tissue sections. The technique allows the visualization of the physical location of nucleic acid sequences homologous to the probe, and is used for the placement of genes on chromosomes and for the spatial and temporal pattern of gene expression of specific mRNA molecules.

Fluorescent dye Molecule that absorbs light at one wavelength and responds by emitting light at another wavelength; the emitted light is of longer wavelength (and hence of lower energy) than the light absorbed.

Fluorescent probe A probe which is labelled with a fluorescent dye, so that the signal emitted can be captured by photometric methods.

Focal contact (adhesion plaque) Small region on the surface of a fibroblast or other cell that is anchored to the extracellular matrix. The attachment is mediated by transmembrane proteins such as integrins, which are linked, through other proteins, to actin filaments in the cytoplasm.

Foetal alcohol syndrome Defects in the growth and development of the foetus resulting from excessive consumption of alcohol by the mother or possibly even the father.

Foetus Pre-natal stage of a viviparous animal, between the embryonic stage and parturition.

Fog Fine particles of liquid suspended in the air, such as of water in a fog chamber used for acclimatizing recent *ex vitro* transplants.

Fold-back The structure formed when a double-stranded DNA molecule containing an inverted repeat sequence is denatured and then allowed to re-anneal at low DNA concentrations. The repeated sequence permits the formation of a double-stranded region within each of the separated strands of the original molecule.

Folded genome The condensed intracellular state of the DNA in the nucleoid of a bacterium. The DNA is segregated into domains, and each domain is independently negatively supercoiled.

Follicle An enclosing cluster of cells that protects and nourishes a cell or structure within. Thus a follicle in the ovary contains a developing egg cell, while a hair follicle envelops the root of hair.

Follicle cell One of the cell types that surround a developing oocyte or egg.

Follicle stimulating hormone (FSH) A hormone, secreted by the anterior pituitary gland in mammals, that stimulates the ripening of the specialized structures in the ovary (Graafian follicles) that produce ova in female mammals; and in males, the formation of sperm in the testis. FSH is a major constituent of fertility drugs.

Follower A component in a cam system that is driven through a pattern of displacements as it rests against a moving contoured surface.

Food and agricultural organisation (FAO) United Nations organisation concerned with promoting agricultural development, improved nutrition, the pursuit of food security, and the access of all people to food.

Food and drug administration (FDA) The U.S. agency responsible for regulation of biotechnology food products. The major laws under which the agency has regulatory powers include the Food, Drug, and Cosmetic Act; and the Public Health Service Act.

Food chain Encompasses the whole process of food production, from the fields of the farmer and the food their animals eat, through to food manufacturing, processing and distribution, all the way to your home.

Food frequency questionnaire A method of dietary assessment in which subjects are asked to recall how frequently certain foods were consumed during a specified period of time.

Food processing enzyme Enzyme used to control food texture, flavour, appearance and, to a certain extent, nutritional value. Amylases break down complex polysaccharides to simplex sugars; proteases tenderize meat proteins. Biotechnology can assist the development of new food enzymes by finding or engineering enzymes which fit better with the other processes which the food must undergo, like cooking or canning.

Food record A method of dietary assessment in which subjects records the foods that they consume.

Food safety objective A government-defined target considered necessary to protect the health of consumers (this may apply to raw materials, a process or finished products).

Food standards australia new Zealand The body responsible for developing, varying and reviewing standards for food available in Australia and New Zealand. They also coordinate national food surveillance and recall systems, conduct research, assess policies about imported food and develop codes of practice with industry.

Foot and mouth disease All species of cloven-hoofed animals are susceptible to FMD, including domestic livestock and wild ungulates such as buffalo, antelope and warthogs. Clinical signs are essentially similar in all species although the severity may vary considerably. The principal signs are pyrexia followed by vesicle formation in the mouth and feet resulting in salivation and lameness. The disease is caused by a virus and is highly contagious but not dangerous to man.

Foot-candle An obsolete photometric measure of light intensity. Now superseded by the lux (symbol: lx) (1 fc » 10.7 lx).

Forced cloning The insertion of foreign DNA into a cloning vector in a predetermined orientation.

Foreign DNA A DNA molecule that is incorporated into either a cloning vector or a chromosomal site.

Formulation For traditional therapeutic agents, this refers to the method by which a therapeutic agent is delivered to its site of action.

Fortify To add strengthening components or beneficial ingredients to a nutrient medium.

Forward mutation A mutation from the wild type to the mutant type.

Fouling The coating or plugging (by materials or micro-organisms) of equipment, thus preventing it from functioning properly.

Founder animal In transgenesis research, an organism that carries a transgene in its germ line and can be used in matings to establish a pure-breeding transgenic line, or one that acts as a breeding stock for transgenic animals.

Founder principle The possibility that a new, small, isolated population may be genetically different from the "parent" population, because the founding individuals (being a small, random sample from the large, "parent" population) could be quite different from typical members of the "parent" population.

Four base cutter; four base pair cutter; four-cutter A type II restriction endonuclease that binds (and subsequently cleaves) DNA at sites that contain a sequence of four nucleotide pairs that is uniquely recognised by that enzyme. Because any sequence of four bases occurs more frequently by chance than any sequence of six bases, four-base cutters cleave more frequently than do six-base cutters. Thus, four-base cutters create smaller fragments than six-base cutters.

Fractal A nested pattern that shows the same symmetry/geometry at any scale a pattern is looked at. It is an example of an infinite pattern in both larger and smaller dimensions. Fractals are geometrical abstractions and sometimes used to explain complexity in living organisms, also the comparison does not hold up on closer inspection. As abstractions, they are thought to continue into infinity, even the infinitely small, although physicists believe that there is a material limit to the what constitutes the smallest dimension. As for complex biological organisms, their structure is hierarchical with higher levels having emergent structures and properties not found at the lower levels. Thus it is not a true fractal.

Fractionation Separation.

Frameshift mutation A mutation that changes the reading frame of a DNA, either by the insertion or the deletion of nucleotides. Because of the triplet nature of codons, this occurs if the number of nucleotides involved is not a multiple of three.

Free energy Free energy is a measure of the ability of a system to do work, such that a reduction in free energy could in principle yield an equivalent quantity of work. The Helmholtz free energy describes the free energy *within* a system; the Gibbs free energy does not.

Free energy change (DG) Change in the free energy during a reaction: the free energy of the product molecules minus the free energy of the starting molecules. A large negative value of DG indicates that the reaction has a strong tendency to occur.

Free living conditions Natural or greenhouse conditions where the plantlets are transferred from *in vitro* conditions to soil mixtures. In such instances, plantlets must manufacture their own food supply for survival.

Free radical A highly reactive chemical species that normally exists for a relatively short time. Some free radicals are formed in the body during processes of oxidation and may be useful, e.g., in killing infectious organisms. Free radicals are also capable of doing extensive damage to tissues unless kept in check by antioxidants. The latter can be enzymes or chemicals, many of which are vitamins obtained from the diet (e.g., vitamins C and E).

Free water Water released by a cell when freezing occurs in intercellular spaces.

Freeze-drying The process of drying a tissue or an organ in a frozen state under vacuum. Tissues are freeze-dried to measure their dry weight or to preserve them for future analysis. Freeze-drying is the standard way of preserving micro-organisms for long periods of time.

Frequency distribution A graph showing either the relative or absolute incidence of classes in a population. The classes may be defined by either a discrete or a continuous variable; in the latter case, each class represents a different interval on the scale of measurement.

Fresh weight The weight, including the water content, of a plant or plant part at the time of harvest.

Friable A term commonly used to describe a crumbling or fragmenting callus. A friable callus is easily dissected and readily dispersed into single cells or clumps of cells in solution.

Functional foods Are food products that confer specific health benefits, in addition to their broader accepted food value. For instance - fish and their ability to provide omega 3 fatty acids to lower the risk of heart disease.

Functional genomics The field of research, that aims to determine patterns of gene expression and interaction in the genome, based on the knowledge of extensive or complete genomic sequence of an organism.

Fungicide A substance or chemical that destroys or inhibits the growth of a fungus.

Fungus/fungi Multinucleate single-celled or multicellular heterotrophic micro-organisms, including yeasts, moulds, and mushrooms. They live as parasites, symbionts, or saprophytes. Lacking any vascular tissues (unlike plants), their cell walls are made of chitin or other non-cellulose compounds.

Furfural; furfuraldehyde Used industrially as a solvent and as a raw material for synthetic resin.

Fusion Joining of the membrane of two cells, thus creating a daughter cell that contains some of the same properties from each parent cells. Used in making hybridomas.

Fusion biopharmaceuticals Biopharmaceutical proteins formed as a result of fusion proteins. Their advantages are:

- synergistic activities in one molecule; thus when the molecule binds to a cell, it does two things at once;

- the adverse effect or poor stability of one molecule may be offset by the properties of the other; and

- one molecule acts as a targeting mechanism to bring the other to the site where it is meant to act.

Fusion gene A hybrid gene created by joining portions of two different genes (to produce a new protein) or by joining a gene to a different promoter (to alter or regulate gene transcription).

Fusion protein A polypeptide made from a recombinant gene that contains portions of two or more different genes. The different genes are joined so that their coding sequences are in the same reading frame: the genetic apparatus reads the gene fusion as a single gene, and so produces a fusion protein. Also known as hybrid protein or chimeric protein. These proteins are used:

- to add an affinity tag to a protein;

- to produce a peptide as part of a larger protein, which is then cut up after it has been made by cloning;

- to produce a protein with the combined characteristics of two natural proteins; or

- to produce a protein where two different activities are physically linked.

G

G Guanine residue in either DNA or RNA.

G proteins Proteins with an important role in relaying signals in mammalian cells. The proteins occur on the inner surface of the plasma membrane and transmit signals from outside the membrane, via transmembrane receptors, to adenylate cyclase, which catalyses the formation of the second messenger, cyclic AMP, inside the cell. G-proteins derive their name from their ability to bind to guanine nucleotides, namely GTP and GDP; the GTP-protein complex is able to activate adenylate cyclase, whereas the GDP-protein complex is unable. G-proteins are activated when the signalling molecule (typically a hormone) binds to the transmembrane receptor.

G cap The 5′-terminal methylated guanine nucleoside that is present on many eukaryotic mRNAs; it is joined, after transcription, to the mRNA.

G0 phase (G "zero" phase) State of withdrawal from the eucaroytic cell-division cycle by entry into a quiescent G1 phase; often seen in differentiated cells.

G1 phase Gap 1 phase of the eucaryotic cell-division cycle, between the end of cytokinesis and the start of DNA synthesis.

G2 phase Gap 2 phase of the eucaryotic cell-division cycle, between the end of DNA synthesis and the beginning of mitosis.

GAG (glycosaminoglycan) Long, linear, highly charged polysaccharide composed of a repeating pair of sugars, one of which is always an amino sugar. Mainly found covalently linked to a protein core in extracellular matrix proteoglycans. Examples include chondroitin sulfate, hyaluronic acid, and heparin.

Gall A tumorous growth in plants.

Gamete A mature reproductive cell which is capable of fusing with a cell of similar origin but of opposite sex to form a zygote from which a new organism can develop. Gametes have a haploid chromosome content. In animals, a gamete is a sperm or egg; in plants, it is pollen, spermatic nucleous, or ovum.

Gamete and embryo storage Storage of ova, sperm or fertilized embryos outside their original source. Almost invariably this means cryopreservation in liquid nitrogen.

Gametic (phase) equilibrium In relation to any two loci, the occurrence of haplotypes (gametes) with a frequency equal to the product of the frequency of the two relevant alleles, e.g., loci A and B are in linkage equilibrium if the frequency of the haplotype (gamete) *AiBi* equals the product of the frequencies of alleles *A*i and *B*i.

Gametoclone A plant regenerated from a tissue culture originating from gametic tissue.

Gametogenesis The process of the formation of gametes.

Gametophyte That phase of the plant life cycle that bears the gamete producing organs; the cells have *n* chromosomes. In angiosperms, the pollen grain is the male gametophyte and the embryo sac is the female gametophyte.

Gametophytic incompatibility A phenomenon controlled by the complex S locus, in which a pollen grain cannot fertilize an ovule produced by a plant that carries the same S allele as the pollen grain. Thus S1 pollen cannot fertilize an ovule of a plant possessing the S1 allele.

Ganglion (plural ganglia) Cluster of nerve cells and associated glial cells located outside the central nervous system.

Ganglioside Any glycolipid having one or more sialic acid residues in its structure. Found in the plasma membrane of eucaryotic cells and especially abundant in nerve cells.

Gap 1. Period of time, during the cell cycle, between M and S phases.

2. A missing section on one of the strands of double-stranded DNA. The DNA will therefore have a single-stranded region.

Gap junction Communicating cell-cell junction that allows ions and small molecules to pass from the cytoplasm of one cell to the cytoplasm of the next.

Gapped DNA A duplex DNA molecule with one or more internal single-stranded regions.

Gas transfer The rate at which gases are transferred from gas into solution. It is an important parameter in fermentation systems because it controls the rate at which the organism can metabolize. Gas transfer can be done by several methods, including use of small gas bubbles, that diffuse faster than larger ones, thanks to their larger surface area per unit of volume; or spreading the liquid out, for example in a thin sheet, or in a thin permeable tube, as in hollow fibre bioreactor.

Gastrointestinal tract The organ along which food travels from the mouth until the undigested remnants emerge as stools. Mixing and some digestion occur in the stomach where the environment is acidic; most digestion and absorption of nutrients occurs in the small intestine; the large intestine, principally the colon, contains very large numbers of micro-organisms capable of fermenting food components that have escaped digestion in the small intestine.GelA complex solid and liquid food system, interconnected to give a network of intermingled particles.

Gastrula Animal embryo at an early stage of development where cells are invaginating to form the rudiment of a gut cavity.

Gate In digital logic, a component that can switch the state of an output dependent on the states of one or more inputs.

GC island A segment of DNA that is rich in G=C base pairs and often precedes a transcribed gene in the genomes of vertebrate organisms.

GDP Guanosine diphosphate.

Gel A lyophilic colloid that has coagulated to a rigid or jelly-like solid. It is used for the electrophoretic separation of nucleic acids or proteins, and for encapsulation.

Gel electrophoresis An analytical method for separating molecules according to their size. Samples are put at one end of a slab of polymer gel; an electric field across the gel pulls the molecules through it; the smaller molecules pass more easily and so move towards the other end faster; the various sizes of molecules end up at different positions according their size. Gels are made from different materials, but common combinations are

Various chemicals can be included in the gel to help separation, such as the detergent sodium dodecyl sulphate (SDS) in protein gels to unfold proteins, or urea in DNA sequencing gels, which unfolds DNA. polyacrylamide gel electrophoresis pulsed field gel electrophoresis (PFGE).

Gel. filtration A method ofprotein or DNA purification, where differences in size are used to separate the components of a complex mixture.

Gelatin A glutinous, proteinaceous gelling and solidifying agent. Gelatin is produced by boiling animal connective tissues, which partially hydrolyses the collagen. Gelatin is used to gel or solidify nutrient solutions for tissue culture.

GelriteTM The brand name of a *Pseudomonas*-derived refined polysaccharide used as a gelling agent and agar substitute.

GEM Genetically engineered micro-organism.

Gene Conceptually, the unit of heredity transmitted from generation to generation during sexual or asexual reproduction. More generally, the term is used in relation to the transmission and inheritance of particular identifiable traits. Since the molecular revolution, it is now known that a gene is a segment of nucleic acid that encodes peptide or RNA.

Gene addition The addition of a functional copy of a gene to the genome of an organism.

Gene amplification The presence of multiple genes. Amplification is one mechanism through which proto-oncogenes are activated in malignant cells.

Gene bank 1. The physical location where collections of genetic material in the form of seeds, tissues or reproductive cells of plants or animals are stored.

2. Field gene bank: A facility established for the *ex situ* storage and maintenance, using horticultural techniques, of individual plants. Used for species whose seeds are recalcitrant, or for clonally propagated species of agricultural importance, such as apple varieties).

3. A collection of cloned DNA fragments from a single genome. Ideally the bank should contain cloned representatives of all the DNA sequences in the genome.

4. A population of micro-organisms, each of which carries a DNA molecule that was inserted into a cloning vector. Ideally, all of the cloned DNA molecules together represent the entire genome of another organism. Also called gene library, clone bank, bank, library. This term is sometimes also used to denote all of the vector molecules, each carrying a piece of the chromosomal DNA of an organism, prior to the insertion of these molecules into a population of host cells.

Gene cloning The process of synthesizing multiple copies of a particular DNA sequence using a bacteria cell or another organism as a host. The gene of interest is inserted into a self-replicating DNA molecule (DNA vector, often a plasmid) and the resulting recombinant DNA molecule is amplified in an appropriate host cell. Used in genetic engineering.

Gene conservation; genetic resources conservation The conservation of species, populations, individuals or parts of individuals, by *in situ* or *ex situ* methods, to provide a diversity of genetic materials for present and future generations.

Gene construct A sequence of genes made by splicing several genes together.

Gene expression Conversion of the genetic information within a gene, into an actual protein (or cell process).

Note that many genes are only expressed at specific times during the lifetime of an organism. Some genes are expressed in a "cascade" of related expressions.

Gene flow The spread of genes from one breeding population to another (usually) related populations by migration, possibly leading to changes in allele frequency.

Gene frequency The percentage of a given allele in a population of organisms.

Gene gun A device for propelling DNA molecules into living cells.

Gene imprinting The differential expression of a single gene according to its parental origin.

Gene insertion The addition of one or more copies of a normal gene into a defective chromosome.

Gene interaction The modification of the action of one gene by another, non-allelic gene.

Gene knockout The replacement of a normal gene with a mutated form of the gene by using homologous recombination. Used to study gene function.

Gene knockout strategy An approach used to determine the function of a specific gene by inactivating (knocking out) that gene in the intact organism and studying the consequences of this modification.

Gene library Collection of cloned DNA fragments that ideally includes all the genetic information of an organism. If the original source of the DNA is the genomic DNA from an organism, then it is called a genomic library; if the DNA is from cDNA made by enzymatic copying of RNA, then the library includes representative fragments from all genes that were being expressed at the time the RNA was sampled, and would be called a cDNA library.

Gene linkage The hereditary association of genes located on the same chromosome.

Gene machine A computerized device for synthesizing genes by combing nucleotides (bases) in the proper order.

Gene mapping The process of determining where genes are located on individual chromosomes, their position in relation to other genes and the distance between them.

Gene modification The chemical repair of a gene's defective DNA sequence.

Gene pair Corresponding genes in each of a pair of matching (homologous) chromosomes.

Gene pool 1. The sum of all genetic information in a breeding population at a given time.

2. In plant genetic resources, use is made of the terms 'primary', 'secondary' and 'tertiary' gene pools. In general, members of the primary gene pool are inter-fertile; those of the secondary can be crossed with those in the primary gene pool under special circumstances; but to introgress variation from the tertiary gene pool, special techniques are required to achieve crossing.

Gene probe A single-stranded DNA or RNA fragment used in genetic engineering to search for a particular gene or other DNA sequence. The probe has a base sequence complementary to the target sequence and will thus attach to it by basepairing. By labelling the probe, it can be identified after subsequent separation and purification.

Gene product Proteins resulting from the transcription of a gene.

Gene replacement The incorporation of a transgene into a chromosome at its normal location by homologous recombination, thus replacing the copy of the gene originally present at the locus.

Gene (DNA) sequencing Determining the exact sequence of nucleotide bases in a strand of DNA to better understand the behavior of a gene.

Gene splicing A stage in the processing of mRNA, occurring only in eukaryote cells, in which intervening sequences (introns) are removed from the primary RNA transcript (hnRNA), and the coding exons are joined together to form the mature mRNA molecule.

Gene stacking Accumulation of several advantageous traits in an organism, usually considered in the context of herbicide tolerance genes in plants (weeds).

Gene technology Branch of modern biotechnology. It includes a range of techniques used by scientists in an attempt to control or modify genes for beneficial outcomes. Examples include utilising genes from within a species (taking a disease resistant gene from one type of pineapple and placing it in another that provides a high level of sweetness) or moving genes between two unrelated species (called recombinant DNA technology) to create a transgenic organism (such as Bt cotton).

Gene testing Methods that identify the presence, absence or mutation of a particular gene in an individual.

Gene therapy The treatment of inherited diseases by introducing into the cells of affected individuals the wild-type copies of the defective gene causing the disorder. If reproductive cells are modified, the procedure is called germ-line or heritable gene therapy. If cells other than reproductive cells are modified, the procedure is called somatic-cell or non-inheritable gene therapy.

Gene tracking Following the inheritance of a particular gene from generation to generation.

Gene translocation The movement of a gene fragment from one chromosomal location to another, which often alters or abolishes expression.

Genera Plural of genus, a group of similar species.

Generally regarded as safe (GRAS) Designation given to foods, drugs, and other materials that have been used for a considerable period of time and have a history of not causing illness to humans, even though extensive toxicity testing has not been conducted. More recently, certain host organisms for recombinant DNA experimentation have been given this status.

Generate To propagate or (mass) proliferate. The process is generation or regeneration.

Generative nucleous One of the two male gametes in the pollen tube of angiosperms.

Genetic Inherited; a genetic disease is one that is inherited via a faulty gene.

Genetic assimilation Botany: Eventual extinction of a natural species as massive pollen flow occurs from another related species and the natural species becomes more like the related species.

Genetic blueprint The blueprints for a living organism are contained in the genes inside all its cells. The genes that an organism is born with define many of the characteristics it has throughout life — they form a 'blueprint'.

Genetic code 1. The set of 64 nucleotide triplets (codons) that specify the 20 amino acids and termination codons (UAA, UAG, UGA).

Second letter

First letter	U	C	A	G	
U	UUU UUC Phenyl-alanine	UCU UCC UCA UCG Serine	UAU UAC Tyrosine	UGU CGC Cystelne	U C
	UUA UUG Leucine		UAA Stop codon UAG Stop codon	UGA Stop codon UGG Tryptophan	A G
C	CUU CUC CUA CUG Leucine	CCU CCC CCA CCG Proline	CAU CAC Histidine	CGU CGC CGA CGG Arginine	U C
			CAA CAG Glutamine		A G
A	AUU AUC AUA Isoleucine	ACU ACC ACA ACG Threonine	AAU AAC Asparagine	AGU AGC Serine	U C
	AUG Methionine initiation codon		AAA AAG Lysine	AGA AGG Arginine	A G
G	CUU CUC CUA CUG Valine	GCU GCC GCA GCG Alanine	CAU CAC Aspartic acid	GGU GGC GGA GGG Glycine	U C
			CAA CAG Glutarnic acid		A G

Fig. Genetic code

2. The relationships between the nucleotide base-pair triplets of an mRNA molecule and the 20 amino acids that are the building blocks of proteins.

Genetic complementation When two DNA molecules that are in the same cell together produce a function that neither DNA molecule can supply on its own.

Genetic counselling The counselling of individuals and prospective parents who are at risk of a particular genetic disease, either themselves or their potential child. It provides them and their families with education and information about genetic-related conditions such as the probabilities, dangers, diagnosis and treatment, and helps them make informed decisions.

Genetic determinism Determinism is the doctrine that all acts, choices and events are the inevitable consequence of antecedent sufficient causes. Genetic determinism is the doctrine that the organism is the inevitable consequence of its genetic makeup, or the sum of its genes.

Genetic disability, disorder or disease Afflictions which can be associated with what some class as defects in a person's genetic inheritance. They may involve changes (mutations) in single genes or in whole chromosomes, parts of which may be lost, duplicated or placed in the wrong position; or from the interaction of multiple genes and external factors in fetal development. Later in life such interactions seem to be the basis of many of the common serious disorders, such as heart disease, diabetes and cancer, although these are not usually thought of as gene disorders.

Genetic distance A measure of the genetic similarity between any pair of populations. Such distance may be based on phenotypic traits, allele frequencies or DNA sequences. For example, genetic distance between two populations having the same allele frequencies at a particular locus, and based solely on that locus, is zero. The distance for one locus is maximum when the two populations are fixed for different alleles. When allele frequencies are estimated for many loci, the genetic distance is obtained by averaging over these loci.

Genetic diversity The heritable variation within and among populations which is created, enhanced or maintained by evolutionary forces.

Genetic drift Change in allele frequency from one generation to another within a population, due to the sampling of finite numbers of genes that is inevitable in all real (finite) populations. The smaller the population, the greater is the genetic drift. Sooner or later (depending on the size of the population), genetic drift results in loss of alleles from a population, and hence leads to a loss of genetic variation. Because of this, the minimization of genetic drift is an important consideration for conservation of genetic resources.

Genetic engineering Changes in the genetic constitution of cells (apart from selective breeding) resulting from the introduction or elimination of specific genes through modern molecular biology techniques. This technology is based on the use of a vector for transferring useful genetic information from a donor organism into a cell or organism that does not possess it.

A broader definition of genetic engineering also includes selective breeding and other means of artificial selection.

Genetic equilibrium Condition in a group of interbreeding organisms in which the allele frequencies remai constant over time.

Genetic fingerprinting A technique which enables genetic relationships between close relatives, or the identity of individuals to be established - usually beyond reasonable doubt.

Genetic heterogeneity The situation in which different mutant genes produce the same phenotype.

Genetic immunization Delivery to a host organism of a cloned gene that encodes an antigen. After the cloned gene is expressed, it elicits an antibody response that protects the organism from infection by a virus, bacterium or other disease-causing organism.

Genetic information Refers to the information content of genes or genomes and is organized in sequences of nucleotides, structures of genes, and arrangement of genes within genomes. Genetic information is the information that is inherited from generation to generation and can be accessed by the cellular machinery with the help of proteins to synthesize all necessary components for the growth maintenance of an organism.

Genetic inheritance The transfer of genetic instructions for different characteristics, such as the colour and size of the cell or organism, to successive generations, via reproduction.

Genetic linkage map A linear map of the relative positions of genes along a chromosome. Distances are established by linkage analysis, which determines the frequency at which two gene loci become separated during chromosomal recombination.

Genetic manipulation Another term for genetic engineering.

Genetic marker A harmless variable inherited change in DNA or protein that can be used to locate a diseased gene on a particular chromosome. Any segment of DNA that can be identified, or whose chromosomal location is known, so that it can be used as a reference point to map or locate other genes. Any gene which has an identifiable phenotype that can be used to track the presence or absence of other genes on the same piece of DNA transferred into a cell.

Genetic modification A technique where individual genes can be copied and transferred to another living organism to alter its genetic make up and thus incorporate or delete specific characteristics into or from the organism. The technology is also referred to as genetic engineering, genetic manipulation and gene technology. Genetic modification can be applied intra species or across species as in the case of transgenics.

Genetic predisposition Susceptibility to disease that is related to a genetic predisposition mutation, which may or may not result in actual development of the disease.

Genetic screening The use of a specific biological test to screen for inherited diseases or medical conditions. Testing can be conducted prenatally to check for metabolic defects and congenital disorders in the developing fetus as well as postnatally to screen for carriers of heritable diseases.

Genetic selection The process of selecting genes, cells, clones, etc., within populations or between populations or species. Genetic selection usually results in differential success rates of different genotypes, reflecting many variables, including selection pressure and genetic variability in populations.

Genetic stability A measure of the resistance to change, with time, of the sequence of genes within a DNA molecule or of the nucleic acid sequence within a gene.

Genetic tags Genes of certain observable traits that are used as markers when introducing foreign genes into an organism. They are usually "tagged" onto the desired genes and indicate whether an organism has successfully incorporated those desired genes into its genome. The gene for antibiotic resistance is often used as a genetic tag. Also known as selectable markers.

Genetic testing The analysis of an individual's genetic material. Genetic testing can be used to gather information on an individual's genetic predisposition to a particular health condition, or to confirm a diagnosis of genetic disease.

Genetic transformation The transfer of extracellular DNA among and between species by using bacterial or viral vectors.

Genetic variation Differences between individuals attributable to differences in genotypes.

Genetically engineered organism (GEO) An organism produced through genetic engineering.

Genetically modified foods Any food containing parts of genetically modified plants, animals or microorganisms. There is some debate whether foods containing products from genetically modified organisms, but not parts of the organism themselves, should be classified as genetically modified foods. (For example, rennet, an enzyme used in cheese, can be produced from genetically modified micro-organisms, but it does not actually contain any of the modified DNA.

Genetically modified organism(s) (GMO) An organism (plant, animal, bacteria, or virus) that has had its genetic material altered, either by the duplication, insertion or deletion of one or more new genes, or by changing the activities of an existing gene.

Genome The genome denotes the full set of genes or genetic information of an organism. It included both coding and non-coding sequences and is physically partitioned into chromosomes. The importance of coding sequences is for protein synthesis and RNA synthesis, while non-coding sequences contain regulatory sequences, sequences of old, inactive genes, repeat sequences that allow recombination of genetic information (DNA pieces) from different chromosomes, locations, or even foreign DNA that is introduces by microbial or viral infection.

Many non-coding sequences are transposable elements meaning that they can copy and insert themselves at many different sites within chromosomes. These rearrangement of physical location of DNA strands affects number, location, and sequence of genes coding for proteins and RNA and thus are vital for generating mutations important for evolutionary fitness of an organism. The human genome has been found to contain only 5% coding sequences (genes that make proteins), while half of all non-coding portions are made of transposable elements reminiscent of viral DNA. This similarity between human and viral DNA indicates that human evolution cannot be thought of as independent but is closely related to the evolution of viruses. The modern existence of pathogenic viruses is an indication of the importance of this co-evolution that likely has helped humans to maintain a heterogeneous genepool important for rapid adaptation to environmental changes.

Genomic library A collection of biomolecules made from DNA fragments of a genome that represent the genetic information of an organism that can be propagated and then systematically screened for particular properties. The DNA may be derived from the genomic DNA of an organism or from DNA copies made from messenger RNA molecules. A computer-based collection of genetic information from these biomolecules can be a "virtual genomic library."

Genotype The genetic description of an organism; often only one or two genes out of thousands are of interest in a genetic comparison between individuals or the analysis of genealogical traits, family history etc.. The genotype is the genetic information underlying a phenotype, the exterior expression of characteristics or traits (e.g. eye color). Most phenotypes that are really physical or functional attributes of an organism are multifactorial, meaning that several genes contribute to its expression. Even at the level of cellular mechanism and metabolism, phenotypes (e.g. photosynthesis) are multifactorial because of the particular composition of enzymes as protein complexes, where individual proteins are coded for by different genes. Photosynthesis, for example, is the concerted action of dozens of proteins (genes) with copy numbers in the hundreds to enable a simple chemical equation: carbon dioxide + water = sugar. In fact this simple overall equation is really performed in many different subsets of reactions.

Genus A category including closely related species. Interbreeding between organisms within the same category can occur.

Geotropism A growth curvature induced by gravity.

Germ 1. In botany, a common name for a plant embryo.

2. Colloquial: A disease-causing micro-organism.

Germ cell A member of a celllineage (the germ line) leading to the production of gametes. In mammals, germ cells are found in the germinal epithelium of the ovaries and testes. Synonym: germ line cell.

Germ cell gene therapy The repair or replacement of a defective gene within the gamete-forming tissues, resulting in a heritable change in an organism's genetic constitution.

Germ layers; primary germ layers The layers of cells in an animal embryo at the gastrula stage, from which the various organs of the animal's body will be derived.

Germ line A lineage of "generative" cells (germ track) ancestral to the gametes (sperm and egg cells proper) which, during the development of an organism (animal or plant), are set aside as potential gamete-forming tissues. These ancestral cells, together with the gametes, are called germ cells, as opposed to somatic cells. Location, nature and time of formation of potential gamete-forming tissues are species specific, and may vary greatly from one species to another.

Germ line gene therapy The delivery of a gene or genes to a fertilized egg or an early embryonic cell. The transferred gene(s) is present in all or some of the nuclei of the cells of the mature individual, including possibly the reproductive cells, and alters the phenotype of the individual that develops.

Germicide Any chemical agent used to control or kill any pathogenic and non-pathogenic micro-organisms.

Germinal epithelium 1. A layer of epithelial cells on the surface of the ovary that are continuous with the mesothelium.

2. The layer of epithelial cells lining the seminiferous tubules of the testis, which gives rise to spermatogonia.

Germination 1. The initial stages in the growth of a seed to form a seedling. The embryonic shoot (plumule) and embryonic root (radicle) emerge and grow upwards and downwards respectively. Food reserves for germination come from endosperm tissue within the seed and/or from the seed leaves (cotyledons).

2. The growth of spores (fungal; algal) and pollen grains.

Germplasm 1. The genetic material that forms the physical basis of hereditary and which is transmitted from one generation to the next by means of the germ cells.

2. An individual or clone representing a type, species or culture, that may be held in a repository for agronomic, historic or other reasons.

Gestation The period in animals bearing live young (especially mammals) from fertilization of the egg to birth of the young (parturition).

Gibberellins Plant growth regulators involved in elongation, enhancement of flower, fruit and leaf size, germination, vernalization and other processes.

Gibbs free energy The Gibbs free energy is the Helmholtz free energy plus the product of the system volume and the external pressure. Changes in the Gibbs free energy at a constant pressure thus include work done against external pressure as a system undergoes volumetric changes. This proves convenient for describing equilibria in gases and liquids at a constant pressure (e.g., at one atmosphere), but is of little use in describing machine-phase chemical processes. Changes in the Gibbs free energy caused by a change in the applied pressure (at constant volume) have no direct physical significance.

Gill irrigation Oxygen transfer from water to the blood in fish.

Gland A group of cells or a single cell in animals or plants that is specialized to secrete a specific substance. In animals, there are two types of glands: endocrine glands discharge their products directly into the blood vessels; exocrine glands secrete through a duct or network of ducts into a body cavity or onto the body surface. In plants, glands are specialized to secrete certain substances produced by the plant. The secretions may be retained within a single cell, or secreted to the outside.

Glaucus A surface with a waxy, white coating. In most cases, this waxy covering can be rubbed off.

Glial cells Cells of the nervous system that aid neurons. They provide nutrition for the neurons, and insulate them from each other. They make up 40 percent of the volume of the brain and spinal cord.

Globulins Common proteins in blood, eggs and milk, and as a reserve protein in seeds. Globulins are insoluble in water, but soluble in salt solutions. Alpha, beta and gamma globulins can be distinguished in blood serum. Gamma globulins are important in developing immunity to diseases.

GLP (good laboratory practice); GMP (good manufacturing practice) They are codes of practice designed to reduce to a minimum the chance of accidents which could affect a research project or a manufactured product. The GLP and GMP prescriptions are quite voluminous, but boil down to a few key points. The essential point of both GLP and GMP is that everything is recorded, and that only established procedures are used and by people who have been trained to use them.

Glucose The major carbohydrate in starch and fruit sugar. The latter is also known as succrose and contains fructose with every glucose molecule.

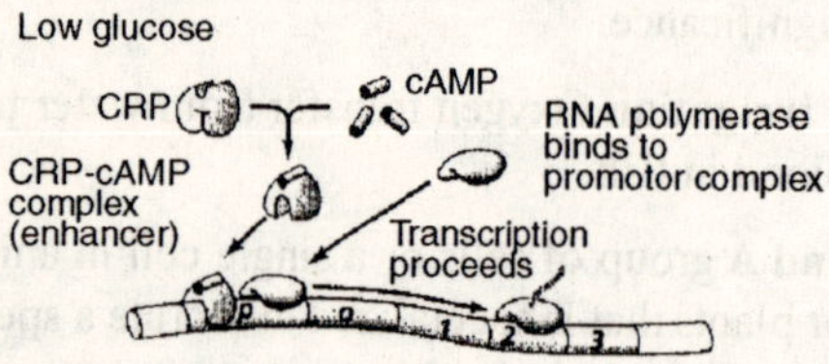

Fig. Glucose

Glucocorticoid A steroid hormone that regulates gene expression in higher animals.

Glucose invertase; glucose isomerase Enzymes that catalyse the interconversion of the two sugars, glucose and fructose. As fructose is chemically more stable than glucose, a mixture of glucose and fructose with the enzyme will end up almost entirely as fructose. This is valuable for the food industry, as fructose is substantially sweeter than glucose, and so more sweetness per gram is achieved using fructose. The usual use for glucose isomerase is to take glucose made by hydrolysis of corn starch and turn it into a mixture of mostly fructose with some glucose. The corn starch is broken down using amylases. The result is called high-fructose corn syrup (HFCS). Invertase takes sucrose and turns it into glucose and fructose.

Glucose tolerance test A test of the body's ability to metabolise carbohydrate by administering a standard dose of glucose under controlled conditions and measuring the blood and urine for sugar at regular intervals thereafter. The glucose tolerance test is usually used to assist in the diagnosis of diabetes or other disorders that affect carbohydrate metabolism.

Glucosinolates A class of molecules produced in the seeds and green tissue of a range of plants, in particular brassicas. Their natural role is thought to be involved in plant-insect interactions. Their importance in plant breeding is largely because of their negative influence on taste and their positive effect on the prevention of cancers of the alimentary tract.

Glutathione A small-molecular-weight antioxidant molecule produced naturally in the human body and present in some foods.

Gluten A mixture of two proteins, gliadin and glutenin, occurring in the endosperm of wheat grain. Their amino acid composition varies, but glutamic acid (33%) and proline (12%) predominate. The composition of wheat glutens determines the strength of the flour and whether or not it is suitable for biscuit or bread making. Sensitivity of the lining of the intestine to gluten occurs in coeliac disease, a condition that must be treated by a gluten-free diet.

Gluthathione peroxidases A family of antioxidant molecule produced naturally in the human body and present in some foods.

Glycaemic index A method for assessing the comparative effects of different carbohydrates on the pattern of changes in the concentration of glucose in the blood following a meal. A dose of 50 g of glucose is assigned a glycaemic index of 100, given by integrating the area under the curve when blood glucose concentration is plotted against time.

Glycoalkaloids Toxic secondary organic compounds found in the potato family.

Glycogen The major complex carbohydrate in animal cells made of glucose. Glycogen is for animals what starch is for plants. Although glycogen is an important long term storage of energy in muscle and liver cells, it is of little nutritional significance, because most glycogen in muscle spontaneously degrades during slaughtering. Thus meats, except liver, have little or no carbohydrate content.

Glycolysis A metabolic pathway found in all organisms. This pathway consists of ten chemical reactions catalyzed by proteins (enzymes) and is responsible for the degradation *and* synthesis of carbohydrates. Glycolysis does not depend on the presence of oxygen and is able to provide the cell with the universal energy currency called ATP, short for adenosine triophosphate. This pathway can degrade glucose only partially and will produce waste products such as lactate (in mammalian muscle causing sour muscle under anaerobic exercise conditions) or ethanol in microorganisms (used for fermentation of wine or beer). In the presence of oxygen, no waste products are formed and instead further degraded to carbon dioxide and water. The latter processes are known as Krebs cycle and oxidative phosphorylation

Glycoprotein A protein conjugated with a carbohydrate group.

Glycosylation The covalent addition of sugar or sugar-related molecules to proteins or polynucleotides.

GM food Abbreviation for genetically modified food. Food that contains above a certain legal minimum content of raw material obtained from genetically modified organisms.

GMAC (genetic manipulation advisory committee) A government expert advisory committee that provided guidance to the government and industry on the safe and responsible development and use of a gene technology in Australia prior to the commencement of the Gene Technology Act 2000 in June 2001.

Golden rice A biotechnology-derived rice, which contains large amounts of beta carotene (a precursor of vitamin A) in its seeds. Achieved by inserting two genes from daffodil and one from the bacterium Erwinia uredovora.

Golgi apparatus An assembly of vesicles and folded membranes within the cytoplasm of plant and animal cells that stores and transports secretory products (such as enzymes and hormones) and plays a role in formation of a cell wall (when this is present). It is named after its discoverer, the Italian cytologist Camillo Golgi (1843-1926).

Gonad Any of the usually paired organs in animals that produce reproductive cells (gametes). The most important gonads are the male testis, which produces spermatozoa, and the female ovary, which produces ova (egg cells). The gonads also produce hormones that control secondary sexual characteristics.

Gonad ridge The area of cells within an embryo which will develop into the gonads of foetus. This usually develops around 32 days after fertilisation

Good laboratory practice (GLP) Written codes of practice designed to reduce to a minimum the chance of procedural or instrument problems which could adversely affect a research project or other laboratory work.

Good manufacturing practice (GMP) Codes of practice designed to reduce to a minimum the chance of procedural or instrument/ manufacturing plant problems which could adversely affect a manufactured product.

Graft 1. To place a detached branch or bud (scion) in close cambial contact with a rooted stem (rootstock) in such a manner that scion and rootstock unite to form a single plant.

2. Colloquial synonym for scion.

Graft inoculation test A test based on the use of a suspected viral carrier which is grafted to an indicator plant. If symptoms appear in the indicator plant, the viral assay is positive.

Graft union The point at which a scion from one plant is joined to a stock from another plant.

Gram negative organism Any prokaryotic organism that does not retain the first stain (crystal violet) used in Gram's staining technique. It does retain the second stain (safranin O) and therefore has a pink colour when viewed under a light microscope. Retention of the stain is due to the structure of the cell wall.

Gram positive organism Any prokaryotic organism that retains the first stain used in the

Gram technique, which gives a purple-black colour when viewed under a light microscope. Retention of the stain is due to the structure of the cell wall.

Gram staining A technique to distinguish between two major bacterial groups, based on stain retention by their cell walls. Bacteria are heat-fixed, then stained with crystal violet, followed by iodine solution, and then rinsed with alcohol or acetone. Gram-positive bacteria are stained bright purple, while Gram-negative bacteria are decolourized.

Grana Structures within chloroplasts, seen as green granules with the light microscope and as a series of parallel lamellae with the electron microscope. Disc- or sac-like structures found in chloroplasts composed of stacked membranes and containing the chlorophyll and carotenoid pigments directly involved in photosynthesis (singular: granum).

Granulocyte One of three types of white blood cells. Granulocytes digest bacteria and other parasites.

Granulocyte-macrophage colony stimulating factor (GMCSF) A natural hormone that stimulates white blood cell production, particularly that of granulocytes and monocytes (the precursors of macrophages).

Gratuitous inducer A substance that can induce transcription of a gene or genes, but is not a substrate for the induced enzyme(s).

Green revolution Name given by William Goud to the dramatic increase in crop productivity during the third quarter of the 20th century, as result of integrated advances in genetics and plant breeding, agronomy, and pest and disease control.

Gro luxTM A wide-spectrum fluorescent lamp suitable for plant growth purposes.

Ground state The lowest-energy state of a system. The electronic ground state of a system cannot reduce its energy by an electronic transition, but may contain vibrational energy (kinetic and potential energy associated with the motions and positions of its atoms); extended systems at ordinary temperatures are always vibrationally excited, and so "ground state" is often taken to mean "electronic ground state."

Group A set of linked atoms in a molecule; a defined substructure. Typically, a set that is usefully regarded as a unit in chemical reactions of interest.

Group velocity In wave propagation, the speed of the waveform (e.g., of a peak) can be different from the speed of a group of waves (e.g., of a set of ripples in water). The latter is the *group velocity*, and is the speed of propagation of information and wave energy. The waveform speed is the *phase velocity*.

Growth An increase in cell size or cell number, or both, resulting in an increase in dry weight.

Growth cabinet A cupboard used for incubating tubes or culture vessels under controlled environmental conditions. The degree of control over temperature, light and humidity is a function of the quality of the cabinet.

Growth factor Any of various chemicals, particularly polypeptides, that have a variety of important roles in the stimulation of new cell growth and cell maintenance. They bind to the cell surface on receptors. Specific growth factors can cause new cell proliferation.

Growth hormone (GH); somatotrophin; somatotropin A hormone, secreted by the mammalian pituitary gland, that stimulates protein synthesis and growth of the long bones in the legs and arms. It also promotes the breakdown and use of fats as an energy source, rather than glucose. Production of growth hormone is greatest during early life. Its secretion is controlled by the opposing actions of two hormones from the hypothalamus: somatocrinin (growth-hormone-releasing hormone), which promotes its release; and somatostatin (growth-hormone-inhibiting hormone), that inhibits the release of growth hormone from the anterior pituitary gland.

Growth inhibitor Any substance inhibiting the growth of an organism. The inhibitory effect can range from mild inhibition (growth retardation) to severe inhibition or death (toxic reaction). Two plant growth regulators that may act as inhibitors are ethylene and abscisic acid. The concentration of the inhibitor, the length of exposure to it, and the relative susceptibility of the organisms exposed to the inhibitor, are all important factors which determine the extent of the inhibitory effect.

Growth phases; growth phase curve The characteristic periods in the growth of a bacterial culture, as indicated by the shape of a graph of viable cell number versus time, namely: lag phase; logarithmic (or exponential) phase ; stationary phase; death phase.

Growth rate Increase in mass per unit of time.

Growth regulator A synthetic or natural compound that at low concentrations elicits and controls growth responses in a manner similar to hormones.

Growth retardant A chemical that selectively interferes with normal hormonal promotion of growth and other physiological processes, but without appreciable toxic effects.

Growth ring Any of the rings that can be seen in a cross-section of a woody stem, such as a tree trunk. It represents the xylem formed in one year as a result of fluctuating activity of the vascular cambium.

Growth substance Any organic substance, other than a nutrient, that is synthesized by plants and regulates growth and development. They are usually made in a particular region, such as the shoot tip, and transported to other regions, where they take effect.

GTAC Gene therapy advisory committee.

Guanine (G) A chemical that is a basic structural unit of a living thing's DNA. It is one of 4 key bases or nucleotides (cytosine (C), adenine (A) and thymine (T)) that repeat throughout the length of a DNA chain. The order of their sequence defines the arrangement of genes.

Guard cell Specialized epidermal cell that occurs as a pair around a stoma and controls opening and closing of the stoma through changes in turgor.

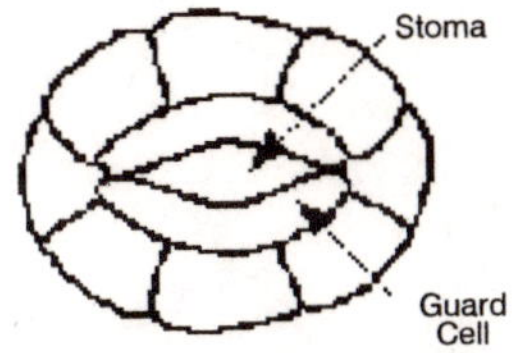

Fig. Guard cell

Guide RNA An RNA molecule that contain sequences that function as a template during RNA editing.

Guide sequence An RNA molecule (or a part of it) which hybridizes with eukaryotic mRNA and aids in the splicing of intron sequences. Guide sequences may be either external (EGS) or internal (IGS) to the RNA being processed and may hybridize with either intron or exon sequences close to the splice junction.

Guidelines Documents produced by the regulatory agencies to accompany the regulations and acts. They provide the steps an individual should follow with respect to a given act or regulation. They are not enforceable by law, but failure to follow them may result in actions contrary to an act or regulation, which is enforced by law.

Guthrie test A bacteria-based diagnostic test used to detect phenylketonuria (PKU) in newborns. The test measures concentrations of amino acids in a sample of blood.

Gymnosperm Any plant whose ovules and the seeds into which they develop are borne unprotected, rather than enclosed in ovaries, as are those of the flowering plants (the term gymnosperm means naked seed).

Gynandromorph An individual in which one part of the body is female and another part is male; a sex mosaic.

Gynogenesis Female parthenogenesis; after fertilization of the ovum, the male nucleus is eliminated and the haploid individual (described as gynogenetic) so produced possesses the maternal genome only.

H

Habituation The phenomena that, after a number of sub-cultures, cells can grow, without the addition of specific factors, such as no longer needing exogenous growth regulators in the tissue culture medium. Such cells are autonomous.

Haematopoietic stem cells Stem cells that make all the blood cells in the body. They are found in the bone marrow - the tissue that fills most bone cavities.

Haemoglobin The red oxygen carrying pigment of the blood. Haemoglobin incorporates two pairs of polypeptides called globins. Various gene mutations which affect the kinds of globins made, or their amounts, can result in diseases called the haemoglobinopathies, notably thalassaemias and sickle cell disease.

Haemolymph The mixture of blood and other fluids in the body cavity of an invertebrate.

Haemophilia A single gene disorder, which affects about 1 in 1000 males, in which there is defective production of a protein that is necessary for blood clotting and thereby prevents excessive bleeding. Its inheritance is X'-linked, i.e. the gene is on the X chromosome.

Hair A single or multicellular, absorptive (root hair) or secretory (glandular hair) and sometimes only a superficial outgrowth (covering hair) of the epidermal cells. The term trichome is often used but includes outgrowths from tissue below the epidermal layer. Distinguishing between hairs and trichomes can be difficult. Trichomes are usually connected to the vascular system, whereas hairs lack a vascular connection.

Hair cell Specialized sensory epithelial cell in the ear with bundles of giant microvilli (stereocilia) protruding from its apical surface. Sound vibrations tilt the stereocilia, evoking an electrical change in the hair cell, which thus acts as a sound detector.

Hairy root culture A fairly recent development in plant culture, consisting of highly branched roots of a plant. A plant tissue is treated with a culture of the bacterium *Agrobacterium rhizogenes*, which transfers part of its own plasmid DNA to the cells of an infected plant. This alters the plant's metabolism, including alterations in hormone levels, which in turn cause the explant to grow highly branched roots from the sites of infection. The roots branch much more frequently than the usual root system of that plant, and are also covered with a mass of tiny root hairs. Their most significant feature is that they produce secondary metabolites at levels similar to those made in the original plant. Thus they can be used as replacement plants for making such compounds as food flavours or fragrances.

Hairy root disease A disease of broad-leaved plants, where a proliferation of root-like tissue is formed from the stem. Hairy root disease is a tumorous state similar to crown-gall, and is induced by the bacterium *Agrobacterium rhizogenes,* containing an Ri plasmid.

Halophyte A plant that can tolerate a high concentration of salt in the growing medium.

Halothane A volatile anaesthetic.

Haploid A cell or organism containing only one representative from each of the pairs of homologous chromosomes found in the normal diploid cell. Having a single complete set of chromosomes, or referring to an individual or generation containing such a single set of chromosomes per cell. Usually a gamete.

Haploid cell A cell containing only one set, or half the usual (diploid) number, of chromosomes.

Haplotype 1. A group of alleles, each from a different locus in the same region of a chromosome, that exist in the same double helix.

Hapten The portion of an antigen that determines its immunological specificity. When coupled to a large protein, a hapten stimulates the formation of antibodies to the two-molecule complex. Also called antigenic determinant.

Haptoglobin A serum protein, alpha globulin, that interacts with haemoglobin during recycling of the iron molecule of haemoglobin.

Hardening off Adapting plants to outdoor conditions by gradually withholding water, lowering the temperature, increasing light intensity, or reducing the nutrient supply. The hardening-off process conditions plants for survival when transplanted outdoors. The term is also used for gradual acclimatization to *in vivo* conditions of plants grown *in vitro*, e.g., gradual decrease in humidity.

Hardy-weinberg equilibrium The frequencies of genotypes at a locus resulting from random mating at that locus; for two alleles, *A*1 and *A*2, with respective frequencies *p* and *q*, the Hardy-Weinberg equilibrium frequencies are *p*2 *A*1*A*1, 2*pq* *A*1*A*2, *q*2 *A*2*A*2. Despite the simplifying assumptions required to predict these frequencies, most loci in most populations are in Hardy-Weinberg equilibrium. Thus the Hardy-Weinberg law, which predicts these frequencies, is one of the great unifying themes of biology.

Harmonic oscillator A system in which a mass is subject to a linear restoring force, like an ideal spring. A harmonic oscillator vibrates at a fixed frequency, independent of amplitude.

Harvesting 1. The process involved in gathering in ripened crops.

2. The collection of cells from cell cultures or of organs from donors for the purpose of transplantation.

Hazard A biological, chemical or physical agent in a food, or condition of a food, with the potential to cause an adverse health effect.

Hazard characterisation The qualitative and/or quantitative evaluation of the nature of the adverse health effects associated with biological, chemical and physical agents which may be present in food. For chemical agents, a close response assessment should be performed. For biological or physical agents, a dose-response assessment should be performed if the data are obtainable.

Hazard identification The identification of biological, chemical and physical agents capable of causing adverse health effects which may be present in a particular food or group of foods.

Hazardous Dangerous.

Heat As defined in thermodynamics, heat is the energy that flows between two systems as a result of temperature differences (a system *contains* neither heat nor work, but can *produce* heat or *do* work). Heat thus differs from thermal energy.

Heat capacity The ratio of the heat input to the temperature increase in a system. Note that this definition does not imply that a system *contains* heat, despite the name *heat capacity*.

Heat pump An apparatus that extracts heat from a fluid or gas that is marginally above ambient temperature. Heat pumps are commonly used to heat (or cool) greenhouses and laboratories.

Heat shock protein (stress-response protein) Protein synthesized in response to an elevated temperature or other stressful treatment; usually helps the cell to survive the stress.

Heat-labile Easily destroyed by heat.

Hedonic Generally involving the psychological range of feelings from pleasant to unpleasant.HepatitisInflammation of the liver, generally accompanied by abdominal discomfort. There may be many causes, including bacterial or viral infections and various drugs, including alcohol.

Hela cell Line of human epithelial cells that grows vigorously in culture. Derived from a human cervical carcinoma.

Helix A structure with a spiral shape. The normal state of double-stranded DNA is in the form of a double helix.

Helmholtz free energy The internal energy of a system minus the product of its entropy and temperature.

Helminth A class of parasitic worms, especially those which are internal parasites of man and animals.

Helper cells T cells that respond to an antigen displayed by a macrophage by stimulating B and T lymphocytes to develop into antibody-producing plasma cells and killer T cells, respectively.

Helper plasmid A plasmid that provides a function or functions to another plasmid in the same cell.

Helper virus A virus that provides a function or functions to another virus in the same cell.

Hemagglutination Clumping (agglutination) of red blood cells.

Heme Cyclic organic molecule containing an iron atom that carries oxygen in hemoglobin and carries an electron in cytochromes.

Hemicellulose Any cellulose-like carbohydrate, but with differing chemical composition. Together with pectin and lignin, hemicelluloses form the cell wall matrix.

Hemidesmosome Specialized cell junction between an epithelial cell and the underlying basal lamina.

Hemizygous; haplozygous The condition in which only one allele of a pair is present, as in sex linkage or as a result of deletion. Genes present only once in the genotype and not in pairs; as in haploids, in differential segments of sex chromosomes, or in diploids as a result of aneuploidy or loss of chromosome segments.

Hemophilia An X-linked recessive genetic disease, caused by a mutation in the gene for clotting factor VIII (hemophilia A) or clotting factor IX (hemophilia B), which leads to abnormal blood clotting.

Hemopoiesis (hematopoiesis) Generation of blood cells, mainly in the bone marrow.

HEPA filter (high efficiency particulate air filter) A filter capable of screening out particles larger than 0.3 µm. HEPA filters are used in laminar air flow cabinets (hoods) for sterile transfer work.

Hepatitis Liver inflammation usually caused by a virus.

Herbicide A substance that kills plants. Herbicides are used in agriculture, horticulture and gardening to control unwanted plants. Herbicides can be selective, and kill selected species, or non-selective (broad spectrum), and kill all plants.

Herbicide resistance The ability of a plant to withstand herbicide. Herbicide resistance has been one of the early targets of plant genetic engineering. If a herbicide is sprayed onto a field planted with such resistant crops, then all the plants except the crop would be killed, thus providing an effective method of weed control without having to develop herbicides specific to each weed type. There is substantial concern in some quarters about the widespread use of this technology, which is essentially giving the plant kingdom the ability to evade man's most effective herbicides. The concern are that, firstly, such engineering will lead to increased use of the herbicides, at a time when it is generally accepted that the use of chemicals should be kept as low as possible, and that, secondly, there is the possibility that resistant crop plants will escape to become weeds, or that their resistant genes could be transferred to other species, including weeds.

Herbicide-tolerant Crops were developed to survive certain herbicides that previously would have destroyed the crop along with the targeted weeds, and allow farmers to use them as postemergent herbicides, providing an effective weed control. The most common herbicide-tolerant crops (cotton, corn, soybeans, and canola) are Roundup Ready (RR) crops resistant to glyphosate, a herbicide effective on many species of grasses, broadleaf weeds, and sedges. Other genetically modified herbicide-tolerant crops include Liberty Link (LL) corn resistant to glufosinate-ammonium, and BXN cotton resistant to bromoxynil.

Hereditary disorders A pathological condition due to changes in individual genes, or groups of genes or in sections of chromosomes or whole chromosomes. These changes may be passed from parents to offspring.

Heredity Resemblance among individuals related by descent; transmission of traits from parents to offspring.

Heritability In the narrow sense: 1. the proportion of phenotypic superiority of parents that is seen in their offspring;

2. the proportion of the total phenotypic variation due to variation in breeding values.

In the broad sense: the proportion of the total phenotypic variation due to genetic variation. The degree to which a given trait is controlled by inheritance.

Herman the bull A Dutch bull engineered to contain the human lactoferrin gene which codes for the protein lactoferrin used in the treatment of sepsis. Human lactoferrin is produced in the milk of cows bred from this bull. (Said to be named after Herman de Boer the managing director of Gene Pharming, Holland)

Hermaphrodite 1. An animal that has both male and female reproductive organs, or a mixture of male and female attributes.

2. A plant whose flowers contain both stamen and carpels. Synonym: intersex.

Heteroalleles Mutations that are functionally allelic but structurally non-allelic; mutations at different sites in a gene.

Heterocaryon Cell with two or more nuclei produced by the fusion of two or more different cells.

Heterochromatin Regions of chromosomes that stain darkly even during interphase; thought to be for the most part genetically inactive.

Heterocyclics Cyclic organic molecules containing one or more hetero atoms (e.g., nitrogen, sulphur, oxygen, etc.).

Heterodimer Protein complex composed of two different polypeptide chains.

Heteroduplex A double-stranded DNA molecule or DNA-RNA hybrid, where each strand is of a different origin, and consequently containing one or more mismatched (non-complementary) base pairs. A DNA duplex is prepared by the hybridization of single-stranded DNA molecules derived from two different sources. Where the two DNAs have identical or very similar sequences, a double-stranded molecule will be established, whereas where the two DNAs differ in sequence, single-stranded regions will remain. A heteroduplex will be revealed as single-strand "bushes" when DNA is observed electron microscopically. A map of homologous and non-homologous regions of the two molecules may thereby be constructed. This process is known as heteroduplex mapping.

Heterogametic sex Producing unlike gametes with regard to the sex chromosomes. In mammals, the XY male is heterogametic, and the XX female is homogametic.

Heterogeneous nuclear RNA (hnRNA) Large RNA molecules, which are unedited mRNA transcripts, or pre-mRNAs found in the nucleous of a eukaryotic cell.

Heterokaryon Cell with two or more different nuclei as a result of cell fusion.

Heterologous probe A DNA probe that is derived from one species and used to screen for a similar DNA sequence from another species.

Heteroplasmy A cellular condition in which two genetically different types of organelles are present.

Heteroploid Term given to a cell culture when the cells comprising the culture possess nuclei containing chromosome numbers other than the diploid number.

Heteropyknosis Property of certain chromosomes, or of their parts, to remain more dense during the cell cycle and to stain more intensely than other chromosomes or parts.

Heterotroph Organisms that receive their nourishment by eating other organisms e.g. plants.

Heterozygous Identifies the two alleles of a diploid organism as being different from each other. If both alleles are genetically identical, the cell or organism is homozygous. Often, one of the two alleles is the dominant, the other the recessive allele. If a dominant allele is present, one copy is enough to establish the corresponding phenotype (e.g. dark eye color). The recessive allele for light eye color cannot be expressed, until two copies of the allele (homozygous) are present.

HFR (high frequency recombination) Strain of *Escherichia coli;* in such strains, the F episome is integrated into the bacterial chromosome.

High density lipoproteins (HDL) Plasma lipoproteins containing relatively low concentrations of cholesterol and other lipids; thought to be beneficial because they cycle cholesterol out of tissues.

Histamine An amine found in all tissues of the body. Large amounts are released by mast cells when the body encounters a substance to which it is sensitive, thereby triggering symptoms of inflammation.

Histocompatibility The degree to which tissue from one organism will be tolerated by the immune system of another organism.

Histocompatibility antigen An antigen that causes the rejection of grafted material from an animal different in genotype from the host animal.

Histocompatibility complex; histocompatibility system The collection of genes coding for peptides present on the surface of nucleated cells; these peptides are responsible for the differences between genetically non-identical individuals that cause rejection of tissue grafts between such individuals. These peptides were originally called histocompatibility antigens. They are now called histoglobulins, reflecting their structural similarity to immunoglobulins.

Histology Science that deals with the microscopic structure of animal and plant tissues.

Histone Group of water-soluble proteins rich in basic amino acids, closely associated with DNA in plant and animal chromatin. Histones are involved in the coiling of DNA in chromosomes and in the regulation of gene activity.

Hollow fibre A tube of a porous material, having an internal diameter of a fraction of a millimetre, and so its ratio of surface area to internal volume is very large. This has had two types of application. Firstly, hollow fibres can be used as filters. Because they have a huge surface area, they take much longer to clog up than normal filters. Secondly, they are used in the hollow fibre bioreactor, in which cells are kept inside the hollow, porous fibres, and the culture medium is circulated outside the reactor. The fibres let nutrients in and products out (as they are in solution), but do not allow the passage of cells. Hollow fibre bioreactors are very effective for maintaining mammalian cells in culture because they have a very large surface area for the cells to grow on without needing a large reactor to hold them, and because the nutrient reaching the cells can be kept fresh. The reactor also provides an easy way of removing the product that the cells are making: such as monoclonal antibodies. Hollow fibre reactors are less use when the cells themselves have to grow, because it is hard to get at the inside of the fibre to remove surplus cells.

Holometabolous An insect that undergoes complete metamorphosis to the adult from a morphologically distinct larval stage.

Homeobox A DNA sequence found in several genes that are involved in the specification of organs in different body parts in animals; they are characteristic of genes that influence segmentation in animals. The homeobox corresponds to an amino acid sequence in the polypeptide encoded by these genes; this sequence is called the homeodomain.

Homeostasis In biology used to describe a condition where an organism maintains a stable structure where in fact a constant flux of molecules occurs. Although many organisms can live for years, all cellular components like proteins, membranes, sugars, and nucleic acids are constantly recycled while never compromising the integrity of the organism as a whole. This turnover processes can be characterized by specific half-life values that for most proteins, membranes, and RNA (but not DNA structures) are measured in hours. In a more narrow sense homeostasis refers to the maintenance of water and salt concentration in cells.

Homeotic mutation A mutation that causes a body part to develop in an inappropriate position in an organism, such as the mutation in *Drosophila* that causes legs to develop on the head in place of antennae.

Homoalleles Mutations that are both functionally and structurally allelic; mutations at the same site in the same gene.

Homodimer A protein with two identical polypeptide chains.

Homogametic sex Producing similar gametes with regard to the sex chromosomes. In mammals, the XY male is heterogametic, and the XX female is homogametic.

Homogenotization A genetic technique used to replace one copy of a gene, or other DNA sequence within a genome, with an altered copy of that sequence. The DNA is first cloned and then altered in some way, e.g., a transposon is inserted into a gene. The mutated gene copy can be used to replace the original gene by recombination *in vivo*. The incorporation of the mutated gene is usually selected, for example, by virtue of its containing a transposon-encoded antibiotic resistance.

Homokaryon Cell with two or more identical nuclei as a result of fusion.

Homologous 1. General definition: from the same source, or having the same evolutionary function or structure.

2. Of chromosomes: identical with respect to gene content and linear ordering. Homologous chromosomes pair and recombine with one another at meiosis.

3. Of DNA/proteins: identical, or nearly identical nucleotide/amino acid sequence.

Homologous chromosomes Chromosomes that occur in pairs and are generally similar in size and shape: one comes from the male parent and the other from the female. Such chromosomes contain the same linear array of genes.

Homologous recombination The exchange of DNA fragments between two DNA molecules or chromatids of paired chromosomes (during crossing over) at the site of identical nucleotide sequences.

Homology The degree of identity between individuals, or characters. The degree of identity between the nucleotide sequences of two nucleic acid molecules or the amino acid sequences of two protein molecules. Although sequence determination is the ultimate test of homology, useful estimates can be provided by either DNA-DNA or DNA-RNA hybridization.

Homoplasmy A cellular condition in which all copies of an organelle are genetically identical.

Homopolymer A nucleic acid strand that is composed of one kind of nucleotide, e.g., GGGGGGGG.

Homozygote An individual that has two copies of the same allele at a particular locus in its two homologous chromosomes.

Honolulu technique A type of nuclear transfer used to clone mammals. It uses donor cells that stay dormant longer so that they do not have to undergo the extra step of being grown in a medium, as in the Roslin technique.

Horizontal gene transfer Transfer of genes from one individual to another, of the same or different species, usually by means other than cross-breeding. Said to have happened naturally in evolution and can now be done artificially.

Hormones Messenger substances synthesized in the body and secreted by the endocrine glands. Hormones regulate the digestive system, growth, hunger, thirst, blood glucose and cholesterol levels, fat burning and storage, absorption and excretion, internal clocks such as day and night cycles, menstrual cycles, and sex drive.

Host An animal or plant on which, or in which, a parasite lives. While the parasite receives nourishment and support from the host, the host does not benefit and is often harmed by the association.

Host cell A cell that houses foreign molecules or viruses and supports them using its own metabolism.

Host specific toxin A metabolite produced by a pathogen which has a host specificity equivalent to that of the pathogen. Such toxins are utilized for *in vitro* selection experiments to screen for tolerance or resistance to the pathogen.

Host vector system Combination of DNA-receiving cells (host) and DNA-transporting substance (vector) used for introducing foreign DNA into a cell.

Human genome project An international research effort aimed at discovering the full sequence of bases in the human genome. Led in the United States by the National Institutes of Health and the Department of Energy.

Human growth hormone (HGH, somatotrophin) A protein produced in the pituitary gland that stimulates the liver to produce somatomedins, which stimulate growth of bone and muscle.

Human immunodeficiency virus (HIV) The virus that causes acquired immune deficiency syndrome (AIDS).

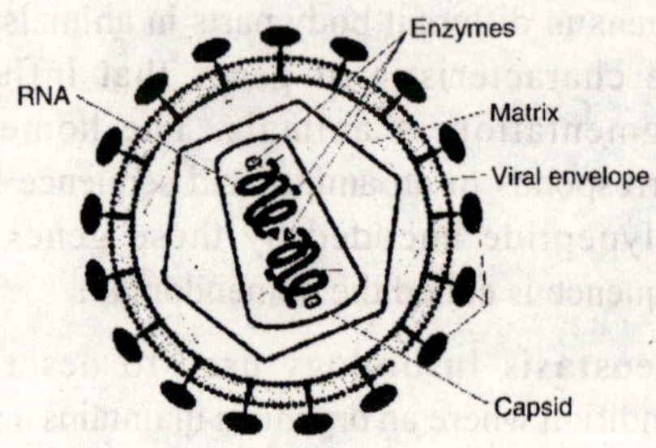

Fig. HIV

Human leukocyte antigen (HLA) The unique proteins on the cells beloning to one's body. They distinguish the cells that blong to the body from those that do not. There are over 100 different HLAs.

Human serum albumin Soluble blood proteins that make up about 55% of plasma proteins. They are involved in maintaining fluid balance in the blood.

Humoral immune response The production of antibody by B cells of the immune system in response to the presence of a foreign antigen.

Huntington disease An inherited disease due to a defective gene on the short arm of chromosome

4. It results in loss of motor control and mental deterioration. Symptoms frequently do not appear until after reproductive age, meaning the defective gene may already have been passed on to offspring when symptoms develop.

Hybrid A plant resulting from pollination between parents that are related, but not genetically identical; or the offspring of two different species. Hybrids can be developed using traditional breeding techniques, or the process can be hastened using gene marker technology to rapidly identify parents with desired genes for certain attributes. Most commercial cereal crops, are hybrids. Seed from a hybrid plant if planted, will not deliver the same benefits as the original seeds and after several offspring, will have lost the desired qualities from the original hybridization.

Hybrid arrested translation A method used to identify the proteins encoded by a cloned DNA sequence. A crude cellular mRNA preparation, composed of many individual types of mRNA, is hybridized with cloned DNA. Only mRNA molecules homologous to the cloned DNA will anneal to it. The rest of the mRNA molecules are put into an *in vitro* translation system and the protein products are compared with the proteins obtained by use of the whole mRNA preparation.

Hybrid cell The mononucleate cell which results from the fusion of two different cells, leading to the formation of a synkaryon.

Hybrid dysgenesis A syndrome of abnormal germ-line traits, including mutation, chromosome breakage, and sterility, which results from activity of transposable elements.

Hybrid released translation A method used to detect the proteins encoded by cloned DNA. The cloned DNA is bound to a nitrocellulose filter and a crude preparation of mRNA is hybridized to the filter-bound DNA. Only mRNA sequences homologous to the cloned DNA will be retained on the filter. These mRNA molecules can then be removed by high temperature or by using formamide. The purified mRNA is then placed in an *in vitro* translation system and the proteins encoded by the message can be analysed by electrophoresis through a polyacrylamide gel.

Hybrid seed 1. Seed produced by crossing genetically dissimilar parents.

2. In plant breeding, used colloquially for seed produced by specific crosses of carefully selected pure lines, such that the F1 crop displays hybrid vigour. As the F1 crop is heterozygous, it does not breed true and so new seed must be purchased each season.

Hybrid selection The process of choosing individuals possessing desired characteristics from among a hybrid population.

Hybrid somaclonal variant cell line The offspring of plant cells from cell cultures. They have enhanced genetic instability.

Hybrid vigour The extent to which a hybrid individual outperforms both its parents with respect to one or many traits. The genetic basis of hybrid vigour is not well understood, but the phenomenon is widespread, particularly ininbreeding plant species. Synonym: heterosis.

Hybridization 1. Interbreeding of species, races, varieties and so on, among plants or animals; a process of forming a hybrid by cross pollination of plants or by mating animals of different types.

2. The production of offspring of genetically different parents, normally from sexual reproduction, but also asexually by the fusion of protoplasts or by transformation.

3. The pairing of two polynucleotide strands, often from different sources, by hydrogen bonding between complementary nucleotides.

Hybridoma A synthetichybrid cell, derived by fusing a B lymphocyte with a tumour cell. The former secretes a single antibody, while the latter confers the property of growing indefinitely in tissue culture. The underlying technology behind the monoclonal antibody.

Hydrate A compound formed by the incorporation of water.

Hydrocarbon A molecule consisting only of H and C.

Hydrogen bond A relatively weak bond formed between a hydrogen atom (which is covalently bound to a nitrogen or oxygen atom) and a nitrogen or oxygen atom with an unshared electron pair. Weak interactions between electronegative atoms and hydrogen atoms (electropositive) that are linked to other electro-negative atoms.

Hydrogen uptake positive (HUP+) A term describing a micro-organism that is capable of assimilating (or taking up) hydrogen gas.

Hydrogenated fats Also known as trans-fatty acids and the result of chemical addition of hydrogen units to polyunsaturated fatty acids to gain optimal texture of fats and oils.

Hydrolysis A chemical reaction in which water is added across a covalent bond, often cleaving the molecule into two. Occurs for example when polynucleotides, polypeptides, and polysaccharides are broken into their component monomers. Thus sucrose can be hydrolysed to glucose and fructose; and proteins to individual amino acids.

Hydrophobic (op. hydrophilic) Means water insoluble and refers to non-polar properties as opposed to polar properties of water. Water soluble, or hydrophilic, molecules do not easily mix with hydrophobic molecules or solutions. Oil is a hydrophobic substance and mixtures of oily and watery molecules tend to minimize their contact surface. Thus the observed separation of oil from water when left standing. However, oil can be partially mixed with water when shaking a container vigorously indicating that the separation or mixing of molecules is determined by the energy of molecular interaction. The energy to overcome these molecular interactions can be provided by mechanical force (shaking) or high temperature. The latter is a well known observation; salts, sugars, and fats normally mix better in hot than cold water, and hot water is better in removing stains form dishes or cloths.

Hydrophobic force Water molecules are linked by a network of hydrogen bonds. A nonpolar, nonwetting, surface (e.g., wax) cannot form hydrogen bonds. To form their full complement of hydrogen bonds, the nearby water molecules must form a more orderly (hence lower entropy) network. This both increases free energy and causes forces that tend to draw hydrophobic surfaces together across distances of several nanometers

Hydrophobic interaction An interaction between a hydrophobic ('water-hating') part of a molecule and an aqueous environment. Particularly significant in establishing the conformation of molecules in solution, and thus their biological activity. Many enzymes have a structure where the polypeptide chain is folded to form a hydrophobic core and a hydrophilic ('water-loving') surface.

Hydroponics The growing of plants in aerated water containing all the essential mineral nutrients, with no soil. Also called soilless gardening or cultivation.

Hypercholesterolaemia Concentrations of cholesterol in the blood higher than normal (or reference) values. Causes include dietary and genetic.

Hyperlipidaemia A condition arising from an increased concentration in the blood of cholesterol, triglycerides or both. These lipids are in the form of lipoproteins.

Hyperploid A genetic condition in which a chromosome or a segment of a chromosome is over-represented in the genotype.

Hyperpolarization A mechanism by which a membrane potential is made more negative inside with respect to the outside of the cell. Hyperpolarization causes neurons and muscle cells to be electrically silent and stabilize at a resting potential. Hyperpolarization is the result of moving positive charges from in to out of a cell which is usually the result of K^+ ions moving out, but can also be achieved by moving Cl^- ions into the cell.

Hypersensitive sites Regions in the DNA that are highly susceptible to digestion with endonucleases.

Hypertension A disorder characterised by elevated blood pressure persistently exceeding systolic\diastolic pressures of 140\90 mm Hg. There may be no visible signs, but the condition poses an increased risk for several diseases such as stroke and coronary heart disease.

Hypertext markup language Used to format information for the World Wide Web.

Hypertriglyceridaemia Concentrations of tryglycerides in the blood higher than normal (or reference) values.

Hypervariable region The parts of both the heavy and light chains of an antibody molecule that enable it to bind to a specific site on an antigen.

Hypervariable segment A region of a protein that varies considerably between strains or individuals.

Hypochlorite Generic term for aqueous solutions of sodium hypochlorite, potassium hypochlorite or calcium hypochlorite, which are oxidizing agents and used for disinfecting surfaces and surface-sterilizing tissues, and for bleaching.

Hypocotyl Portion of an embryo or seedling below the cotyledons, which is a transitional area between stem and root.

Hypomorph A mutation that reduces but does not completely abolish gene expression.

Hypoplastic Reduction in plant growth or development (dwarfing, stunting) resulting from an abnormal condition associated with a disease or nutritional stress.

Hypoploid A genetic condition in which a chromosome or segment of a chromosome is underrepresented in the genotype.

Hypothalamic peptides Peptides generated in the vertebrate forebrain and concerned with regulating the body's physiological state.

Hypothalamus Part of the brain that is responsible for regulating body temperature, water and food intake, and sleep. It is also responsible for regulating the release of hormones.

Hypothesis A testable scientific idea that can be proved right or wrong with experiments. A hypothesis is a formulation of a question that lends itself to a prediction. This prediction can be verified or falsified. A question can only be use as scientific hypothesis, if their is an experimental approach or observational study that can be designed to check the outcome of a prediction.

Hypotonic Osmotic potential less than that of living cells. Cells placed in a hypotonic solution display swelling and turgidity.

I

Iatrogenic Created by a medical doctor.

Ichthyosis A genetic skin disorder that causes abnormal scaling of the skin. There are several different forms of the disorder, each caused by a different genetic defect related to skin shedding.

Identity preservation The segregation of one crop type from another at every stage from production and processing to distribution. This process is usually performed through audits and site visits and provides independent third-party verification of the segregation.

Idiopathic Describes any disorder with unknown causes.

Illuminate To supply or brighten with light. Illumination is an absolute requirement for tissue cultures. Fluorescent lights are commonly employed. The light intensity is dependent on the light source and the requirements of the culture.

Imaginal disc A mass of cells in the larvae of *Drosophila* and other holometabolous insects that gives rise to particular adult organs, such as antennae, eyes or wings.

Imbibition 1. The absorption of liquids or vapours into the ultramicroscopic spaces or pores found in materials. 2. The initial water uptake by seeds starting germination.

Immediate early gene A viral gene that is expressed promptly after infection.

Immobilized cells Cells entrapped in matrices such as alginate, polyacrylamide and agarose designed for use in membrane and filter bioreactors.

Immortalization Production of a cell line capable of an unlimited number of cell divisions. Can be the result of a chemical or viral transformation or of fusion with cells of a tumor line.

Immortalizing oncogene A gene that upon transfection enables a primary cell to grow indefinitely in culture.

Immune machines Medical nanomachines designed for internal use, especially in the bloodstream and digestive tract, able to identify and disable intruders such as bacteria and viruses.

Immune response The immune response is the reaction of the body to substances that are foreign or treated as foreign. The response is in a variety of forms from the recognition of antigens in the body, the production of antibodies against the foreign substance and the response of lymphocytes (white blood cells, T cell and B cells).

Immune serum Blood serum containing antibodies.

Immune system The immune response is the reaction of the body to substances that are foreign or treated as foreign. The response is in a variety of forms from the recognition of antigens in the body, the production of antibodies against the foreign substance and the response of lymphocytes (white blood cells, T cell and B cells).

Immunity Nonsusceptibility to a disease or to the toxic effects of antigenic material.

Immunization The production of immunity in an individual by artificial means. Active immunization involves the introduction, either orally or by infection, of specially treated bacteria, viruses or their toxins so as to stimulate the production of antibodies.

Immuno-affinity chromatography A purification technique in which an antibody is bound to a matrix and is subsequently used to bind and separate a protein from a complex mixture.

Immunoassay; immunodiagnostics An assay system which detects proteins by using an antibody specific to that protein. A positive result is seen as a precipitate of an antibody-protein complex. The antibody can be linked to a radioactive atom or to an enzyme which catalyses an easily monitored reaction such as a change in colour.

Immunochemical control Use of immune agents to combat infections.

Immunocontraception A method of reducing fertility of a pest species by controlling or preventing conception and pregnancy. Used in rabbits, it depends on the insertion of genes using the myxoma virus.

Immunodiagnostic The use of specific antibodies to measure a substance. This tool is useful in diagnosing infectious diseases and the presence of foreign substances in a variety of human and animal fluids (blood, urine, etc.). The approach is currently being investigated as a way of locating tumor cells in the body.

Immunofluorescence Technique for identifying antigenic material that uses an antibody labeled with fluorescent material. Specific binding of the antibody and antigen can be seen under a microscope by applying ultraviolet light rays and noting the visible light that is produced.

Immunogen Any substance that can elicit an immune response.

Immunogenicity The ability to elicit an immune response.

Immunoglobulin General name for proteins that function as antibodies. These proteins differ somewhat in structure and are grouped into five categories on the basis of these differences; immunoglobulin G (IgG), IgM, IgA, IgE and IgD.

Immunology Immunology is the science of molecular self-defense of organisms against infections. It deals with the immune system, a complex organ that produces both cells and proteins involved in detecting and destroying foreign molecules and microorganisms (the 'non-self'). It is most elaborate in mammals. The immune cells are white blood cells, originate from the bone marrow and mature in the lymph system. Some of these cells produce antibodies (proteins) that circulate in the blood as a result of a detected infection. These antibodies can be produced for a very long time after the initial infection and are the basis of immunity against further infection by the same microorganism (e.g. flu vaccine). If the microorganism mutates, as if often the case, immunity is no longer given and a new response must be provided by the immune system. In order to avoid costly development of antibodies, the innate immune system provides a broad and less specific protection against a large class of pathogens. Sometimes, the immune system overreacts and destroys cells of the body (the 'self'). The result is an autoimmune disease. Common autoimmune diseases are arthritis, lupus, and type I or juvenile diabetes. There are no known cures for autoimmune disease except for immune system suppressant and pain relievers.

Immunomodulators A diverse class of proteins that boost the immune system. Many are cell growth factors that accelerate the production of specific cells that are important in mounting an immune response in the body. These proteins are being investigated for use in possible treatments for cancer.

Immunosensor A biosensor having an antibody as biological part.

Immunosuppression The suppression of immune response. Immunosuppression is necessary following organ transplants in order to prevent the host rejecting the grafted organ.

Immunotherapy The use of an antibody or a fusion protein containing the antigen binding site of an antibody to cure a disease or enhance the well-being of a patient.

Immunotoxin Protein drugs consisting of an antibody joined to a toxin molecule. Immunotoxins can be made by linking toxin and antibody molecules chemically, or by fusing the genes for the toxin and the antibody. The antibody portion of the molecule facilitates binding to a target molecule or cell, and the toxin inactivates or kills the target molecule or cell.

Impeller An agitator that is used for mixing the contents of a bioreactor.

Imprinting This is when genes are suppressed or silenced depending on which parent they were received from. When DNA is passed to daughter cells after fertilization of an egg by a sperm, certain alleles can become active only if they were received from the mother, others only if they came from the father. If a gene is suppressed through imprinting from one parent, and the allele from the other parent is not expressed because of mutation, neither can act and the child will be deficient. A healthy child cannot be produced when both sets of chromosomes come from the same parent. Imprinting of the same areas will occur and all these genes will be suppressed.

In situ (Latin for "in place") Meaning in the natural place or in the original place. 1. Experimental treatments performed on cells or tissue rather than on extracts from them.

2. Assays or manipulations performed with intact tissues.

In situ colony; in situ plaque hybridization A procedure for screening colonies or plaques growing on plates or membranes for the presence of specific DNA sequences by the hybridization of nucleic acid probes to the DNA molecules present in these colonies or plaques.

In situ conservation of farm animal genetic diversity All measures to maintain live animal breeding populations, including those involved in active breeding programmes in the agro-ecosystem where they either developed or are now normally found, together with husbandry activities that are undertaken to ensure the continued contribution of these resources to sustainable food and agricultural production, now and in the future.

In situ hybridization The visualization of in vivo location of macromolecules (particularly polynucleotides and polypeptides) by the histological staining of tissue sections or cytological preparations via labelled probes/antibodies.

In utero Within the uterus.

In vitro From the Latin meaning "in glass". The term is applied to biological processes studied experimentally in isolation from the organism, as distinct from in vivo, which refers to the study of processes in the living organism.

In vitro embryo production (IVEP) The combination of ovum pickup, *in vitro* maturation of ova, and *in vitro* fertilization. A potential means of overcoming the variability between donors in number of ova collected in embryo-transfer programmes.

In vitro fertilization (IVF) A widely used technique in human and animal science, whereby the egg is fertilized with sperm outside the body. Usually, the fertilized egg is cultured outside the body for a few days (to confirm that fertilization has occurred) before re-implantation into a female.

In vitro maturation (IVM) Culture of immature ova in the laboratory, usually until they are ready for *in vitro* fertilization.

In vitro translation; cell free translation The synthesis of proteins in the test-tube from purified mRNA molecules using cell extracts containing ribosomal sub-units, the necessary protein factors, tRNA molecules and aminoacyl tRNA synthetases, ATP, GTP, amino acids and an enzyme system for re-generating the nucleóside triphosphates. Prokaryotic translation systems are usually prepared from *E. coli* or the thermophilic bacterium, *Bacillus stearothermophilus.* Eukaryotic systems usually employ rabbit reticulocyte lysates or wheat germ.

In vivo The natural conditions in which organisms reside. Refers to biological processes that take place within a living organism or cell under normal conditions.

In vivo gene therapy The delivery of a gene or genes to a tissue or organ of an individual to alleviate a genetic disorder.

Inactivated agent A virus, bacterium or other organism that has been treated to prevent it from causing a disease.

Inbred line The product of inbreeding, i.e., the mating of individuals that have ancestors in common; in plants and laboratory animals, it refers to populations resulting from at least 6 generations of selfing or 20 generations of brother-sister mating, that are for all practical purposes, completely homozygous. In farm animals, the term is sometimes used to describe populations that have resulted from several generations of the mating of close relatives, without having reached complete homozygosity.

Inbreeding Matings between individuals that have one or more ancestors in common, the extreme condition being self-fertilization, which occurs in many plants and some primitive animals.

Inbreeding depression Reduction in vigour, yield, etc., of a population that is commonly seen as the level of inbreeding increases. The traits that show greatest inbreeding depression are those that are most closely associated with viability and reproductive ability.

Inclusion body Protein that is overproduced in a recombinant bacterium and forms a crystalline array inside the bacterial cell.

Incompatibility 1. Selectively-restricted mating competence, which limits fertilization, such as lack of proper functions by otherwise normal pollen grains or certain pistils, a condition that may be caused by a variety of factors.

2. A physiological interaction resulting in graft rejection or failure.

3. A function of a related group of plasmids. Plasmids which are closely related share similar replication functions and this leads to the exclusion of one or the other plasmid if they are present in the same cell; thus such plasmids are

incompatible. Plasmids are placed in incompatibility groups by this simple reaction and, in general, plasmids belonging to one incompatibility group are very closely related.

Incompatibility group A classification scheme indicating which plasmids can co-exist within a single cell. Plasmids must belong to different incompatibility groups to co-exist within the same cell. Plasmids that belong to the same incompatibility group are unstable when placed in the same cell. A plasmid cloning vector should always belong to an incompatibility group different from that of the host bacterium's endogenous plasmids.

Incomplete dominance A type of gene action in which heterozygotes have a phenotype that is distinctly different from, and intermediate to, the homozygous phenotypes.

Incomplete penetrance When some individuals in a population have a specific genotype that causes an abnormality but are not affected.

Incubation 1. The hatching of eggs by means of heat, either natural or artificial.

2. Period between infection and appearance of symptoms induced by parasitic organisms.

3. The culture of cells and organisms.

Incubator An apparatus in which environmental conditions (light, photoperiod, temperature, humidity, etc.) are fully controlled, and used for hatching eggs, multiplying micro-organisms, culturing plants, etc.

Indehiscent Describing a fruit or fruiting body that does not open to release its seeds or spores when ripe.

Independent assortment The random distribution of alleles (from different loci) to the gametes that occurs when genes are located in different chromosomes or far apart on large chromosomes. The distribution of alleles at one locus is independent of other the distribution of alleles at another locus.

Indeterminate growth 1. Unlimited growth potential for a definite or indefinite period. Some apical meristems can produce unrestricted numbers of lateral organs.

2. In legumes, used to describe plant architecture.

Indirect embryogenesis Embryo formation on callus tissues derived from zygotic or somatic embryos, seedling plant or other tissues in culture.

Indirect organogenesis Organ formation on callus tissues derived from explants.

Inducer A low-molecular-weight compound or a physical agent that is bound by a repressor so as to produce a complex that can no longer bind to the operator; thus, the presence of the inducer turns on the expression of the gene(s) controlled by the operator.

Inducible A gene or gene-product is said to be inducible if its transcription or synthesis is increased by exposure of the cells to an effector. Effectors are usually small molecules whose effects are specific to particular operons or groups of genes.

Inducible enzyme An enzyme that is synthesized only in the presence of the substrate that acts as an inducer.

Inducible gene A gene that is expressed only in the presence of a specific metabolite, the inducer.

Induction Act or process of causing to occur; process whereby a cell or tissue influences neighbouring cells or tissues. Turning on transcription of a specific gene or operon. Getting an organism to make a protein by exposing it to some stimulus.

Induction media 1. Media used to induce the formation of organs or other structures.

2. Media causing variation or mutation in the tissues exposed to it.

Industrial corn A variety of corn that is processed and used in livestock feeds, processed foods and industrial products.

Inert A support structure that makes no chemical contribution, and whose only function is support. Physiologically it is a neutral or immobile unit.

Infection The invasion of any living organisms by disease-causing micro-organisms, which proceed to establish themselves, multiply, and produce various clinical signs in their host.

Infectious agent Typically, a proliferating virus, bacterium or parasite that causes a disease in plants or animals.

Infectivity The potential for a given micro-organism to cause an infection.

Infertile Incapable of initiating, sustaining, or supporting reproduction, alternatively: not fertilised and therefore incapable of growing and developing.

Infiltrate To force the passage of liquid into tissue pores or space, such as by applying a vacuum, then releasing it, during the disinfectation procedure.

Inflammatory response Local response of a tissue to injury or infection. Caused by invasion of white blood cells, which release various local mediators such as histamine.

Inflorescence 1. The way the flowers are arranged on the stalk of a plant.

2. The flowers of a plant collectively.

3. The flowering or blooming process.

Informed consent A term used to describe the responsibility of doctors or researchers to ensure that patients or people being researched have an understanding of the relevant facts regarding their care or participation in research. We can also say that consumers of have a right to practice informed consent when they buy particular foods. Informed consent relies on our having access to reliable, truthful, and complete information.

Inheritance The transmission of particular characteristics and/or genes from generation to generation.

Inherited Traits or characteristics that come from one's ancestors and are transmitted from parents to offspring through genes. The traits will therefore be present at birth.

Initial Cells in a meristem that remain permanently meristematic and form tissues of particular structure and function.

Initiation 1. Early steps or stages of a tissue culture process (culture growth, organogenesis, embryogenesis)

2. Early stages of biosynthesis.

Initiation codon; initiator The codon AUG which specifies the first amino acid methionine of a polypeptide chain. In bacteria, the initiation codon is either AUG, which is translated as *n*-formyl methionine or, rarely, GUG (valine). In eukaryotes, the initiation codon is always AUG and is translated as methionine. The term is also used to describe the corresponding sequence in DNA, namely ATG.

Initiation factors Soluble proteins required for the initiation of translation.

Inner cell mass Within a blastocyst - the hollow ball of cells that forms soon after an egg is fertilised - there is a mass of cells on one side which will form the body of the embryo. This is referred to as the inner cell mass. This is where embryonic stem cells are taken from.

Inoculants Disease-causing organisms or infective agents that promote immunity against the disease when introduced (injected) into an organism or medium.

Inoculate Deliberately introduce, in contrast to contamination. 1. In bacteriology, tissue culture, etc., placing an inoculum into (or onto) medium to initiate a culture.

2. In immunology, to carry out immunization.

3. In plant pathology, application of pathogen spores etc. on to plants under conditions where infection should result in the absence of resistance.

Inoculation cabinet Small room or cabinet for inoculation (of tissue or micro-organism cultures) operations, often with a current of sterile air to carry contaminants away from the work area.

Inoculum (pl: inocula) 1. A small piece of tissue cut from callus, or an explant from a tissue or organ, or a small amount of cell material from a suspension culture, transferred into fresh medium for continued growth of the culture.

2. Microbial spores or parts (such as mycelium).

3. Vaccine.

Inorganic This term has several meanings, including: Chemicals which are not organic, that is, not manufactured within living organisms.

Any chemical compound which is not based on carbon chains or rings (except oxides, sulphides of carbon and metallic carbides which are also inorganic).

Inorganic compound A chemical compound that generally is not derived from living processes; compounds that do not contain carbon.

Inositol (hexahydroxycyclohexane; $C_6H_6(OH)_6$) 1. A cyclic acid that is constituent of certain cell phosphoglycerides.

2. A water-soluble nutrient frequently referred to as a "vitamin" in plant tissue culture. Also acts as a growth factor in some animals and micro-organisms.

Inositol lipid A membrane-anchored phospholipid that transduces hormonal signals by stimulating the release of any of several chemical messengers.

Input traits Traits introduced into crop plants with the aim of lowering the cost of production and improving the performance of the crop in the field. For example: pesticide resistance, herbicide tolerance and disease resistance. This is in comparison to traits introduced into the crop to produce products with enhanced value. These are referred to as output traits.

Insect resistant plants Are resistant to certain insect pests because of naturally occurring or introduced factors. Some crops are engineered to produce toxin proteins through the introduction of genes, which come from a soil bacterium, Bacillus thuringiensis (Bt). In their natural form, Bt toxins are lethal to the larvae of certain pests but not to other living things.

Insecticide A substance used to control certain population s of insect pests. In agriculture, insecticides are used to control insect pests that feed on crops or carry plant disease.

Insecticide resistance The development or selection of heritable traits (genes) in an insect population that allow individuals expressing the trait to survive in the presence of levels of an insecticide (biological or chemical control agent) that would otherwise debilitate or kill this species of insect. The presence of such resistant insects makes the insecticide less useful for managing pest populations.

Insect-resistance management A strategy for delaying the development of pesticide resistance by maintaining a portion of the pest population in a refuge that is free from contact with the insecticide. For Bt crops this allows the insects feeding on the Bt toxin to mate with insects not exposed to the toxin produced in the plants.

Insect-resistant crops Plants with the ability to withstand, deter or repel insects and thereby prevent them from feeding on the plant. The traits (genes) determining resistance may be selected by plant breeders through cross-pollination with other varieties of this crop or through the introduction of novel genes such as Bt genes through genetic engineering.

Insert A DNA molecule that is incorporated into a cloning vector.

Inserted genes Genes introduced into the DNA of a recombinant organism which are not present at the same position in the DNA of the organism before genetic modification.

Insertion element Generic term for DNA sequences found in bacteria capable of genome insertion. Postulated to be responsible for site-specific phage and plasmid integration.

Insertion mutation Changes in the base sequence of aDNA molecule resulting from the random integration of DNA from another source.

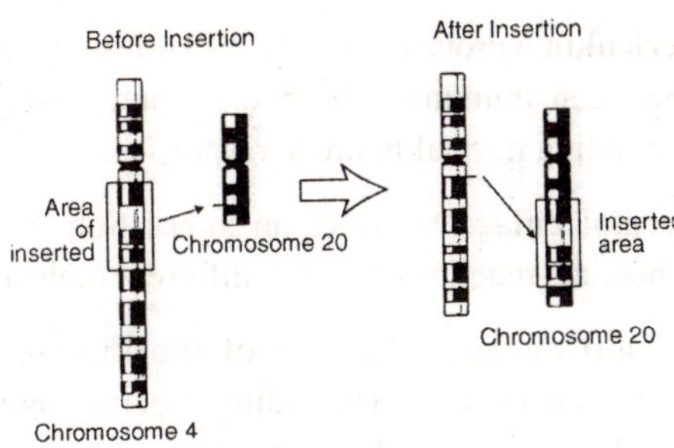

Fig. Insertion

Insertion site 1. A unique restriction site in a vector DNA molecule into which foreign DNA can readily be inserted. This is achieved by treating both the vector and the insert with the relevant restriction endonuclease and then ligating the two different molecules, both having the same sticky ends. Synonym: cloning site.

2. The position of integration of a transposon.

Instability A random-type variation or a lack of steadiness. Due to genetic instability, cell lines lose certain characteristics or functions in culture.

Insulin A protein hormone that regulates the use of glucose after a carbohydrate rich meal stimulating the degradation of glucose to extract energy and the storage of excess glucose in glycogen or metabolic conversion to fatty acids and cholesterol.

Integrated crop management (ICM) A system which integrates a number of ways to control pests and increase yields (including natural, biological controls and careful use of chemicals).

Integrated pest management (IPM) A sustainable, ecological approach to pest control that includes biological, mechanical and chemical means. The goal of IPM is to produce a healthy crop in an economically efficient and environmentally sound manner.

Integrating vector A vector that is designed to integrate cloned DNA into the host cell chromosomal DNA.

Integration The recombination process which inserts a small DNA molecule (usually by homologous recombination) into a larger one. If the molecules are circular, integration involves only a single crossing-over; if linear, then two crossings-over are required. A well known example is the integration of phage l (lambda) DNA into the *E. coli* genome.

Integration-excision (I/E) region The portion of bacteriophage lambda (l) DNA that enables bacteriophage lambda (l) DNA to be inserted into a specific site in the *E. coli* chromosome and excised from this site.

Integument One of the layers that enclosed the ovule, and is the precursor of the seed coat.

Intellectual property (IP) A term that refers to the content of the human intellect, or the result of intellectual effort, which is considered to be unique and original and o have value in the marketplace, and therefore requires legal protection and ownership. This includes copyrighted material such as literary or artistic works, industrial processes, and trademarks and patents.

Intellectual property rights The legal protection for inventions, including new technologies or new organisms (such as new plant varieties). The owner of these rights can control their use and earn the rewards for their use. This encourages further innovation and creativity for the benefit of us all. Intellectual property rights protection includes various types of patents, trademarks, and copyrights.

Intensifying screen A plastic sheet impregnated with a rare-earth compound, such as calcium tungstate, which absorbs ί radiation and emits light. When placed on one side of a piece of X-ray film with a radioactive sample on the other side, the intensifying screen will capture some of the ί emissions which pass through the film, blackening the X-ray film and so greatly enhancing the sensitivity of the detection. An intensifying screen is used in Southern and northern blotting procedures.

Interaction In statistics, an effect that cannot be explained by the additive action of contributing factors; a departure from strict additivity.

Intercalary Meristematic tissue or growth not restricted to the apex of an organ, i.e., growth at nodes.

Intercalary growth A pattern of stem elongation typical of grasses. Elongation proceeds from the lower internodes to the upper internodes through the differentiation of meristematic tissue at the base of each internode.

Intercalating agent A chemical capable of inserting between adjacent base pairs in a DNA molecule.

Intercellular space Pore space between cells, especially typical of leaf tissues.

Interfascicular cambium Cambium that arises between vascular bundles.

Interference Crossing over at one point that alters the chance of another crossing-over nearby; detected by studying the pattern of crossings-over with three or more linked genes. Interference is positive or negative depending on whether the chance of another crossing-over nearby is reduced or increased, respectively.

Interferon A class of lymphokine proteins important in the immune response. There are three major types of interferon: alpha (leukocyte), beta (fibroblast) and gamma (immune). Interferons inhibit viral infections and may have anticancer properties.

Intergeneric A cross between two different genera.

Intergenic regions DNA sequences located between genes; they comprise a large percentage of a genome and have no known function.

Interleukin A group of proteins that transmit signals between immune cells and are necessary for mounting normal immune responses.

Intermolecular Describes an interaction (e.g., a chemical reaction) between different molecules.

Internal energy The sum of the kinetic and potential energies (including electromagnetic field energies) of the particles that make up a system.

Internal validity The accuracy of a study's findings with regard to the study subjects.

International regulatory status The regulations that govern the acceptable uses of, and the tests performed on, a particular product in worldwide markets.

Internode The region of a stem between two successive nodes.

Interphase The stage in the cell cycle when the cell is not dividing and during part of which DNA replication occurs; it follows telophase of one mitotic division and extends to the beginning of prophase in the next division.

Interrupted genes Genes whose coding sequence is interrupted at intervals by long stretches of non-coding sequences. The coding regions came to be known as *exons* and the non-coding regions as *introns.* This structure is now found to be characteristic of most eukaryotic genes. The number and size of introns vary greatly, and they are often much longer than the coding sequences. After transcription, the intron regions are removed, or *spliced* out form the RNA transcript before it is translated into protein.

Intersex An organism displaying a mixture of male and female attributes.

Interspecific Between two different species, e.g., an interspecific cross is a cross between two species.

Intervention trials Trials in which one or more factors that may affect health are altered, with the aim of demonstrating beneficial effects compared with a control group not receiving the intervention.

Intracellular Within cells; the intracellular compartment refers to the space, mainly water, inside cells.

Intracytoplasmic Into the cytoplasm, the part of the cell outside the nucleus.

Intracytoplasmic sperm injection (ICSI) The micro-injection of a single sperm into the cytoplasm of a mature oocyte.

Intrageneric Within a genus, such as a hybrid resulting from a cross between species within one genus.

Intragenic complementation Complementation that occurs between two mutant alleles of a gene; common only when the product of the gene functions as a homomultimer.

Intramolecular Describes an interaction (e.g., a chemical reaction) within a single molecule. Intramolecular interactions between widely separated parts of a molecule resemble intermolecular interactions in most respects.

Intraspecific Within a species or its populations, including subspecies, such as an intraspecific cross, or variation.

Intravenous Describes fluids or medication administered directly into the vein.

Introgression The introduction of new gene(s) into a population by crossing between two populations, followed by repeated backcrossing to that population while retaining the new gene(s).

Intron; intervening sequence A segment of DNA sequence of a eukaryotic gene, not represented in the mature (final) mRNA transcript, because

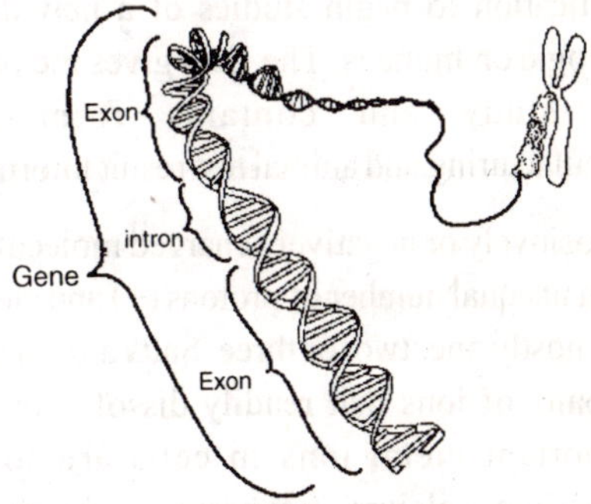

Fig. Gene

it is spliced out of the primary transcript before it can be translated; a process known as intron splicing. Some genes of higher eukaryotes contain a large number of introns, which make up the bulk of the DNA sequence of the gene. Introns are also found in genes whose RNA transcripts are not translated, namely eukaryotic rRNA and tRNA genes. In these cases the intron sequence does not appear in the functional RNA molecule.

Invariant Constant, unchanging; usually referring to the portion of a molecule that is the same across species.

Invasiveness Ability of a plant to spread beyond its introduction site and become established in new locations where it may provide a deliterious effect on organisms already existing there.

Inversion A chromosome re-arrangement that reverses the order of a linear array of genes in it.

Inverted repeat Two regions of a nucleic acid molecule which have the same nucleotide sequence but in an inverted orientation, such as

5′ GCACTTG... ...CAAGTGC 3′3′ CGTGAAC ... ... GTTCACG 5′

Because they contain exactly the same message when read in either direction, inverted repeats are said to be palindromes.

Investigational new drug application (IND) An application to begin studies of a new drug or biologic on humans. The IND gives the plan for the study and contains formulation, manufacturing and animal test result information.

Ions Positively or negatively charged molecules due to an unequal number of protons (+) and electrons (-), mostly one, two, or three. Salts are composed of pairs of ions that readily dissolve in water. Important metal ions in cells are sodium, potassium, calcium, magnesium, zinc and iron. The positively charged metal ions are always found with an equal number of negatively charged counter ions like chloride or phosphate.

Ion channel A protein integral to a cell membrane, through which selective ion transport occurs.

Ionic bond A chemical bond resulting chiefly from the electrostatic attraction between positive and negative ions.

Ionizing radiation The portion of the electromagnetic spectrum that results in the production of positive and negative charges (ion pairs) in molecules. X-rays and gamma rays are examples of ionizing radiation.

IPTG (Isopropyl b d thiogalactopyranoside) An inducer of the *lac* (lactose) operon. In recombinant DNA technology, IPTG is often used to induce cloned genes that are under the control of the *lac* repressor-*lac* promoter system.

Irradiation 1. Exposure to any form of radiation.

2. Treatment with a ray, such as ultraviolet rays, etc.

3. Apparent enlargement of objects, due to difference in illumination.

4. In food technology: exposure of food to controlled low levels of ionizing radiation to sterilize or disinfest it, without inducing radioactivity in the target. If in sealed containers, the product is sterile and can be stored under ambient conditions. Used for continuous flow disinfestation of grain at unloading facilities.

IS element (insertion sequence) A short (800-1400 nucleotide pairs) DNA sequence found in bacteria that is capable of transposing to a new genomic location; other DNA sequences that are bounded by IS elements may also be transposed.

Ischaemia Reduced or inadequate blood, and thus oxygen, supply to a part of the body.

Islet cells Insulin-producing cells in the pancreas. If these cells are destroyed, Type I diabetes develops. Also known as beta cells.

Iso electric focusing gels Variant of DNA gel electrophoresis, which separates macromolecules on the basis of their iso-electric point rather than their size.

Isoalleles Different forms of a gene that produce the same phenotype or very similar phenotypes.

Isochromosome A chromosome with two identical arms and identical genes. The arms are mirror images of each other.

Isodiametric Term commonly used to describe cells with equal diameters.

Isoelectronic Two molecules are described as isoelectronic if they have the same number of valence electrons in similar orbitals, although they may differ in their distribution of nuclear charges (e.g., H-C N and $H\text{-}N^+ C^-$.

Isoenzyme One of the several forms that a given enzyme can take. The forms may differ in certain physical properties, but function similarly as biocatalysts.

Isoform A member of a family of closely related proteins that have some amino acid sequences in common and some different.

Isogamy Sexual reproduction involving the production fusion of gametes that are similar in size and structure.

Isogenic stocks Strains of organisms that are genetically identical; completely homozygous.

Isolating mechanism Any of the biological properties of organisms that prevent interbreeding (and therefore exchange of genetic material) between members of different species that inhabit the same geographical area.

Isolation medium An optimum medium suitable for explant survival, growth and development.

Isomer 1. Structural isomers have the same chemical formula but different structures; e.g. leucine and isoleucine.

2. Stereoisomers are different topological forms of an otherwise single chemical structure, due to changes in bond configurations about some axis or plane of symmetry; eg, D- and L-glucose or cis- and trans-cinnamic acid.

Isopentenyltransferase (IPT) Selection system An alternative genetic tag to antibiotic resistance genes. IPT is an enzyme involved in plant growth hormone biosynthesis.

Isotonic; iso osmotic Solutions having the same osmotic potential; the same molar concentration. For protoplasts to survive, the medium they are suspended in must be isotonic with them.

Isotope One of two or more forms of an element that have the same number of protons (atomic number) but differing numbers of neutrons (mass numbers). Radioactive isotopes are commonly used to make DNA probes and metabolic tracers.

Isozyme A variant of a particular enzyme. In general, all the isozymes of a particular enzyme have the same function and sometimes the same activity, but they differ in amino-acid sequence. With the help of isozyme analysis, based on electrophoresis techniques that can separate the different variants, related species or cultivars can be distinguished.

J

J 1. Joining segment

2. Joule.

J/m2; J m-2 (joules per square metre) A unit of light measurement.

Jiffy potTM Pots made from wood pulp and peat, commonly used for transplanting tissue-culture-derived plants into soil medium.

Joining (J) segment A small DNA segment that links genes to yield a functional gene encoding an immunogobulin.

Joule Standard unit of energy in the meter-kilogram-screened system. One joule is the energy delivered in one second by a one-watt power source. Approximately equal to 0.24 calories.

Jumping genes More correectly called transposons or mobile genetic elements which can move round genomes and, assisted by suitable carriers, transfer between organisms.

Junk DNA Portions of DNA that don't have genes with an obvious function.

Juvenile hormone A hormone secreted by insects from a pair of endocrine glands close to the brain. It inhibits metamorphosis and maintains the larval features.

Juvenile *in vitro* embryo technology (JIVT; JIVET) A technology involving collection of immature ova from young animals, *in vitro* maturation of those ova, *in vitro* fertilization, and then transfer of the resultant embryos into recipient females. A method of achieving rapid generation turnover.

Juvenility Early phase of development in which an organism is juvenile and incapable of sexual reproduction.

K

Kanamycin An antibiotic of the aminoglycoside family that inhibits translation by binding to the ribosomes. Important as a substrate for selection of plant transformants.

Kanr Kanamycin-resistance gene.

Kappa chain One of two classes of antibody light chains.

Kartagener's syndrome A genetic disorder in which the heart and other internal organs are on the opposite side of the body to the normal location for the organs.

Karyogamy The fusion of nuclei or nuclear material that occurs at fertilization during sexual reproduction.

Karyogram A diagrammatic representation of the full chromosome set of a species, highlighting characteristic physical features of individual chromosomes.

Karyokinesis The division of a cell nucleus.

Karyotype The number and ordering of eukaryotic chromosomes according to size and appearance. The karyotype of an organism is a conserved feature and any changes in number and size of chromosomes are mutations that cause severe diseases and birth defects and are commonly lethal. The importance of chromosome structure is found in the spatial organization of genes on chromosomes, which has been found to be influencing when genes can be expressed, i.e., used to make a protein or functional RNA.

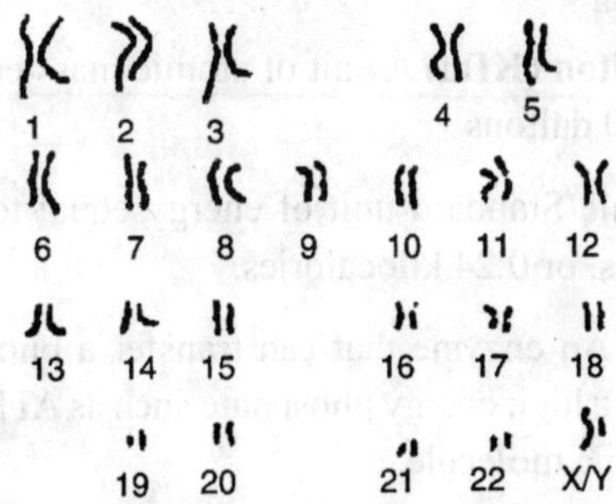

Fig. Karyotype

K*cat* /K*m* The catalytic efficiency of an enzyme-catalysed reaction. The greater the value of *kcat*/K*m*, the more rapidly and efficiently the substrate is converted into product.

Keratin (cytokeratin) Member of the family of proteins that form keratin intermediate filaments, mainly in epithelial cells. Some specialized keratins are found in hair, nails, and feathers.

Keratinocytes A type of skin cell that makes up 95 percent of the epidermis. They contain the keratin proteins that make up a large part of nail, hair, and skin.

Ketone Organic molecule containing a carbonyl group linked to two alkyl groups.

Kidney plasminogen activator A precursor to the enzyme urokinase, which has blood-clotting properties.

Killer t cells T cells that carry T-cell receptors and kill cells displaying the recognized antigens.

Kilo Prefix denoting 1000.

Kilobase (KB) A length unit equal to 1 000 base pairs of a double-stranded nucleic acid molecule. One kilobase of double-stranded DNA has a mass of about 660 kilodalton.

Kilocalorie (kcal) Unit of heat energy equal to 1000 calories. Often used to express the energy content of food or molecules: bond strengths, for example, are measured in kcal/mole. An alternative unit in wide use is the kilojoule, equal to 0.24 kcal.

Kilodalton (KDa) A unit of atomic mass equal to 1 000 daltons.

Kilojoule Standard unit of energy equal to 1000 joules, or 0.24 kilocalories.

Kinase An enzyme that can transfer a phosphate from a high energy phosphate such as ATP, to an organic molecule.

Kinesin One type of motor protein that uses the energy of ATP hydrolysis to move along a microtubule.

Kinetics Kinetics is the science of measuring changes, of assessing rates of movements and flow. In biology, kinetics is concerned with *enzyme kinetics*, the rate of how proteins help catalyze a chemical reactions. Another application of kinetics is the rate of flow of molecules in solution by diffusion or in an energy field (such as charges in an electric field, or mass in a gravitational field). Flux rates of molecules across biological membranes are also studied by kinetics.

Kinetic energy Energy resulting from the motion of masses.

Kinetin One of the cytokinins, a group of growth regulators that characteristically promote cell division in plants.

Kinetochore Structure at the centromere of eukaryotic chromosomes. The kinetochore consists of inner and outer electron dense plates and a central zone containing repetitive DNA elements. Kinetochores are involved in the control of chromosome movement in cell division.

Kinetosome Granular cytoplasmic structure which forms the base of a cilium or flagellum.

Kinin A substance promoting cell division. In plant systems, the prefix cyto- has been added (cytokinin) to distinguish it from kinin in animal systems.

Klenow fragment A product of proteolytic digestion of the DNA polymerase I from *E. coli*; it has both polymerase and 3′-exonuclease activities but not 5′-exonuclease activity.

Klinefelter's syndrome A genetic growth disorder in males, resulting in small testes, enlarged breasts, and failure to produce sperm. It is caused by the presence of an extra sex chromosome.

K*m* A dissociation constant that characterizes the binding of an enzyme to a substrate. The smaller the value of K*m*, the tighter the binding of the enzyme to the substrate. Also called the Michaelis constant.

Knockout An animal resulting from an embryonic stem cell in which a normal functional gene has been replaced by a non-functional form of the gene. This technique is used extensively in mice - much can be learned about the function of a gene by studying the phenotype of animals that lack the peptide product of that gene.

L

Labelled probe Molecules with special properties, like radioactivity, used to search for, and highlight, other specific molecules in a sample.

Labelling The process of replacing a stable atom in a compound with a radioactive isotope of the same element to enable it to be detected by autoradiography or other techniques. Increasingly, radioactive labelling is being replaced by fluorescent labelling. The method is used to trace the path of the labelled compound through a biological or chemical system.

Lactoferrin A protein in breast milk that promotes infant growth.

Lactose Milk sugar, a disaccharide with one unit each of glucose and galactose.

Lag phase 1. A state of apparent inactivity preceding a response; called also a latent phase.

2. The initial growth phase, during which cell number remains relatively constant, prior to rapid growth.

3. The first of five growth phases of most batch-propagated cell suspension cultures, being the phase in which inoculated cells in fresh medium adapt to the new environment and prepare to divide.

Lagging strand One of the two newly made strands of DNA found at a replication fork. The lagging strand is made in discontinuous lengths that are later joined covalently.

Lambda bacteriophage (l bacteriophage) Virus that infects *E. coli;* widely used as a DNA cloning vector.

Lambda chain One of two classes of antibody light chains.

Lamella A double-membrane structure, plate or vesicle that is formed by two membranes lying parallel to each other.

Lamins Intermediate filament proteins that form the fibrous matrix (nuclear lamina) on the inner surface of the nuclear envelope.

Laminar air flow cabinet; laminar air flow hood Cabinet for inoculation of cultures. The working area is kept sterile by a continuous, non-turbulent flow of sterilized air through a HEPA filter.

Laminarin A storage polysaccharide of the brown algae.

Lampbrush chromosomes Large diplotene chromosomes present in oocyte nuclei, and particularly conspicuous in amphibians. These chromosomes have extended regions called loops, which are active sites of transcription.

Land based grow out systems (in aquaculture) Refers to a land rearing facility for aquatic organisms.

Landrace In plant genetic resources, an early, cultivated form of a crop species, evolved from a wild population, and generally composed of a heterogeneous mixture of genotypes.

Latent agent Something, usually a virus, that is present in a host organism without producing any symptoms.

Latent bud An inactive bud not held back by rest or dormant period, but which may start growth if stimulated.

Lateral meristem A meristem giving rise to secondary plant tissues, such as the vascular and cork cambia. The term is sometimes used to refer to an axillary meristem.

Lawn In biotechnology: a uniform and uninterrupted layer of bacterial growth, in which individual colonies cannot be observed.

Layering Technique for vegetative propagation, in which new plants produce adventitious roots before being severed from the parent plant.

LD50 (lethal dose50%) The amount of a chemical required to kill 50% of the test population. The higher the LD50, the lower the presumed toxicity of the chemical.

Leader sequence A sequence of nucleotides at the 5′ end of an mRNA that is not translated into protein.

Leading strand One of the two newly made strands of DNA found at a replication fork. The leading strand is made by continuous synthesis in the 5′-to-3′ direction.

Leaf blade The usually flattened portion of the leaf.

Leaf bud cutting A cutting that includes a short section of stem with attached leaf.

Leaf margin The edge of a leaf.

Leaf primordium A lateral outgrowth from the apical meristem, which will become a leaf when fully developed and expanded.

Leaf roll Virus diseases characterized by symptomatic curling of the host's leaves.

Leaf scar Mark left on a stem after leaf abscission.

Leaflet Expanded leaflike part of a compound leaf.

Lecithin A major component of cell membranes containing equal amounts of saturated and monounsaturated fatty acids, phosphate, and choline. Lecithin is a member of the lipid group called phospholipids. Its biochemical name is phosphatidylcholine (PC).

Lectin Protein that binds tightly to a specific sugar. Abundant lectins derived from plant seeds are often used as affinity reagents to purify glycoproteins or to detect them on the surface of cells.

Legume A member of the pea family that possesses root nodules containing nitrogen-fixing bacteria.

Leptokurtic A normal statistical curve that is quite steep or sharp.

Leptonema Stage in meiosis immediately preceding synapsis, in which the chromosomes appear as single, fine, threadlike structures (but they are really double because DNA replication has already taken place).

Lethal allele; lethal gene A mutant form of a gene that eventually results in the death of an organism if expressed in the phenotype.

Leucine zipper Structural motif seen in many DNA-binding proteins in which two a helices from separate proteins are joined together in a coiled-coil, forming a protein dimer.

Leukaemia An increase in the number of ineffective and immature white blood cells causing a weakened immune system which leaves the body susceptible to infection.

Leukocytes White blood cell, up to 0.02 mm in diameter, of which there are normally 4-11 million per millilitre of human blood. There are several kinds, all involved in the body's defence mechanisms. Granulocytes have granules in their cytoplasm; monocytes ingest and feed on bacteria and other micro-organisms that cause infection; lymphocytes include the B cells that are involved with the production of antibodies.

Library (gene library) A collection of cells, usually bacteria or yeast, that have been transformed with recombinant vectors carrying DNA inserts from a single species.

Life-cycle The sequence of events from a given developmental stage in one generation to the same stage in the following generation. In sexually reproduced organisms, the starting point is the fusion of gametes to form the zygote.

Life expectancy Expected further number of years of life measured at any age.

Ligand In protein chemistry, a small molecule that is (or can be) bound by a larger molecule is termed a ligand. In organometallic chemistry, a moiety bonded to a central metal atom is also termed a ligand; the latter definition is more common in general chemistry.

Ligase (DNA ligase) An enzyme that catalyzes a condensation reaction that links two DNA molecules via the formation of a phosphodiester bond between the 3' hydroxyl and 5' phosphate of adjacent nucleotides.

Ligase chain reaction (LCR) A technique for determining the presence or absence of a specific nucleotide pair within a target gene.

Lignans A small molecule (e.g. activators, substrates and inhibitors of enzyme activity) bound to a protein by non-covalent forces; an ion or a molecule that binds to another chemical entity to form a larger complex.

Ligation The joining of two linear nucleic acid molecules by the formation of phospho-diester bonds. In cloning experiments, a restriction fragment is often ligated to a linearized vector molecule using T4 DNA ligase.

Lignin A group of high-molecular-weight amorphous materials, comprising polymers of phenylpropanoid compounds, giving strength to certain tissues. Wood is composed of lignified xylem cells (about 15 to 30% by weight).

Lignocellulose The combination of lignin, hemicellulose and cellulose that forms the structural framework of plant cell walls.

Limited assembler Assembler capable of making only certain products; faster, more efficiently and less liable to abuse, than a general purpose assembler.

Lineage Several individuals originating from a common descent, such as the production of a cell line from a single cell plated *in vitro*.

Linear Aside from its geometric meaning, *linear* describes systems in which an output is directly proportional to an input. In particular, a linear elastic system is one in which the internal displacements are (at equilibrium) directly proportional to applied forces.

Linear phase The growth phase during cell culture when cell number increases arithmetically. The linear phase follows a period of exponential growth.

Linearized vector A covalently closed circular DNA vector (typically a plasmid) which has been opened by restriction digestion to convert it to a linear molecule. In molecular cloning, DNA to be cloned is mixed with the linearized vector, and treated with ligase to join and recircularize the resulting hybrid molecule.

LINEs (long interspersed nuclear elements) Families of long (average length = 6 500 bp), moderately repetitive (about 10,000 copies). LINEs are cDNA copies of functional genes present in the same genome; also known as processed pseudo-genes.

Linkage The tendency of a set of genes to be inherited together more often than would be expected if they were assorting independently. exists between two genes when they are located sufficiently close to one another on the same chromosome that a proportion of gametes is produced without crossing-over occurring between them.

Linkage group Includes all loci (in DNA molecule) that can be connected (directly or indirectly) by linkage relationships; equivalent to a chromosome

Linkage map Depending on how often certain characteristics are seen to appear at the same time in an organism, it's possible to predict where on a chromosome their genes may be found.

Linked genes/markers Genes and/or markers that are so closely associated on the chromosome that they are coinherited in 80% or more of cases.

Linker A short, double-stranded oligonucleotide containing a restriction endonuclease recognition site, which is ligated to the ends of a DNA fragment.

Lipases Enzymes which break down lipids into their component fatty acid and "head group" moieties. The lipases used in biotechnology are almost invariably digestive lipases, meant to break down the fats in food. They can be used to break down complex fats into their components, which are then used to make other materials.

Lipid A lipid is a water insoluble (hydrophobic) substance and is the name of a large class of structurally and functionally diverse molecules. Important lipids include fatty acids (saturated and unsaturated), they are a component of phospholipids and fats

Phospholipids, main component of biological membranes composed of glycerol phosphate backbone, fatty acid substituents, and hydrophobic headgroups.

- sterols, cholesterol derived membrane components, (sex) hormones, and bile salts (intestinal detergents)

Lipid peroxidation A process in which unsaturated fat-soluble substances (lipids) are oxidised to form radicals and therefore capable of causing extensive tissue damage. The polyunsaturated fatty acid components of lipids are particularly prone to oxidation in this way. The process can occur in foods before they are eaten or can take place in the body.

Lipofection Delivery into eukaryotic cells of DNA, RNA or other compounds that have been encapsulated in an artificial phospholipid vesicle.

Lipogenesis The conversion of carbohydrates and organic acids to fat.

Lipopolysaccharide (LPS) A compound containing lipid bound to a polysaccharide; often a component of microbial cells walls.

Lipoproteins A class of serum proteins that transport lipids and cholesterol in the bloodstream. Abnormalities in lipoprotein metabolism have been implicated in certain heart diseases.

Lipoprotein particle These are protein based carriers of triglycerides (fats) and cholesterol in the blood circulation. The low-density lipoprotein particle LDL is also known as 'bad cholesterol' as it is the major carrier of blood plasma cholesterol and high levels of LDL particles are associated to increased risk of heart disease.

Liposome A synthetic microscopic spherical structure consisting of a phospholipid bilayer membrane containing a user-defined aqueous solution. Liposomes can be used to transport relatively toxic drugs into diseased cells, where they can exert their maximum effect. DNA molecules may be entrapped in, or bound to the surface of, the vesicles, and subsequent fusion of the liposome with the cell membrane will deliver the DNA into the cell. Liposomes have been used to develop an efficient transfection procedure for Streptomyces bacteria.

Liquefaction Enzymatic digestion (often by a-amylase) of gelatinized starch to form lower molecular weight polysaccharides.

Liquid media Media without a solidifying agent.

Liquid membranes Thin films made up of liquids (as opposed to solids) which are stable in another liquid (usually water). Thus the liquid must not dissolve in the water, but nevertheless must be prevented from collapsing into a lot of small droplets.

Liquid nitrogen Nitrogen gas condensed to a liquid with a boiling point of about -196°C. Very commonly the medium in which containers of genetic material are stored.

Litmus A pH indicator paper (range 4.5-8.3) impregnated with an extracted lichen pigment. It turns red in acidic and blue in alkaline solutions. However, the use of litmus paper as an indicator is not a precise method of pH measurement.

Live vaccine A living, non-virulent form of a micro-organism or virus that is used to elicit an antibody response that will protect the inoculated organism against infection by a virulent form of the micro-organism or virus. Also a living, non-virulent micro-organism or virus that express a foreign antigenic protein and is used to inoculate humans or animals. The latter organisms are also called a live recombinant vaccines.

Living modified organism (LMO) "Living organism that possess a novel combination of genetic material obtained through the use of modern biotechnology" (Convention on Biological Diversity). Synonym of GMO, but restricted to organisms that can endanger biological diversity.

Locus In genetics the place on a chromosome which is occupied by a gene. Plural: loci.

Lod score The logarithm of the odds of linkage between two loci. Calculated from pedigree data, as the log (to base 10) of the ratio of the probability of the observed pedigree assuming linkage with a specified recombination fraction q, to the probability of the observed pedigree assuming no linkage, i.e., recombination fraction = 0.5. A lod score (also called a z value) is thus calculated as z = log10 {Prob(data|q)/ Prob(data|0.5)}. A lod score of +3.00 (which is odds of 1000:1) or greater is regarded as acceptable evidence for linkage; -2.00 (which is the log of 1:100) or less indicates that no linkage exists.

Logarithmic phase (log or exponential growth phase) The steepest slope of the growth curve—the phase of vigorous growth during which cell number doubles every 20-30 minutes.

London dispersion force An attractive force caused by quantum-mechanical electron correlation. For example, a neutral spherical molecule (such as a single argon atom) has no charge and produces no external electric field, yet a pair of molecules has a distribution of electron configurations weighted toward those with lesser electron-electron repulsions; this creates a small net attraction.

Lone pair Two valence electrons of an atom that share an orbital but do not participate in a bond.

Long day plant Plant requiring short nights before flowering is initiated.

Long template A DNA strand that is synthesized during the polymerase chain reaction and has a primer sequence at one end but is extended beyond the site that is complementary to the second primer at the other end.

Long terminal repeat (LTR) A string of bases that occurs at each end of the "genome" of a retrovirus that has become integrated into the host genome. Involved in the integration process.

Loop bioreactors Fermenters in which the fermenting material is cycled between a bulk tank and a smaller tank or loop of pipes. The circulation helps to mix the materials and to ensure that gas injected into the fermenter is well distributed in the liquid. The reactors are also very useful for photosynthetic fermentations, where they allow the photosynthesizing organisms to be passed along a large number of small pipes, where the light can get to them easily, rather than inside a single volume, where only the organisms near the edges get much light.

Low density lipoproteins (LDL) Plasma lipoproteins containing high concentrations of lipids (which are low in density compared to that of water), including cholesterol. Increased concentrations are a risk factor for coronary heart disease.

Lumen The channel within a tubular organ; for example, the lumen of the intestine.

Luteinizing hormone A pituitary hormone which causes growth of the yellow body of the ovary and also stimulates activity of the interstitial cells of the testis.

Lux (SI symbol: lx) The unit of measurement for illuminance (i.e., the amount of illumination) impinging upon a surface. 1 lx is the illuminance impinging upon a surface of 1 m2, each point of which is at a distance of 1 m away from a uniform point source of light of 1 cd (candela). It supersedes the foot-candle.

Lyase Any of a class of enzymes that catalyse either the cleavage of a double bond and the addition of new groups to a substrate, or the formation of a double bond.

Lycopene A carotenoid found primarily in tomatoes. Lycopene has antioxidant activity but does not act as a precursor of vitamin A.

Lymphocyte A type of leukocyte found in lymphatic tissue in the blood, lymph nodes and organs. Lymphocytes are continuously made in the bone marrow and mature into antibody-forming cells.

Lymphokine Generic name for proteins that are released by lymphocytes to act on other cells involved in the immune response. The term includes interleukins and interferons. A sub-class of cytokines.

Lymphoma Cancer originating in the lymph nodes, spleen and other lympho-reticular sites.

Lyophilize Rapid freezing followed by a dehydration under high vacuum. The process is lyophilization.

Lysis The destruction or breakage of cells either by viruses or by chemical or physical treatment.

Lysogen A bacterial cell whose chromosome contains integrated viral DNA.

Lysogenic A type or phase of the virus life cycle during which the virus integrates into the host chromosome of the infected cell, often remaining essentially dormant for some period of time.

Lysogenic bacteria Bacteria harbouring temperate (non-virulent, lysogenic) bacteriophages.

Lysogeny A condition in which a bacteriophage genome (pro-phage) survives within a host bacterium, either as part of the host chromosome or as part of an extrachromosomal element, and does not initiate lytic functions.

Lysosome A membrane-bound sac within the cytoplasm of animal cells that contains enzymes responsible for the digestion of material in food vacuoles, the dissolution of foreign particles entering the cell and, on the death of the cell, the breaking down of all cell structures. The digestive system of the cell.

Lysozyme A naturally occurring enzyme extracted from egg white protein and other animal and plant sources, which attacks the cell wall of gram-positive bacteria leading to cell lysis and death.

Lytic A phase of the virus life cycle during which the vi- rus replicates within the host cell, releasing a new generation of viruses when the infected cell lyses.

Lytic cycle The steps in viral production that usually lead to cell lysis.

M

M phase Period of the eucaryotic cell cycle during which the nucleus and cytoplasm divide.

M13 strand The single-stranded DNA molecule that is present in the infective form of bacteriophage M13.

Macerate To disintegrate tissues to obtain a cell dissociation. Cutting, soaking or enzymatic actions are commonly used.

Machine-phase chemistry The chemistry of systems in which all potentially reactive moieties follow controlled trajectories (e.g., guided by molecular machines working in vacuum).

Macromolecule Molecule such as a protein, nucleic acid, or polysaccharide with a molecular mass greater than a few thousand daltons. (Macro from Greek *makros*, large.)

Macro-nutrients Also called caloric nutrients including proteins, carbohydrates, and fats. Some definitions include water and alcohol. The term is used by gardeners and in agriculture referring to common minerals needed for proper plant growth.

Macrophage A type of white blood cell produced in blood vessels and loose connective tissues that can ingest dead tissues and cells and is involved in producing interleukin-1. When exposed to the lymphokine macrophage-activating factor, macrophages also kill tumor cells.

Macrophage colony stimulating factor (M CSF) A natural hormone that stimulates the production of white blood cells, particularly monocytes (the precursors of macrophages).

Macropropagation Production of plant clones from growing parts.

Macroscopic Used in science to describe large scale processes like the temperature, volume, pressure,and energy of a system characterizing the behavior of a very large number of molecules. The macroscopic values tend to be predictable and represent the average behavior of a system. They give no detailed information about the behavior of individual molecules or units of a system.

Mad cow disease Colloquial term for bovine spongiform encephalopathy.

Major histocompatibility complex A group of genes that control several aspects of the immune response. They code for markers which go on the surface of all body cells and are recognise by the body as 'self', belonging to the body. These genes define a person's tissue type and are used to determine whether a transplant would be compatible or not

Malformation A fault in the primary development of an organ or tissue, and therefore present at birth.

Malignant Describes tumors and tumor cells that are invasive and/or able to undergo metasis; a malignant tumor is a cancer.

Malt extract A mixture of organic compounds from malt, used as a culture medium adjunct.

Malting A process of generating starch-degrading enzymes in grain by allowing it to germinate in a humid atmosphere.

Mammalian Pertaining to the group of vertebrates that have: internal development of the embryo;

mammary glands that can produce milk;

live -born young, a body covering of hair or fur;

a four-chambered heart;

a well developed cerebral cortex;

the ability to maintain a constant body temperature; and

a permanent set of teeth.

Mammary glands The milk-producing organs of female mammals, which provide food for the young.

Mammary tumours Tumours of the milk glands.

Management of farm animal genetic resources The sum total of technical, policy and logistical operations involved in understanding (characterization), using and developing (utilization), maintaining (conservation), accessing, and sharing the benefits of animal genetic resources.

Mannitol ($C_6H_{14}O_6$) A sugar alcohol widely distributed in plants. Mannitol is commonly used as a nutrient and osmoticum in suspension medium for plant protoplasts.

Mannose ($C_6H_{12}O_6$) A hexose component of many polysaccharides and mannitol. Mannose is occasionally used as a carbohydrate source in plant tissue culture media.

Map 1. To determine the relative positions of loci on a DNA molecule. Linkage mapping is done by estimating the recombination fraction between loci, from the genotypes of offspring of particular matings. The further apart two loci are on a chromosome, the greater will be the frequency of recombination between them up to a maximum of 50%, the situation observed when they are sufficiently far apart on a chromosome that recombinant gametes are as frequent as non-recombinant gametes, or when they are on different chromosomes. Physical mapping is usually performed by the use of *in situ* hybridisation of cloned DNA fragments to metaphase chromosomes.

2. A diagram showing the relative positions of, and distances between, loci.

Map distance The standard measure of distance between loci, expressed in centiMorgans (cM). Estimated from recombination fraction via a mapping function. For small recombination fractions, map distance equals the percentage of recombination (recombination frequency) betwen two genes. 1% recombination = 1 cM. Sometimes called a map unit.

MAP kinase (mitogen-activated protein kinase) A protein kinase that performs a crucial step in relaying signals from the plasma membrane to the nucleus. Turned on by a wide range of proliferation- or differentiation-inducing signals.

Mapping Determining the location of a locus (gene or genetic marker) on a chromosome.

Mapping function A mathematical expression relating observed recombination fraction to map distance expressed in centiMorgans. Two common mapping functions are those developed by Haldane (1919; *J. Genet.*, 8: 299-309) and Kosambi (1944; *Ann. Eugen.*, 12: 172-175). In both functions, the relationship between recombination fraction and map distance is approximately linear for recombination fractions less than 10%; as recombination fraction increases above 10% (up to its maximum of 50%), map distance is increasingly greater than recombination fraction.

Marker An identifiable DNA sequence that facilitates the study of inheritance of a trait or a gene. Such markers are used in mapping the order of genes along chromosomes and in following the inheritance of particular genes: genes closely linked to the marker will generally be inherited with it. Markers must be readily identifiable in the phenotype, for instance by controlling an easily observable feature (such as eye colour) or by being readily detectable by molecular means, e.g., microsatellite markers.

Marker assisted selection (MAS) The use of DNA markers to increase the response to selection in a population. The markers will be closely linked to one or more quantitative trait loci.

Marker gene A gene of known function or known location, used for marker-assisted selection or genetic studies.

Marker peptide A portion of fusion protein that facilitates its identification or purification.

Marker-assisted introgression The use of DNA markers to increase the speed and efficiency of introgression of a new gene or genes into a population. The markers will be closely linked to the gene(s) in question.

Marsupial A mammal whose distinguishing features include the birth of young at an early foetal stage of development, and generally, a pouch (marsupium) in which further development of the foetus occurs.

Mash Mixture of boiled grain, bran or other substances (including bacteria) for brewing or production purposes.

Mass selection As practised in plant and animal breeding, the choosing of individuals for reproduction from the entire population on the basis of individual phenotypes.

Mass spectrometry A technique for determining the composition of a molecule and its fragments.

Mast cell Widely distributed tissue cell that releases histamine as part of an inflammatory response. Closely related to blood basophils.

Maternal effect An effect attributable to some aspect of performance of the mother of the individual being evaluated.

Maternal inheritance Inheritance controlled by extrachromosomal (cytoplasmic) factors that are transmitted through the egg.

Mating system The mode of transmission of genes from one generation to the next through sexual reproduction.

Matric potential A water potential component, always of negative value, resulting from capillary, imbibitional and adsorptive forces.

Maturation The formation of gametes or spores.

Maximum residue levels Criteria that govern the safe maximum levels of pesticide residues on goods produced for consumption.

Mean In statistics, the arithmetic average; the sum of all measurements or values in a sample divided by the sample size.

Mechanical Pertaining to the positions and motions of atoms, as defined by the positions of their nuclei. A purely mechanical device can be described in terms of atomic positions and motions without reference to electronic properties, save through their effect on the potential energy function.

Mechanochemistry In this volume, the chemistry of processes in which mechanical systems operating with atomic-scale precision either guide, drive, or are driven by chemical transformations. In general usage, the chemistry of processes in which energy is converted from mechanical to chemical form, or vice versa.

Mechanosynthesis Chemical synthesis controlled by mechanical systems operating with atomic-scale precision, enabling direct positional selection of reaction sites; synthetic applications of mechanochemistry. Suitable mechanical systems include AFM mechanisms, molecular manipulators, and molecular mill systems. Processes that fall outside the intended scope of this definition include reactions guided by the incorporation of reactive moieties into a shared covalent framework (i.e., conventional intramolecular reactions), or by the binding of reagents to enzymes or enzymelike catalysts.

Median In a set of measurements, the central value above and below which there are an equal number of measurements.

Medium (pl: media) 1. In plant tissue culture, a term for the liquid or solidified formulation upon which plant cells, tissues or organs develop.

2. In general terms, it could also means a substrate for plant growth, such as nutrient solution, soil, sand, etc., e.g., potting medium.

Medium formulation In tissue culture, the particular formula for the culture medium. It commonly contains macro-elements and micro-elements (high and low salt), some vitamins (B vitamins, inositol), plant growth regulators (auxin, cytokinin and sometimes gibberellin), a carbohydrate source (usually sucrose or glucose) and often other substances, such as amino acids or complex growth factors. Media may be liquid or solidified with agar; the pH is adjusted (ca. 5-6) and the solution is sterilized (usually by filtration or autoclaving). Some formulations are very specific in the kind of explant or plant species that can be maintained; some are very general.

Mega Prefix denoting 10^6.

Megabase (Mb) A length of DNA consisting of 106 base pairs (if double-stranded) or 10^6 bases (if single-stranded). 1 Mb = 10^3 kb = 10^6 bp.

Megabase cloning The cloning of very large DNA fragments.

Megadalton (MDa) One megadalton is equal to 106 daltons.

Megaspore; macrospore A haploid (n) spore developing into a female gametophyte in heterosporous plants.

Meiosis The special cell division process by which the chromosome number of a reproductive cell becomes reduced to half (n) the diploid (2n) or somatic number. Two consecutive divisions occur. In the first division, homologous chromosomes became paired and may exchange genetic material (via crossing over) before moving away from each other into separate daughter nuclei (reduction division). These new nuclei divide by mitosis to produce four haploid nuclei. Meiosis results in the formation of gametes in animals or of spores in plants. It is an important source of variability through recombination.

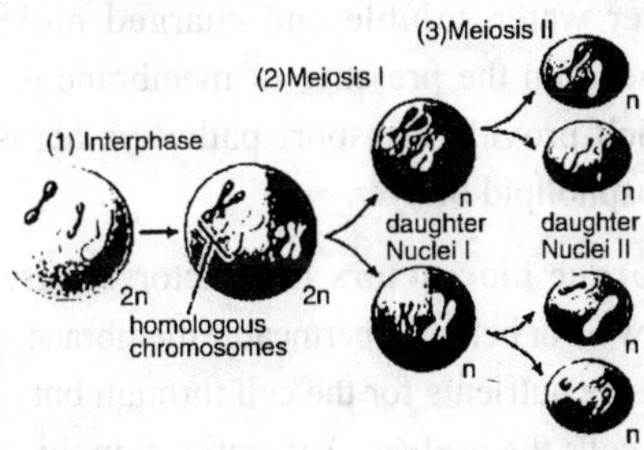

Fig. Meiosis

Meiotic analysis A technique used to analyse chromosome-pairing relationships.

Meiotic drive Any mechanism that causes alleles to be recovered unequally in the gametes of a heterozygote.

Melanin Pigment, as typically produced by specialised epidermal cells called melanocytes.

Melanoma A type of cancer that begins in the melanocytes - the skin cells that produce pigments. It can spread to other areas of the body if not detected and treated early.

Melting temperature (Tm) The temperature at which a double-stranded DNA or RNA molecule denatures into separate single strands. The Tm is characteristic of each DNA species and gives an indication of its base composition. DNAs rich in G:C base pairs are more resistant to thermal denaturation than A:T rich DNA since three hydrogen bonds are formed between G and C, but only two between A and T.

Membrane (cell membrane; phospholipid bilayer) A planar structure surrounding cells and organelles within eukaryotic cells (e.g. membranes of cell nucleus of mitochondrion) separating aqueous compartments which carry out different metabolic processes. Cell membranes are electrical insulators but permeable to hydrophobic molecules such as steroidal hormones and small gases (carbon dioxide, molecular oxygen, nitric oxide). All other water soluble and charged molecules depend on the presence of membrane proteins which provide transport pathways across the phospholipid bilayer.

Membrane bioreactors Bioreactors where cells grow on or behind a permeable membrane, which lets the nutrients for the cell through but retains the cells themselves. A variations on this theme is the hollow-fibre reactor.

Memory cell Long-lived B cells and T cells that mediate rapid secondary immune responses to a previously encountered antigen.

Mendel's laws Two laws summarizing Gregor Mendel's theory of inheritance. The Law of Segregation states that each hereditary characteristic is controlled by two 'factors' (now called alleles), which segregate and pass into separate germ cells. The Law of Independent Assortment states that pairs of 'factors' segregate independently of each other when germ cells are formed.

Mendelian inheritance A hereditary process where genetic traits are passed from parents to offspring and are explained in terms of chromosomes separating, independent assortment of genes and the homologous exchange of segments of DNA. There are three modes of Mendelian inheritance: autosomal dominant, autosomal recessive and X-linked inheritance. Named after Gregor Mendel, who first studied and recognised the existence of genes and this method of inheritance by experimenting with and breeding different varieties of peas.

Mendelism The theory of heredity that forms the basis of classical genetics, proposed by Gregor Mendel in 1866 and formulated in two laws.

Mendel's laws Two laws summarizing Gregor Mendel's theory of inheritance. The Law of Segregation states that each hereditary characteristic is controlled by two 'factors' (now called alleles), which segregate and pass into separate germ cells. The Law of Independent Assortment states that pairs of 'factors' segregate independently of each other when germ cells are formed.

Mericloning A propagation method using shoot tips in culture to proliferate multiple buds, which can then be separated, rooted and planted out.

Meristele The vascular cylinder tissue in the stem.

Meristem Undifferentiated but determined tissue, the cells of which are capable of active cell division and differentiation into specialized and permanent tissue such as shoots and roots.

Meristem culture A tissue culture containing meristematic dome tissue without adjacent leaf primordia or stem tissue. The term may also imply the culture of meristemoidal regions of plants, or meristematic growth in culture.

Meristem tip An explant comprising the meristem (meristematic dome) and usually one pair of leaf primordia. Also refers to explants originating from apical meristem tip or lateral or axillary meristem tip. Do not confuse the meristem tip with the term "shoot tip," which is much larger and usually has more immature leaves and stem tissue.

Meristem tip culture Cultures derived from meristem tip explants. The use of meristem tip culture is for virus elimination or axillary shoot proliferation purposes, but less commonly for callus production.

Meristemoid A localized group of cells in callus tissue, characterized by an accumulation of starch, RNA and protein, and giving rise to adventitious shoots or roots.

Merozygote Partial zygote produced by a process of partial genetic exchange, such as transformation in bacteria.

Mesoderm The middle germ layer that forms in the early animal embryo and gives rise to parts such as bone and connective tissue.

Mesophile A micro-organism able to grow in the temperature range 20 to 50°C; optimal growth often occurs at about 37°C.

Mesophyll Leaf parenchyma tissue occurring between epidermal layers.

Messenger RNA (mRNA) An RNA molecule that specifies the amino acid sequence of a protein. It is based on the sequence of a complementary strand of DNA and is translated into protein by ribosomes.

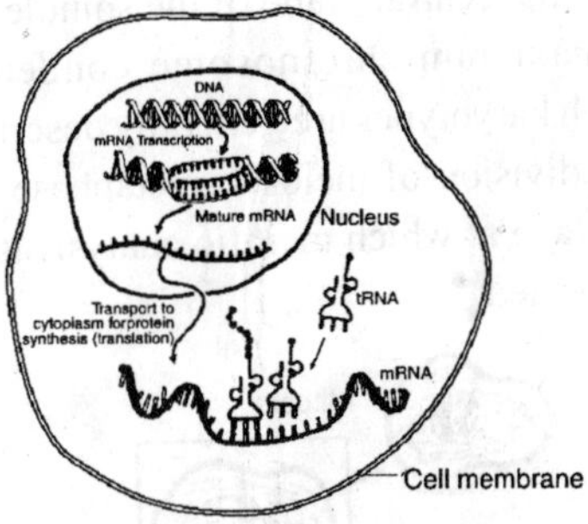

Fig. Messenger RNA (mRNA)

Metabolism, metabolic The totality of all chemical processes in cells and all living organisms. Metabolism is the chemistry of energy extraction from nutrients and the biosynthesis of the building blocks of life (amino acids, sugars, lipids). A process by which absorbed nutrients are converted to chemical energy and building blocks for vital processes or cellular structures. Metabolism regulates the rate at which you burn Calories.

Metabolic cell A cell that is not dividing.

Metabolite 1. A low-molecular-weight biological compound that is usually synthesized by an enzyme.

2. A compound that is essential for a metabolic process. A substance synthesized by the organism, or taken in from the environment. Autotrophic organisms take in inorganic metabolites, such as water, CO_2, nitrates and some trace elements.

Metacentric chromosome A chromosome in which thecentromere is located in the middle and, consequently, the chromosome arms are of about equal length.

Metallothionein A protective protein that binds heavy metals, such as cadmium and lead.

Metaphase Stage of mitosis or meiosis (following prophase and preceding anaphase) during which the chromosomes, or at least the kinetochores, lie in the central plane of the spindle. The stage of maximum chromosome condensation, at which karyotypes are generally described. In the first division of meiosis, metaphase represents the stage at which meiotic analysis is generally performed.

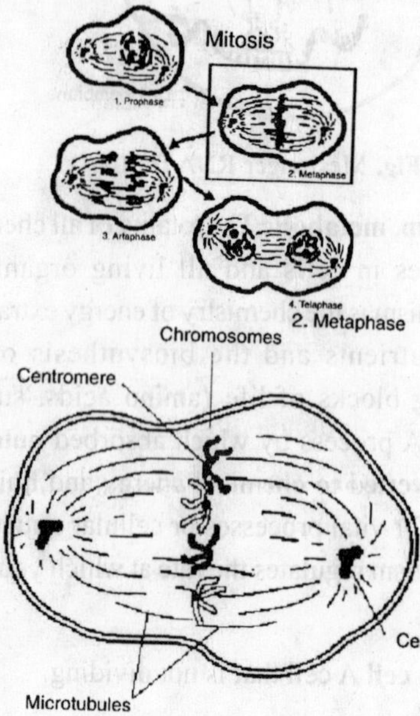

Fig. Metaphase

Metastable A classical system is metastable if it is above its minimum-energy state, but requires an energy input before it can reach a lower-energy state; accordingly, a metastable system can act like a stable system, provided that energy inputs (e.g., thermal fluctuations) remain below some threshold. Systems with strong metastability are commonly described as stable. Quantum mechanical effects can permit metastable states to reach lower energies by tunneling, without an energy input; an associated, broader definition of *metastable* embraces all systems that have a long lifetime (by some standard) in a state above the minimum-energy state.

Metastasis The spread of cancer cells to previously unaffected organs.

Methanogenesis The process of creating methane gas during metabolism.

Methionine A naturally occurring amino acid which is essential in the mammalian diet, providing both methyl groups and sulfur needed for normal metabolism.

Methylation The addition of a methyl group ($-CH_3$) to a macromolecule, such as the addition of a methyl group to specific cytosine and, occasionally, adenine residues in DNA.

Micro carriers Small particles used as a support material for cells, and particularly mammalian cells, which are too fragile to be pumped and stirred as bacterial cells are in a large-scale culture.

Microalgae An ancient life form that produces oxygen. Almost all microalgae are unicellular.

Microalgal culture Culture in bioreactors of microalgae; microalgae include seaweeds.

Micro-array A large set of cloned DNA molecules immobilized as a compact and orderly pattern of sub-microlitre spots onto a solid matrix (typically a glass slide). Used to analyse patterns of gene expression, presence of markers, or nucleotide sequence. The major advantage of micro-arrays is the extent to which the process of genotyping can be automated, thereby enabling large numbers of individuals to be simultaneously genotyped at many loci. A similar approach may be used with other immobilized components for other purposes.

Microbe An organism that can only be seen using a microscope and not with the naked eye e.g. bacteria.

Microbial herbicides and pesticides Microorganisms that are toxic to specific plants or insects. Because of their narrow host range and limited toxicity, these microorganisms may be preferable to their chemical counterparts for certain pest-control applications.

Microbial mats (biofilms) Layered groups or communities of microbial populations.

Microbiology Study of living organisms that can be seen only under a microscope.

Microbody A cellular organelle always bound by a single membrane, frequently spherical, from 20 to 60 nm in diameter, containing a variety of enzymes.

Microdroplet array; multiple drop array (MDA); hanging droplet technique Introduced by Kao and Konstabel (1970), this technique is used to evaluate large numbers of media modifications, employing small quantities of medium into which are placed small numbers of cells. Droplets of liquid are arranged on the lid of a Petri dish, inverted over the bottom half of the dish containing a solution with a lower osmotic pressure, and the dish is sealed. The cells or protoplasts form a monolayer at the droplet meniscus and can easily be examined.

Micro-element An element required in very small quantities.

Micro-encapsulation A process of enclosing a substance in very small sealed capsules from which material is released by heat, solution or other means.

Micro-environment The environment close enough to the surface of a living or non-living object to be influenced by it.

Microfibrils Microfibrils are exceedingly small fibres visible only at the high magnification of the electron microscope.

Microflora The bacterial colonies found in the large intestine. These bacteria are important for proper digestion and fecal excretion of waste products.

Micro-injection The introduction of small amounts of material (DNA, RNA, enzymes, cytotoxic agents) into a single eukaryotic cell with a fine, microscopic needle, penetrating the cell membrane.

Micro-isolating system Mechanical separation of single cells or protoplasts thus allowing them to proliferate individually.

Micrón; micrometre A unit of distance: 10^{-6} m; 0.001 mm. Symbol: mm.

Micronutrient For growth media: An essential element normally required in concentrations < 0.5 millimole/litre.

Micro-organisms Organisms that can be seen only with the aid of a microscope. They are also known as microbes.

Microplasts Vesicles produced by subdivision and fragmentation of protoplasts or thin-walled cells.

Microprojectile bombardment A technique for introducing new genes into cells. The gene is attached to tiny "bullets" made of tungsten or other metal. The tiny particles are then literally "shot" through the plasma membrane into plant cells by a "gene gun."

Micropropagation Miniaturized *in vitro* multiplication and/or regeneration of plant material under aseptic and controlled environmental conditions on specially prepared media that contain substances necessary for growth; used for three general types of tissue: excised embryos (= embryo culture); shoot-tips (= meristem culture or mericloning); and pieces of tissue that range from bits of stems to roots. Four stages of plant tissue culture have been defined by Murashige: Stage I. Establishment of an aseptic culture.

Stage II. The multiplication of propagules.

Stage III. Preparation of propagules for successful transfer to soil (rooting and hardening).

Stage IV. Establishment in soil.

Micropyle 1. A small opening in the surface of a plant ovule through which the pollen tube passes prior to fertilization.

2. A small pore in some animal cells or tissues.

Microsatellite A form of VNTR. Specifically, a segment of DNA characterized by the occurrence of a variable number of copies (from a few up to 30 or so) of a sequence of around 5 or fewer bases (called a repeat unit). A typical microsatellite is the repeat unit AC, which occurs at approximately 100 000 different sites in a typical mammalian genome. At any one site (locus), there are usually several different "alleles," each identifiable according to the number of repeat units. These alleles can be detected by PCR using primers designed from the unique sequence that is located on either side of the microsatellite. When the PCR product is run on an electrophoretic gel, alleles are seen to differ in length in units equal to the size of the repeat unit, e.g., if the primers correspond to the unique sequence immediately on either side of the microsatellite and are each 20 bases long, and an individual is heterozygous for an AC microsatellite with one allele comprising 5 repeats and the other comprising 6 repeats, the heterozygote will exhibit two bands on the gel, one band being 20 + (2 X 5) + 20 = 50 bases long, and the other allele being 20 + (2 X 6) + 20 = 52 bases long. Microsatellites have been the standard DNA marker: they are easily detectable by PCR, and they tend to be evenly located throughout the genome. Thousands have been mapped in many different species.

Microscopic The description of the behavior of individual molecules (at the molecular level). The behavior of an individual molecule may be very different from other identical molecules. In systems with a very large number of identical molecules, these individual deviations are negligible and contribute to the noise (random fluctuation) of a property of this system. The decay of a radioactive material is a good example of such an internal fluctuation. While the decay of isotopes can be predicted for a large clump of radioactive material (e.g. its half-life time, which is a macroscopic property), the actual time point of decay of an individual isotope cannot be predicted with accuracy and is random. However, we can assign a probability that it will decay within a certain amount of time (similar arguments can be made for individuals who have risk factors indicating the chance of developing a disease like cancer or a heart attack).

Microspore The smaller of the two kinds of meiospores produced by heterosporous plants in the course of microsporogenesis; in seed plants, microspores give rise to the pollen grain, the male gametophyte.

Microtubules A minute filament in living cells that is composed of the protein tubulin and occurs singly, in pairs, triplets or bundles. Microtubules help cells to maintain their shape; they also occur in cilia, flagella and the centrioles, and form the spindle during nuclear division.

Mid parent value In quantitative genetics, the average of the phenotypes of two mates.

Middle lamella Original thin membrane separating two adjacent protoplasts and remaining as a distinct cementing layer between adjacent cell walls.

Mini prep A small-scale (mini-) preparation of plasmid or phage DNA. Used to analyse DNA in a cloning vector after a cloning experiment.

Minimum effective cell density The inoculum density below which the culture fails to give reproducible cell growth. The minimum density is a function of the tissue (species, explant, cell line) and the culture phase of the inoculum suspension. Minimum density decreases inversely to the aggregate size and division rate of the stock culture.

Minimum inoculum size The critical volume of inoculum required to initiate culture growth, due to the diffusive loss of cell materials into the medium. The subsequent culture growth cycle is dependent on the inoculum size, which is determined by the volume of medium and size of the culture vessel.

Minisatellite A form of VNTR in which the repeat units typically range from 10 to 100 bases. They are usually detected by Southern hybridization, using a probe comprising a clone of the repeat unit. The first DNA fingerprints were minisatellites detected in this way. Minisatellites tend to be located at the ends of chromosomes and in regions with a high frequency of recombination.

Minitubers Small tubers (5-15 mm in diameter) formed on shoot cultures or cuttings of tuber-forming crops, such as potato.

Mismatch The lack of a complementary pair of bases in a double helix of DNA, e.g., A:C, G:T.

Mismatch repair DNA repair processes that correct mismatched base pairs.

Misreaction A chemical reaction that fails by yielding an unwanted product.

Missense mutation A mutation that changes a codon for one amino acid into a codon specifying another amino acid.

Mist propagation Application of fine droplets of water to leafy cuttings in the rooting stage to reduce transpiration.

Mites Free-living and parasitic animals belonging to the order Acarina, class Arachnida (with spiders). Mites may infest plant crops, reducing their harvest. They may also infest plant tissue culture work areas and incubation facilities in search of sugars, and so contaminate culture vessels and spread bacteria and fungi.

Mitochondria Small structures in living cells which are found outside the nucleus and which contain a small number of the cell's genes. Mitochondria are often described as the 'power house of the cell'. They carry their own complement of DNA and are replicated independently so that when the cell divides, each daughter cell will receive half of the mitochondria.

Mitochondrial DNA The genetic material of the mitochondria - the organelle that generates energy for the cell. The DNA in mitochondria is different from that in the nucleus. Many scientists believe that this DNA is the remnant of a bacterium that invaded the cell in very early evolution. Mitochondrial DNA (mtDNA) is typically passed on only from the mother during sexual reproduction as it is only the nucleus of the sperm that enters the egg upon fertilisation. This means there is little change in the mtDNA from generation to generation.

Mitochondrion (pl.: mitochondria) Organelle possessing its own DNA which appear in all eukaryotic cells (and never in prokaryotic cells) and produce adenosine triphosphate as an energy source for the cell via oxidative phosphorylation. Mitochondria contain many enzymes of the respiratory cycle, although most of these proteins are nuclear encoded.

Mitosis Disjunction of replicated chromosomes and division of the cytoplasm to produce two genetically identical daughter cells. The division involves the appearance of chromosomes, their longitudinal duplication, and equal distribution of newly formed parts to daughter nuclei. It is separated into five stages: interphase, prophase, metaphase, anaphase and telophase.

Mixed bud A bud containing both rudimentary leaves and flowers.

Mixoploid Cells with variable (euploid, aneuploid) chromosome numbers. Mosaics or chimeras differ in chromosome number as a result of a variety of mitotic irregularities.

MM2/CSC A molecular mechanics program developed by Cambridge Scientific Computing that closely follows the MM2 model, adding a graphical user interface and other features.

Mobilization 1. The transfer between bacteria of a non-conjugative plasmid by a conjugative plasmid.

2. The transfer between bacteria of chromosomal genes by a conjugative plasmid.

Mobilizing functions The genes on a plasmid that give it the ability to facilitate the transfer of either a non-conjugative or a conjugative plasmid from one bacterium to another.

Modal class; mode In a frequency distribution, the class having the greatest frequency.

Model A mathematical description of a biological phenomenon.

Modern biotechnology This is the term used to describe the range of processes and techniques that have been made possible since the discovery of DNA in the 1950s.

Modification 1. Enzymatic methylation of a restriction enzyme DNA recognition site.

2. Specific nucleotide changes in DNA or RNA molecules.

Modifier; modifying gene A gene that affects the expression of some other gene.

Modulus Any of several measures of strain versus applied stress.

Moiety A portion of a molecule structure having some property of interest.

Molarity The number of moles of a substance contained in a kilogram of solution.

Mole (M) Amount of substance that has a weight in grams numerically equal to the molecular weight of the substance. Also called gram molecular weight. A mole contains 6.023×10^{23} molecules or atoms of a substance.

Molecular biology The science of studying the genetic composition and mechanism of living organisms at the molecular level. It historically refers to the understanding and manipulation of genes (DNA). The molecular studies of all other organic molecules like proteins, fats, and carbohydrates is called biochemistry.

Molecular cloning The biological amplification of a DNA sequence via the mitotic division of a host cell into which it has been transformed or transfected.

Molecular electronics Any system with atomically precise electronic devices of nanometer dimensions, especially if made of discrete molecular parts, rather than the continuous materials found in today's semiconductor devices.

Molecular genetics The area of knowledge concerned with the genetic aspects of molecular biology, especially with DNA, RNA and protein molecules.

Molecular manipulator A device combining a proximal-probe mechanism for atomically precise positioning with a molecule binding site on the tip; can serve as the basis for building complex structures by positional synthesis.

Molecular markers Genetic markers are specific sets of base sequences (strings of C, G, A and T) along a living thing's chromosomes.

Molecular mechanics models Many of the properties of molecular systems are determined by the molecular potential energy function. Molecular mechanics models approximate this function as a sum of 2-atom, 3-atom, and 4-atom terms, each determined by the geometries and bonds of the component atoms. The 2-atom and 3-atom terms describing bonded interactions roughly correspond to linear springs.

Molecular medicine A variety of pharmaceutical techniques and therapies in use today.

Molecular mill A mechanochemical processing system characterized by limited motions and repetitive operations without programmable flexibility.

Molecular nanotechnology Thorough, inexpensive control of the structure of matter based on molecule-by-molecule control of products and by-products; the products and processes of molecular manufacturing, including molecular machinery.

Molecular recognition A chemical term referring to processes in which molecules adhere in a specific way, forming a large structure; an enabling technology for nanotechnology.

Molecular surgery / molecular repair Analysis and physical correction of molecular structures in the body using medical nanomachines.

Molecular systems engineering Design, analysis and construction of systems of molecular parts working together to carry out a useful purpose.

Molecular weight The sum of the atomic weights of all atoms in a molecule.

Molecule A unit of matter, the smallest portion of an element or a compound that retains chemical identity with the substance in mass. The molecule usually consist of a union of two or more atoms; some organic molecules containing a very large number of atoms.

Monoclonal antibody (mAB) A single type of antibody that is directed against a specific epitope (antigen, antigenic determinant) and is produced by a single clone of B cells or a single hybridoma cell line, which is formed by the fusion of a lymphocyte cell with a myeloma cell. Some myeloma cells synthesize single antibodies naturally.

Monocotyledon A plant whose embryo has one seed leaf (cotyledon). Examples are cereal grains (corn, wheat, rice), asparagus, and lily. Colloquially called a monocot.

Monoculture The agricultural practice of cultivating a single crop on a whole farm or area.

Monocytes One of three types of white blood cells. Monocytes are precursors to macrophages.

Monoecious Denoting plant species that have separate male and female flowers on the same plant (e.g., maize).

Monogastric animals Animals with simple stomachs that do not ruminate.

Monogenic Controlled by a single gene, as opposed to multigenic.

Monohybrid The offspring of two homozygous parents that differ from one another by the alleles present at only one locus.

Monohybrid cross A cross between parents differing in only one trait or in which only one trait is being considered.

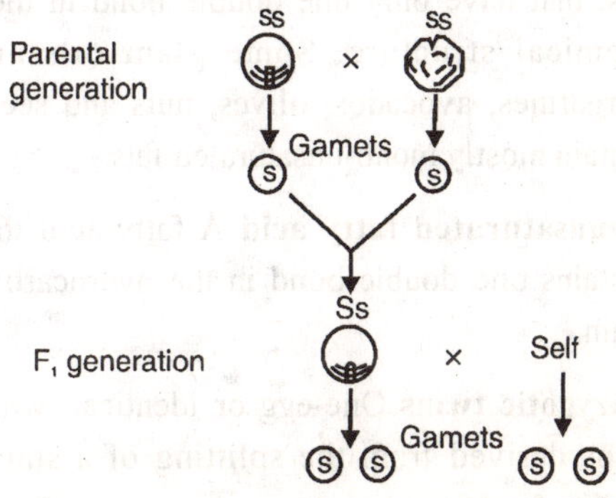

Fig. Monohybrid cross

Monolayer A single layer of cells growing on a surface.

Monomer A single molecular entity that may combine with others to form more complex structures.

Monophyletic Describing any group of organisms that are assumed to have originated from the same ancestor.

Monosaccharide A single sugar.

Monosomic (n: monosomy) Describing a diploid organism lacking one chromosome (2n -1) of its proper (disomic) complement; a form of aneuploidy.

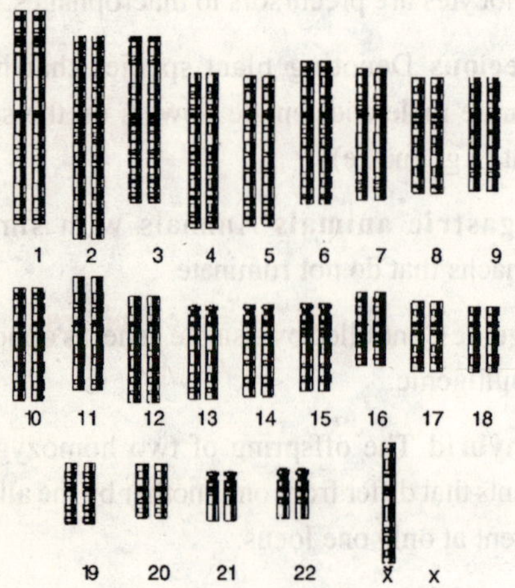

Fig. Monosomic

Mono-unsaturated Refers to molecules, such as fats, that have only one double bond in their chemical structure. Some plant oils and margarines, avocados, olives, nuts and seeds contain mostly mono-unsaturated fats.

Monounsaturated fatty acid A fatty acid that contains one double bond in the hydrocarbon chain.

Monozygotic twins One-egg or identical twins; twins derived from the splitting of a single fertilized ovum.

Monsanto (trade name) A transnational corporation based in St. Louis, Missouri. Monsanto markets a huge range of products, with a particular focus on agricultural biotechnology.

Moral standing To say that a group of organisms has moral standing is to say that their wellbeing must be given some consideration. It does not decide the question of whether they have the same moral standing as people (and thus have 'human' rights).

Morphogen A substance that stimulates the development of form or structure in an organism.

Morphogenesis The development, through growth and differentiation, of form and structure in an organism.

Morphogenic response The effect on the developmental history of a plant or its parts exposed to a given set of growth conditions or to a change in the environment.

Morphology 1.The science of studying form and its development.

2. General: Shape, form, external structure or arrangement.

Mosaic An organism or part of an organism that is composed of cells with different origin.

Mosaic viruses Viruses that cause leaf-mottling in plants.

MRUs Minimum recognition units.

Multi copy Describing plasmids which replicate to produce many plasmid molecules per host genome, e.g., pBR322 is a multi-copy plasmid, there are usually 50 pBR322 molecules (or copies) per *E. coli* genome.

Multi locus probe A probe that hybridizes to a number of different sites in the genome of an organism.

Multigene families Genes that exist in multiple copies in the genome. from several copies to many thousands or hundreds of thousands of copies.

Multimer; multimeric A protein made up of more than one peptide chain.

Multinational corporation Corporation operating (with offices, factories, headquarters) in more than one country. Frequently multinational companies split different parts of a production process between different countries i.e. manufacturing may occur where labour is cheaper, and marketing where their are consumers with money to spend.

Multiple alleles The existence of more than two alleles at a locus in a population.

Multiple ovulation and embryo transfer (MOET) A technology by which a single female that usually produces only one or two offspring can produce a litter of offspring. Involves stimulation of a female to shed large numbers of ova; natural mating or artificial insemination; collection of fertilized ova (either surgically, or non-surgically through the cervix); and transfer (usually non-surgical, through the cervix) of these fertilized ova to recipient females.

Multipotent The potential to make a few cell types in the body. Usually in reference to adult stem cells.

Multivalent vaccine A single vaccine that is designed to elicit an immune response either to more than one infectious agent or to several different epitopes of a molecule.

Muscular dystrophy A group of hereditary diseases that cause progressive muscle wastage due to defects in the biochemistry of a muscle tissue. The most common type is Duchenne muscular dystrophy, which is due to a defective gene on the X chromosome. Because the condition is sex-linked, it usually only affect males. It is usually lethal by the early 20's.

Mutable gene A gene which has an unusually high rate of mutation.

Mutagen Something that is capable of producing a genetic mutation (change), by causing changes in the DNA of living organisms.

Mutagenesis Induction of heritable change(s) in the genetic constitution of a cell through alterations to its DNA.

Mutation, mutant Mutations are changes in the nucleotide sequence of the genome (affecting genes) and the resulting changes in the amino acid sequence of proteins. The amino acid sequence of proteins determines their structure and function, the latter being subject to natural selection. Mutation occur in different forms, from single nucleotide mutations to insertions and deletions of longer sequences as well as gene duplications, deletions and chromosomal rearrangements in higher organisms.

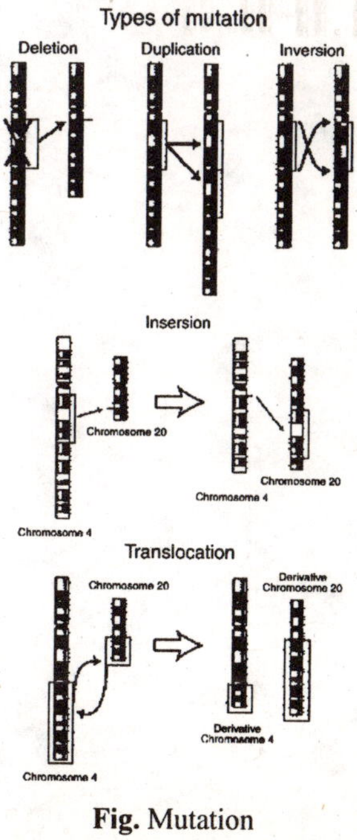

Fig. Mutation

Mutation breeding A technique used to cause mutations to take place in crops to introduce the desirable genes and therefore the trait into the plant.

Mutation pressure A constant mutation rate that adds mutant genes to a population; repeated occurrences of mutations in a population.

Mycelium (pl: mycelia) Threadlike filament making up the vegetative portion of thallus fungi.

Mycoprotein Fungal protein.

Mycorrhiza Fungi that form an association with or have a symbiotic relationship with roots of more developed plants.

Mycotoxin Toxic substance of fungal origin, such as aflatoxin.

Myeloma A type of cancer cell (plasma cell) that is used in monoclonal antibody technology to form hybridomas.

Myxomatosis A disease of rabbits caused by the myxoma virus. It is an early form of biological control.

N

N_2 **free nitrogen gas** In liquid form, used as a cryopreservant.

NAD⁺ (nicotine adenine dinucleotide) Coenzyme that participates in an oxidation reaction by accepting a hydride ion (H^-) from a donor molecule. The NADH formed is an important carrier of eléctrons for oxidative phosphorylation.

Naked bud A bud not protected by bud scales.

Nano A prefix meaning one billionth (1/ 1,000,000,000).

Nanocell surgery Modifying cellular structures by the use of medical nanomachines.

Nanocomputer A computer with parts built on a molecular scale.

Nanoelectronics Electronics on a nanometer scale, whether by current techniques or nanotechnology; includes both molecular electronics and nanoscale devices resembling today's semiconductor devices.

Nanomachine An artificial eutactic mechanical device that relies on nanometer-scale components.

Nanometer (NM) Unit of length commonly used to measure molecules and cell organelles. 1 nm = 10^{-3} µm = 10^{-9} m.

Nanoscale On a scale of nanometers, from atomic dimensions to ~100 nm.

Nanosurgery A generic term including molecular repair and cell surgery.

Nanosystem A eutactic set of nanoscale components working together to serve a set of purposes; complex nanosystems can be of macroscopic size.

Nanotechnology A technology that creates small materials at the scale of molecules by manipulating single atoms. The name nano comes from the size of molecules which is measured in nanometers - or one billionth of a meter (0.000000001 meter). The dimension of single atoms is ten fold smaller. The molecular processes of life, particularly the activity of proteins (enzymes) and the self-organizing behavior of many biological molecules has greatly inspired nanotechnology and molecular motors (i.e. protein complexes) could be considered the result of natures nanotechnology.

Narrow host range plasmid A plasmid that can replicate in one, or at most a few, different bacterial species.

Narrow sense heritability In quantitative genetics, the proportion of the phenotypic variance that is due to variation in breeding values.

National institutions of health (NIH) A nonregulatory agency which has oversight of research activities that the agency funds.

Native Refers to organisms that have not been recently introduced into an ecosystem.

Native protein The naturally occurring form of a protein.

Natural active immunity Immunity that is established after the occurrence of a disease.

Natural killer (NK) cell A type of leukocyte that attacks cancerous or virus-infected cells without previous exposure to the antigen. NK cell activity is stimulated by interferon.

Natural passive immunity Immunity conferred by the mother on the fetus or newborn.

Natural selection The process described by Darwin's theory of evolution that favors certain genotypes and disfavors others. This process is entirely guided by the interaction of an organism with its environment.

Necrosis Death associated with discoloration and dehydration of all or some parts of organs.

Negative autogenous regulation; negative self-regulation Inhibition of the expression of a gene or set of coordinately regulated genes by the product of the gene or the product of one of the genes.

Negative control system A mechanism in which the regulatory protein(s) is required to turn off gene expression.

Negative selection Selection of individuals that do not possess a certain character. A method by which growing cells that do not carry a DNA insert integrated at a specific chromosomal location are selected.

Nematode Slender, unsegmented worms, often parasitic. Also known as eelworm, especially when phytoparasitic.

Neo formation Organogenesis; production of newly formed structures, such as tissues, meristems and embryos.

Neoplasm Localized cell multiplication. Generally it designates a collection of cells which have undergone genetic transformation, forming a tumour. Neoplasmic cells differ in structure and function from the original cell type.

Neoteny The retention of juvenile body characters in the adult state, or the occurrence of adult characters in the juvenile state.

Net photosynthesis Photosynthetic activity minus respiratory activity, as measured by carbon dioxide exchange.

Neurodegenerative disease Diseases caused by the irreversible deterioration of essential cell and tissue components of the nervous system.

Neuroepithelial Forming connections of neurons at the epithelium.

Neurons Cells of the nervous system, which transmit nerve impulses (electrical signals) between cells.

Neurotransmitter A chemical substance released from neurons in synapses that binds to corresponding receptors on nearby cell surfaces (post synaptic membrane) causing a physiological stimulus in form of a membrane current (triggering action potentials) or second messenger cascade activating channels, pumps, kinases, or proteases. The molecular mechanisms of activation are similar to those of hormones.

Neutral mutation A mutation that changes the nucleotide sequence of a gene but has negligible effect on the fitness of the organism.

Neutral theory The theory that much of evolution has been primarily due to random drift of neutral mutations.

Nick To break a phospho-diester bond in the backbone of one of the strands of a duplex DNA molecule.

Nick translation A procedure for labelling DNA. A DNA fragment is treated with DNase to produce single-stranded nicks. The nick is moved along the DNA molecule in the presence of labelled deoxyribonucleoside triphosphates by the concerted action of the 5′-> 3′ exonuclease and 5′-> 3′ polymerase activities of *E. coli* DNA polymerase I.

Nicked circle (relaxed circle) During extraction of plasmid DNA from the bacterial cell, one strand of the DNA becomes nicked. This relaxes the torsional strain needed to maintain supercoiling, producing the familiar form of plasmid.

Nitrate The only form in which nitrogen can be used directly by plants; a component of chemical fertilizers.

Nitrification A chemical process in which nitrogen in plant and animal wastes and dead remains is oxidized, first to nitrites and then to nitrates.

Nitrile herbicide Chemical class of herbicides used for post-emergent control of annual broadleaf weeds. Broken down by soil bacteria.

Nitrite A salt or ester of nitrous acid.

Nitrocellulose; cellulose nitrate A nitrated derivative of cellulose. It is made into membrane filters of defined porosity, used to immobilize DNA, RNA or protein, which can then be probed with a labelled sequence or antibody. These filters have a variety of uses in molecular biology, particularly in nucleic acid hybridization experiments. Used extensively in the Southern and northern blotting procedures involving DNA and RNA.

Nitrogen assimilation The incorporation of nitrogen into organic cell substances by living organisms.

Nitrogen fixation The conversion of atmospheric nitrogen (N_2) into oxidized forms that can be assimilated by plants. Biological nitrogen fixation is catalysed by the enzyme nitrogenase, which is found only in prokaryotes. Certain blue-green algae and some genera of bacteria (e.g., *Rhizobium* spp.; *Azotobacter* spp.) are capable of biochemically fixing nitrogen. Such bacteria are very important symbionts for plants growing in nitrogen-poor soils.

Nitrogenous bases The purines (adenine and guanine) and pyrimidines (thymine, cytosine and uracil) that form DNA and RNA molecules.

NMOS An acronym for n-*channel metal-oxide-semiconductor*, as in NMOS transistor and NMOS logic.

NMP Ribonucleoside monophosphate.

No til A method of farming without tillage that helps conserve soil.

Nod box A DNA sequence that controls the transcriptional regulation of *Rhizobium* nodulation genes.

Nodal culture The culture of a lateral bud and a section of adjacent stem tissue.

Node Slightly enlarged portion of the stem where leaves and buds arise and where branches originate. Stems have nodes but roots do not.

Nodular Term commonly used to describe a pebbly (rough) texture of a callus.

Nodule The enlargement or swelling on roots of nitrogen-fixing plants. The nodules contain symbiotic nitrogen-fixing bacteria.

Non autonomous A term referring to biological units that cannot function by themselves; such units require the assistance of another unit, or "helper".

Non coding linker sequences Short pieces of DNA between genes which do not direct protein synthesis or perform a regulatory function.

Non disjunction Failure of disjunction or separation of homologous chromosomes or chromatids in mitosis or meiosis, resulting in too many chromosomes in some daughter cells and too few in others. Examples: In meiosis, both members of a pair of chromosomes go to one pole so that the other pole does not receive either of them; in mitosis, both sister chromatids go to the same pole.

Non histone chromosomal proteins In chromosomes, all of the proteins except the histones.

Non sense mutation A mutation which converts an amino-acid-specifying codon into a stop codon, e.g., a change from UAU (tyr) to UAG (amber) would lead to the premature termination of a polypeptide chain at the place where a tyrosine was inserted in the wild-type.

Non selective (in herbicides) Lacking specificity. Lethal to many plant varieties, whether or not they are considered weeds.

Non target organism/species Organisms that may be exposed to a given product or event, even though they are not the intended target.

Non template strand In transcription, the non-transcribed strand of DNA. It will have the same sequence as the RNA transcript, except that T is present at positions where U is present in the RNA transcript.

Non transformed Those cells which failed to incorporate foreign DNA into their genomes during an engineering attempt to add a gene(s).

Nontarget organism An organism which is affected by an interaction for which it was not the intended recipient.

Nopaline synthase An enzyme used as a genetic tag.

Northern blot A cellulose or nylon membrane to which RNA molecules have been attached by capillary action. The transferred RNA is hybridized to single-stranded DNA probes. northern blot technique is often used to measure expression (transcription) of a gene for which a specific cDNA is available for use as a probe.

Northern hybridization (northern blotting) A procedure in which RNA fragments are transferred from an agarose gel to a nitrocellulose filter, where the RNA is then hybridized to a radioactive probe.

Novel/novel food New. In the context of food, the term often refers to genetically modified foods, foods developed using a new process (including the use of genetically modified organisms), or substances that have never been used as foods before.

Novel organism or food A term used to describe an organism or food that has been genetically engineered or modified.

Novel pollination mechanisms A process that involves the use of genetic modification to create a unique pollination environment where intentionally selected varieties of a crop must pollinate each other.

Novel traits A characteristic that has been intentionally selected, created or introduced into a distinct, stable population of the same species (animal or plant) through a specific genetic change.

Nucellar embryo An embryo which has developed vegetatively from somatic tissue surrounding the embryo sac, rather than by fertilization of the egg cell.

Nuclear transfer A technology by which animals are created by cloning a single diploid somatic cell. It involves taking a single diploid cell from a culture of cells, and inserting it into an enucleated ovum, i.e., an ovum from which the haploid nucleus has been removed. The resultant diploid ovum develops into an embryo that is placed in a recipient female, which gives birth to the cloned animal in the normal manner. Note that the term is somewhat of a misnomer, since it is a whole cell that is transferred, not just the nucleus.

Nuclease A class of enzymes that degrade DNA or RNA molecules by cleaving the phospho-diester bonds that link adjacent nucleotides. In deoxyribonuclease (DNase), the substrate is DNA. In endonuclease, it cleaves at internal sites in the substrate molecule. Exonuclease progressively cleaves from the end of the substrate molecule. In ribonuclease (RNase), the substrate is RNA. In the S1 nuclease, the substrate is single-stranded DNA or RNA. Nucleases have varying degrees of base-sequence specificity, the most specific being the restriction endonucleases.

Nucleic acid A macromolecule composed of phosphoric acid, pentose sugar, and organic bases. The two nucleic acids, deoxyribonucleic acid (DNA) and ribonucleic acid (RNA), are made up of long chains of molecules called nucleotides. They were first isolated as part of a protein complex in 1871, and were separated from the protein moiety in 1889.

Nuclein The term used by Friedrich Miescher to describe the nuclear material he discovered in 1869, which today is known as DNA.

Nucleo-cytoplasmic ratio In a cell, the ratio of nuclear to cytoplasmic volume. This ratio is high in meristematic cells and low in differentiated cells.

Nucleolar organizer (NO); nucleolar organizer region (NOR) A chromosomal segment containing genes that encode ribosomal RNA; located at the secondary constriction of some chromosomes.

Nucleolar zone Any chromosome region, irrespective of whether or not it is a secondary-constriction, that is associated with the formation of the nucleolus during telophase.

Nucleolus An RNA-rich intranuclear organelle in the nucleus of eukaryotic cells, produced by a nucleolar organizer. It represents the storage place for ribosomes and ribosome precursors. The nucleolus consists primarily of ribosomal precursor RNA, ribosomal RNA, their associated proteins, and some, perhaps all, of the enzymatic equipment (RNA polymerase, RNA methylase, RNA cleavage enzymes) required for synthesis, conversion and assembly of ribosomes. Subsequently the ribosomes are transported to the cytoplasm.

Nucleoplasm The non-staining or slightly chromophilic, liquid or semi-liquid, ground substance of the interphase nucleus and which fills the nuclear space around the chromosomes and the nucleoli. Little is known of the chemical composition of this ground substance, which is not easily defined. Sometimes called "karyoplasm" when it is gel-like, and "karyolymph" when it is a colloidal fluid.

Nucleoprotein Conjugated protein composed of nucleic acid and protein; the material of which the chromosomes are made.

Nucleoside A base (purine or pyrimidine) that is covalently linked to a 5-carbon (pentose) sugar. When the sugar is ribose, the nucleoside is a ribonucleoside; when it is deoxyribose, the nucleoside is a deoxyribonucleoside. Adenine, guanine and cytosine occur in both DNA and RNA; thymine occurs in DNA; and uracil in RNA. They are the building blocks of DNA and RNA.

Nucleoside analog A synthetic molecule that resembles a naturally occuring nucleoside, but that lacks a bond site needed to link it to an adjacent nucleotide.

Nucleosome Spherical sub-units of eukaryotic chromatin that are composed of a core particle consisting of an octamer of histones (two molecules each of histones H2a, H2b, H3 and H4) and 146 nucleotide pairs.

Nucleotide A nucleoside with one or more phosphate groups linked at the 3'- or 5'-hydroxyl

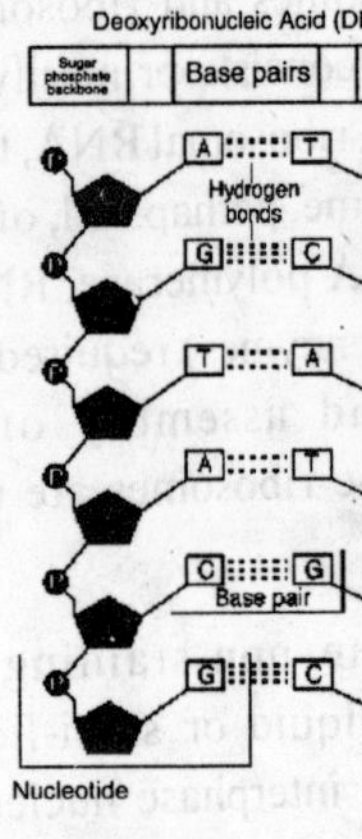

Fig. Nucleotide

of a pentose sugar. When the sugar is ribose, the nucleotide is a ribonucleotide; when it is 2-deoxyribose, the nucleotide is a deoxyribonucleotide. RNA and DNA are polymers of, respectively, ribonucleoside 5'-monophosphates and deoxyribonucleoside 5'-monophosphates. Nucleotides containing the bases adenine, guanine and cytosine (A, G, C) occur in both DNA and RNA; thymine (T) occurs only in DNA, and uracil (U) only in RNA. Ribonucleoside mono-, di-, and triphosphates for which a specific base is not assigned are abbreviated NMP, NDP, and NTP, while deoxyribonucleoside mono-, di, and tri-phosphates are abbreviated dNMP, dNDP, and dNTP. Otherwise, the "N" is replaced by the base letter abbreviation.

Nucleus A dense protoplasmic-membrane-bound region of a eukaryotic cell that contains the chromosomes separated from the cytoplasm by a membrane; present in all eukaryotic cells except mature sieve-tube elements.

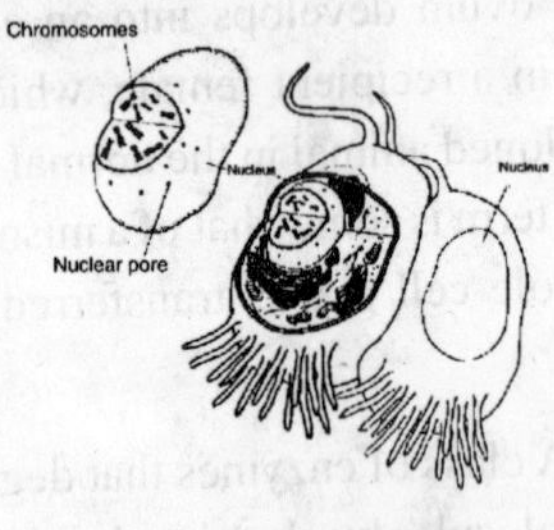

Fig. Nucleus

Nullisomy An otherwise diploid cell or organism lacking both members of a chromosome pair (chromosome formula *2n -2)*.

Nurse culture Planting a cell from a suspension culture on a raft of filter paper above a callus tissue piece (nurse tissue). The filter paper serves to prevent tissue union but allows the flow of essential substances from the nurse to the isolated cell. In such culture, a piece of callus is first placed on nutrient agar. Over this tissue is laid a strip of filter paper.

Nutraceuticals Chemicals that have beneficial effects (pharmacological effects) on our physiology if taken in appropriate amounts with food. Plants are the major source of nutraceuticals also known as phytochemicals. Plants produce those molecules for self-defense or to attract insects and animals to facilitate pollen distribution. For the latter purpose, they often are colorants while tasting bitter to ward of animals.

Nutrient Molecules that can be used by cells or living organism to extract energy through metabolic processes. Although nutrients are often sought off only as energy providers, they can also be used as molecular building block for the biosynthesis of cellular structures.

Nutrient cycle The passage of a nutrient or element through an ecosystem, including its assimilation and release by various organisms and its transformation into various organic or inorganic chemical forms.

Nutrient deficiency Absence or insufficiency of an essential factor for normal growth and development.

Nutrient density Nutrient content of a food expressed in relation to energy content (e.g. Mg per 1000 kcal).

Nutrient film technique (NFT) A hydroponic technique used to grow plants. NFT delivers a thin film of water or nutrient solution either continuously or through on-off cycles.

Nutrient gradient A diffusion gradient of nutrients and gases that develops in tissues where only a portion of the tissue is in contact with the medium. Gradients are less likely to form in liquid media than in callus cultures.

Nutrient medium (pl: nutrient media) A solid, semi-solid or liquid combination of: major and minor salts; an energy source (sucrose); vitamins; plant growth regulators; and occasionally other defined or undefined supplements. Often made from stock solutions, then sterilized by autoclaving or filtering through a micropore filter.

O

Obesity An excessive accumulation of body fat, often defined as a body mass index (BMI) greater than 30. BMI is the ratio of body weight in kilograms to height in metres squared.

Observable expression The observable properties of an organism, thought to be a result of the interaction of the organism's genetic constitution with its environment.

Occupational safety and health administration (OSHA) One of the U.S. agencies responsible for regulation of biotechnology. The major law under which the agency has regulatory powers is the Occupational Safety and Health Act.

Ocean pout Marcrozoarces americanus. A type of fish found in western North Atlantic Ocean.

Octoploid Cell or organism with eight sets of chromosomes, i.e., chromosome number $2n = 8x$.

Oestrogen (estrogen) One of a group of hormonal steroid compounds that promote the development of female secondary sex characteristics; sometimes used in the prevention of osteoporosis.

Oestrous cycle The cycle of reproductive activity shown by most sexually mature non-pregnant female mammals.

Oestrus In female mammals, the period of sexual excitement and acceptance of the male.

Office of the gene technology regulator (OGTR) A part of the Australian Department of Health and Aging that assists the Gene Technology Regulator, a statutory office holder, to administer the Gene Technology Act 200 (the Act). The objective of the Act is to protect the health and safety of people and to protect the environment by identifying risks posed by, or resulting from, gene technology and by managing those risks through regulating certain dealings with genetically modified organisms.

Offset Young plant produced at the base of a mature plant.

Offshoot Short, usually horizontal, stem produced near the crown of a plant.

Offspring; progeny (both same in plural) New individual organisms that result from the process of sexual or asexual reproduction.

Oil Triglycerides (fats) that are liquid at room temperature.

Oilseeds Plant species whose seeds produce high oil content (20 - 50%). They include canola, soybeans, flax, cottonseed and corn.

Okazaki fragment Since DNA can be replicated in only one direction, i.e., nucleotides can be added only at the 3′ end, only one of the two strands of a double helix can be replicated continuously. The other strand is replicated in small segments (Okazaki fragments) that are subsequently joined together by DNA ligase.

Oligodendrocytes A class of glial cells. They provide insulating sheaths around neurons to prevent them from transmitting impulses to the wrong neuron.

Oligomer Short polymer, usually consisting (in a cell) of amino acids (oligopeptides), sugars (oligosaccharides), or nucleotides (oligonucleotides).

Oligonucleotide A short molecule (usually 6 to 100 nucleotides) of single-stranded DNA.

Oligonucleotide ligation assay (OLA) A diagnostic technique for determining the presence or absence of a specific nucleotide pair within a target gene, often indicating whether the gene is wild type (normal) or mutant (defective).

Oligosaccharide Carbohydrate consisting of several linked sugar units.

Omega 3 fatty acids An unsaturated fatty acid that may reduce blood concentrations of triglycerides.

Omitted reaction A chemical reaction that fails by not occurring.

Oncogene A gene that causes cells to grow in an uncontrolled manner, i.e., that causes cancer. Oncogenes are mutant forms of normal functional genes (called proto-oncogenes) that have a role in normal cell proliferation. Oncogenes are found in tumours and in retroviruses; in the latter case, having been picked up along with retroviral genes during retroviral replication in the host. Once incorporated as part of the retroviral genome, they are activated at inappropriate times and places by retroviral promoters, thereby becoming oncogenes.

Oncology Study of cancer.

Onco mouse A mouse that has been genetically modified to incorporate an oncogene, which acts as an animal model for studies of human cancer.

Oncovirus A virus associated with cancer.

Ontogenetic delay A delay in the course of growth and development to maturity.

Ontogeny The course of growth and development of an individual to maturity.

Oocyte The egg mother cell; it undergoes two meiotic divisions (oogenesis) to form the egg cell. The primary oocyte is before completion of the first meiotic division; the secondary oocyte is after completion of the first meiotic division.

Oogenesis The formation and growth of the egg or ovum in an animal ovary.

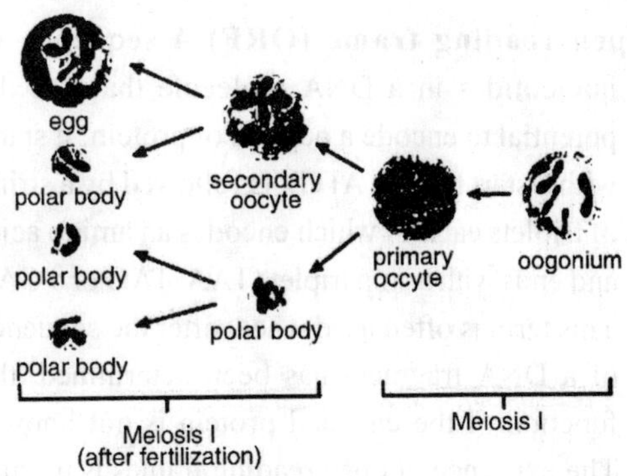

Fig. Oogenesis

Oogonium (pl: oogonia) 1. A germ cell of the female animal, that gives rise to oocytes by mitotic division.

2. The female sex organ of algae and fungi.

Ooplasmic transplantation Adding some of the cytoplasm of an ovum from a donor to the ovum of another woman so that the recipient will receive healthy mitochondria. As the mitochondria contain some 30 or so genes the child resulting from the fertilised treated ovum would strictly have 3 genetic parents. By May 2001 some 30 children were reported to have been born by this IVF variant.

Oosphere The non-motile female gamete in plants and some algae.

Oospore A resistant spore developing from a zygote, resulting from the fusion of heterogametes in certain algae and fungi.

Open continuous culture A continuous culture in which inflow of fresh medium is balanced by outflow of a corresponding volume of culture. Cells are constantly washed out with the outflowing liquid. In a steady state, the rate of cell wash-out equals the rate of formation of new cells in the system.

Open pollination Pollination by wind, insects, or other natural mechanisms.

Open reading frame (ORF) A sequence of nucleotides in a DNA molecule that has the potential to encode a peptide or protein: it starts with a start triplet (ATG), is followed by a string of triplets each of which encodes an amino acid, and ends with a stop triplet (TAA, TAG or TGA). This term is often used when, after the sequence of a DNA fragment has been determined, the function of the encoded protein is not known. The existence of open reading frames is usually inferred from the DNA (rather than the RNA) sequence.

Operational definition An operation or procedure that can be carried out to define or delimit something.

Operator A prokaryotic regulatory element that interacts with a repressor to control the transcription of adjacent structural genes.

Operator gene A region of the chromosome, adjacent to the operon, where a repressor protein binds to prevent transcription of the operon.

Operon A functionally integrated genetic unit for the control of gene expression in bacteria. It consists of one or more genes that encode one or more polypeptide(s) and the adjacent site (promoter and operator) that controls their expression by regulating the transcription of the structural genes.

Opine The condensation product of an amino acid with either a keto-acid or a sugar, produced by the crown gall tissues. Opine synthesis is a unique characteristic of tumour cells.

Orbital In the approximation that each electron in a molecule has a distinct, independent wave function, the spatial distribution of an electron wave function corresponds to a molecular orbital. These, in turn, can be approximated as sums of contributions from the orbitals characteristic of the isolated atoms. An electron added to a molecule—or, similarly, one excited to a higher-energy state within a molecule—would occupy a state with a different wave function from the rest; an unoccupied state of this kind corresponds to an unoccupied molecular orbital. Orbital-symmetry effects on reaction rates arise when a reaction requires overlap between two lobes of the orbitals on each of two reagents: if the algebraic signs of the wave functions in the facing lobes do not match, bond formation between those orbitals is prohibited.

Ordinate The vertical axis of a graph.

Organ A tissue or group of tissues that constitute a morphologically and functionally distinct part of an organism.

Organ culture The aseptic culture of complete living organs of animals and plants outside the body in a suitable culture medium. Animal organs must be small enough to allow the nutrients in the culture medium to penetrate all the cells.

Organelle An organelle is a structure within a cell that performs a particular function, for example, mitochondria, the endoplasmic reticulum, vacuoles, chloroplasts and lysosomes. Organelles are like smaller versions of the organs in your body, each performing a particular function to keep the whole cell alive.

Organellar genes Genes located on organelles outside the nucleus.

Organic Compounds that contain carbon, such as vitamins, carbohydrates, proteins and fats, but not minerals. Organic is a chemical term designating compounds containing a carbon skeleton plus hydrogens, oxygen, and in smaller and variable amounts nitrogen, phosphorus, or sulfur.

Organic agriculture A concept and practice of agricultural production that focuses on production without the use of synthetic inputs and does not allow the use of transgenic organisms. USDA's National Organic Program has established a set of national standards for certified organic production which are available online.

Organic complex A chemically undefined compound added to nutrient media to stimulate growth, e.g., coconut milk; yeast extract; casein hydrolysate.

Organic compound A compound containing carbon.

Organic co solvent A compound used to dissolve some neutral organic substances, such as in media preparation. Organic co-solvents include alcohols (usually ethanol), acetone and dimethylsulphoxide (DMSO).

Organic crops Organic crop includes those that: 1. prohibit the use of ionizing radiation in the preservation of food. 2. Prohibit the use of genetically engineered or modified organisms

3. Encourages the maximum use of recycling

4. Encourages the maximum rotation of crops and promotion of biodiversity.

Organic evolution The process by which changes in the genetic composition of populations of organisms occur in response to environmental changes.

Organic farming A farming method free that does not use synthetic chemicals or genetically modified crops. Organic farming means produced in soils of enhanced biological activity, determined by the humus level, crumb structure and feeder root development, so that plants are fed through the soil ecosystem and not primarily through soluble fertilisers added to the soil. Organic farming relies strategies such as on crop rotations, crop residues, animal manures, mechanical cultivation, approved mineral-bearing rocks and aspects of biological pest control to maintain soil productivity, to supply plant nutrients and to control insects, weeds and other pests.

Organic toxin Toxins derived from bacteria and viruses, killed by chemical exposure, heat, or irradiation, which are incapable of causing disease. Also known as inactivated toxins.

Organism The individual member of a species; can be a single cell or a multicellular organism. Organisms are the biological unit of reproduction and while cells of single cell organism are autonomous (bacteria, archaea), individual cells of multicellular organisms (fungi, plants, animals) are not.

Organized growth The development under tissue culture conditions of organized explants (meristem tips or shoot tips, floral buds or organ primordia).

Organized tissue Composed of regularly differentiated cells.

Organizer An inductor; a chemical substance in a living system that determines the fate in development of certain cells or groups of cells.

Organogenesis The initiation of adventitious or *de novo* shoots or roots from callus, meristem or suspension cultures.

Organoid An organ-like structure produced in culture, such as leaves, roots or callus.

Organoleptic Having an effect on one of the organs of sense, such as taste or smell.

Origin of replication The nucleotide sequence at which DNA synthesis is initiated.

Orphan receptor A receptor for which a cellular function or ligand has yet to be identified.

Ortet The plant from which a clone is obtained.

Osmic acid (osmium tetroxide) (OsO_4) A fixing agent commonly used to prepare tissue samples for electron microscopy.

Osmolarity The total molar concentration of the solutes. Osmolarity affects the osmotic potential of solution or nutrient medium.

Osmosis Diffusion from areas of high concentration to areas of lower concentration of a solvent through a differentially permeable membrane.

Osmotic pressure The pressure exerted when ions flow across a membrane from one cellular compartment to another to equalise the ionic concentration, which is normally close to what is obtained by adding 9 g salt (Na^+Cl^-) to 1 litre distilled water.

Osmoticum An agent, such as PEG, mannitol, glucose or sucrose, employed to maintain the osmotic potential of a nutrient medium equivalent to that of the cultured cells (isotonic). Because of this osmotic equilibrium, cells are not damaged *in vitro*.

Osteoporosis A disorder characterised by an abnormally low density of the bone mineral. The condition predisposes to fracture of certain vulnerable bones and occurs most frequently in women after menopause.

Outbreeding A mating system characterized by the breeding of genetically unrelated or dissimilar individuals. Since genetic diversity tends to be enhanced and since vigour or fitness of individuals can be increased by this process, it is often used to counter the detrimental effects of continuous inbreeding.

Outcrossing Mating between different populations or individuals of the same species that are not closely related. The term "outcrossing" can be used to describe unintended pollination by an outside source of the same crop during hybrid seed production.

Outflow The volume of growing cells that is removed from a bioreactor during a continuous fermentation process.

Output traits Traits produced in genetically modified crops that are beneficial or of direct value to the consumer. For example, enhancing the quality of food, fibre, lowering the fat content or increasing anti-oxidant levels.

Ovary 1. Enlarged basal portion of the pistil of a plant flower that contains the ovules.

2. The reproduction organ in female animals in which eggs (ova) are produced.

Overdominance A condition in which heterozygotes are superior (on some scale of measurement) to either of the associated homozygotes.

Overlap Orbitals lack sharply defined surfaces, declining in amplitude exponentially in their surface regions. When two orbitals are brought together, regions of substantial amplitude overlap. The resulting system can be described as two new orbitals, one formed by joining the two original orbitals without introducing a node in the wave function, and the other formed with a node between them. The nodeless joining reduces the energy of the electrons relative to the separate orbitals, resulting in a bonding interaction; joining with a node raises the energy, producing an antibonding interaction. If both new orbitals are occupied, antibonding forces dominate, resulting in overlap repulsion. Molecular mechanics models give an approximate description of overlap (and other) forces for a certain range of atoms and geometries.

Overlapping reading frames Start codons in different reading frames generate different polypeptides from the same DNA sequence.

Ovulation In mammals, the process of escape of the ovum (egg cell) from the ovary.

Ovule The part of the reproductive organs in seed plants that consists of the nucellus, the embryo sac and integuments.

Ovum 1. A gamete of female animals, produced by the ovary.

2. The oosphere in plants.

Ovum pickup (OPU) The non-surgical collection of ova from a female.

Oxidation A chemical reaction that involves the loss of electrons; it usually but not always involves

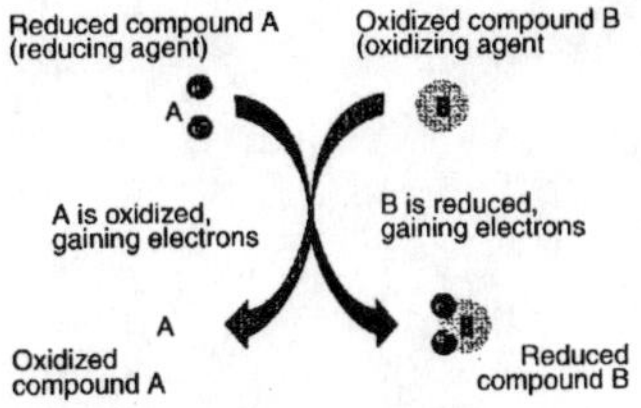

Fig. Oxidation

direct participation of oxygen and is an important process by which food components are utilised by the body's metabolism.

Oxidative stress A condition in which the production of oxidants and free radicals exceeds the body's ability to inactivate them.

Oxygen electrode based sensor Sensor in which an oxygen electrode - a standard electrochemical cell which measures the amount of oxygen in a solution - is coated with a biological material which generates or absorbs oxygen. When the biological coating is active, the amount of oxygen next to the electrode changes and the signal from the electrode changes.

P

P1 Symbol for the parental generation or parents of a given individual.

P450 metabolism genes A specific group of chemical metabolism genes that produce the group of P450 enzymes.

Pachynema A mid-prophase stage in meiosis, immediately following zygonema and preceding diplonema. In microscopic preparations, the chromosomes are visible as long, paired threads. Rarely, four chromatids are detectable.

Packaging cell line A cell line that is designed to produce viral particles that do not contain nucleic acid. After transfection of these cells with a full-size viral genome, fully infective viral particles are assembled and released.

Packed cell volume (PCV) The volume of cells in a set volume of culture expressed as a percentage of that set volume after sedimentation (packing) by means of low speed centrifugation.

Pairing The pairing of homologous chromosomes during the prophase of the first meiotic division. Pairing is the first prerequisite before crossing over and recombination can occur.

Pair-rule gene A gene that influences the formation of body segments in *Drosophila.*

Palaeontology The study of the fossil record of past geological periods and of the phylogenetic relationships between extinct and contemporary plant and animal species.

Palindrome Originally a word, sentence or verse that reads the same from right to left as it does from left to right. In biotechnology, the term is applied to a segment of DNA in which the base-pair sequence reads the same in both directions from a central point of symmetry.

Palindromic sequence A segment of duplex DNA whose 5′-to-3′ sequence is identical on each DNA strand. The sequence is the same when one strand is read left-to-right and the other strand is read right-to-left. Typically, recognition sites for type II restriction endonucleases are palindromes.

Palisade parenchyma Elongated cells found just beneath the upper epidermis of leaves, and containing many chloroplasts.

PAMP Ampicillin-resistant plasmid developed for this laboratory course.

Panicle An inflorescence, the main axis of which is branched; the branches bear loose racemose flower clusters.

Panicle culture Aseptic culture of grain panicle segments to induce microspore germination and development.

Panmictic population A population in which mating occurs at random.

Panmixis Random mating in a population.

Papillae Minute elevations of the upper surface of the tongue in which nerves end.

Par gene A gene found in bacterial and plant cells, and involved in partition of plasmids in cells.

Paracentric inversion An inversion that is entirely within one arm of a chromosome and does not include the centromere.

ParafilmTM A stretchable film based on paraffin wax; used to seal tubes and Petridishes. ParafilmTM is a proprietary name which is applied colloquially to similar products.

Parahormone A substance with hormone-like properties that is not a secretory product (e.g., ethylene; carbon dioxide).

Parallel evolution The development of different organisms along similar evolutionary paths due to similar selection pressures acting on them.

Parameter A value or constant pertaining to an entire population.

Parasexual cycle A sexual cycle involving changes in chromosome number but differing in time and place from the usual sexual cycle; occurring in those fungi in which the normal cycle is suppressed or apparently absent.

Parasite An organism deriving its food from the living body of another organism.

Parasitism The close association of two or more dissimilar organisms where the association is harmful to at least one.

Parasporal crystal Tightly packaged insect pro-toxin molecules that are produced by strains of *Bacillus thuringiensis* during the formation of resting spores.

Parenchyma 1. A plant tissue consisting of spherical, undifferentiated cells, frequently with air spaces between them.

2. Loose connective tissue formed of large cells.

Fig. Parenchyma

Parenchymatous Adjective used to describe spherical and undifferentiated cells with primary cell walls, capable of both cell division and differentiation.

Parthenocarpy The development of fruit without fertilization.

Parthenogenesis Production of an embryo from an unfertilized egg.

Partial digest Addition of a restriction enzyme to a DNA sample under particular conditions or for a limited period, such that only a proportion of the target sites in any individual molecule are cleaved. Partial digests are often performed to give an overlapping collection of DNA fragments for use in the construction of a gene bank.

Particle radiation Refers to gamma (g) particles (positively charged) and beta (i) particles (negatively charged), electrons, protons and neutrons. In plant tissue culture, these particles are used to produce mutant cells or organisms.

Partition coefficient K The partition coefficient most often refers to the oil-water or air-water partition coefficient expressing the concentration ratio of a solute in a two-phasic system (e.g. oil and water). After thoroughly mixing an oil-water solution with a particular solute (an amino acid, lipid, hormone, gas etc.) the concentration of the solute are measured in each phase after the system comes to rest and the oil (gas) phase is well separated from the water phase. The ratio thus is an expression of the relative solubility of a molecule in oil vs water quantifying its hydrophobicity, or its ability to be soluble in oil. The oil-water partition coefficient of molecules is a good indicator if a molecule can easily diffuse across cell membranes or not. The more soluble it is in oil, the better its permeation across membranes. However, very high oil solubility is also an indicator of small molecules to function as general anesthetics, because they tend to stick in cell membranes rather than diffuse across it, thus altering the composition and physical properties of membranes and some of its proteins. If membrane solubility affects neuronal membranes, a loss of sensation or consciousness can be the result.

Partition function A function determined by the probability distribution (over phase space in the classical treatment; over quantum states in the quantum treatment) describing a thermally equilibrated system; many thermodynamic quantities can be expressed in terms of the partition function and its derivatives.

Parts per million (PPM) Units of any given substance per one million equivalent units, such as the weight units of solute per million weight units of solution (i.e., 1 ppm = 1 mg/l).

Parturition The process of giving birth.

Passage The transfer or transplantation of cells from one culture vessel to another.

Passage number The number of times the cells in the culture have been sub-cultured. In descriptions of this process, the dilution ratio of the cells should be stated so that the relative cultural "age" can be ascertained.

Passage time Interval between successive sub-cultures.

Passive immunity 1. Natural acquisition of antibodies by the foetus or neonate (newborn) from the mother.

2. The artificial introduction of specific antibodies by the injection of serum from an immune animal.

In both cases, temporary protection is conferred on the recipient.

Patch-clamp recording Electrophysiological technique in which a tiny electrode tip is sealed onto a patch of cell membrane, thereby making it possible to record the flow of current through individual ion channels in the patch.

Patent Intellectual property protection which gives the owner exclusive right to exploit an invention for a fixed period (e.g 17-20 years) in exchange for full disclosure of how the invention is made.

Pathogen A disease-causing organism (generally microbial: bacteria, fungi, viruses; but can extend to other organisms: e.g. nematodes etc.). Synonym: infectious agent.

Pathogen free Freedom from disease-causing organisms (bacteria, fungi, viruses, etc.).

Pathogenicity The ability to cause disease.

Pathogenecity potential The potential to cause disease.

Pathotoxin Very dilute substance synthesized and released by some pathogen, and which interacts with the host metabolism.

Pathovar In plant pathology, strains of bacteria causing disease in specific plant cultivars.

Pattern of clearance The method by which the body breaks down or disposes of infectious organisms or substances such as drugs (i.e. what major organs are involved, how long it stays in the system).

PBR322 One of the first plasmid vectors widely used; especially used for cloning DNA in *E. coli.*

PCR (polymerase chain reaction) Technique for amplifying specific regions of DNA by multiple cycles of DNA polymerization, each followed by a brief heat treatment to separate complementary strands.

Pectin A white amorphous substance which, when combined with acid and sugar, yields a jelly substance cementing cells together (the middle lamella).

Pedicel Stalk or stem of the individual flowers of an inflorescence.

Pedigree A diagram mapping the genetic history of a par- ticular family.

Peduncle Stalk or stem of a flower that is born singly; the main stem of an inflorescence.

Pehenotype The outward appearance (structure) or other visible characteristics of an organism (which of course, is determined by the DNA of its genotype). This also includes (and/or determines) how that organism's body responds to a given physical agent (e.g., a pharmaceutical, a toxin, sunlight, etc.). For example, genetically fair-skinned people tend to get sunburned easier/ faster than other people do.

Penetrance The percentage of individuals in population that show a particular phenotype among those capable of showing it, i.e., among those that have the genotype normally associated with that phenotype.

Peptide A sequence of amino acids linked by peptide bonds; a breakdown or build-up unit in protein metabolism. Typically used to describe low molecular weight species.

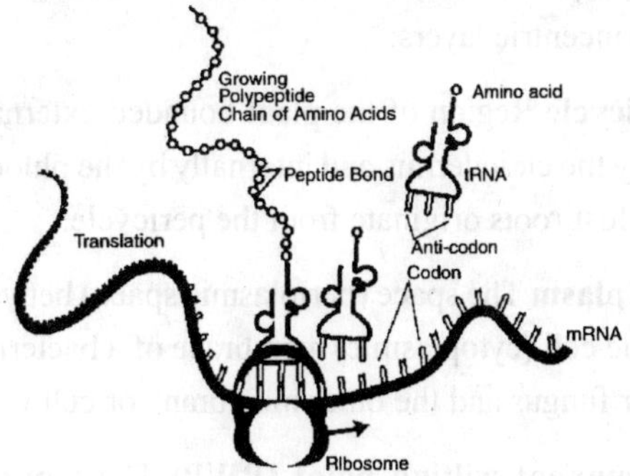

Fig. Peptide

Peptide bond The chemical bond holding amino acid residues together in peptides and proteins. The (CO-NH) bond is formed by the condensation, with loss of a water molecule, between the carboxyl (-COOH) group of one amino acid and the amino (-NH_2) of the next amino acid.

Peptide map Characteristic two-dimensional pattern (on paper or gel) formed by the separation of the mixture of peptides produced by the partial digestion of a protein.

Peptide vaccine A short chain of amino acids that can induce antibodies against a specific infectious agent.

Peptidyl transferase An enzyme bound tightly to the large sub-unit of the ribosome that catalyses the formation of peptide bonds between amino acids during translation.

Perennial A plant that grows more or less indefinitely from year to year and, once mature, usually produces seed each year.

Pericentric inversion An inversion that includes the centromere, hence involving both arms of a chromosome.

Periclinal The plane of cell wall orientation or cell division parallel to the surface of the organ.

Periclinal chimera Genotypically or cytoplasmically different tissues arranged in concentric layers.

Pericycle Region of the plant bounded externally by the endodermis and internally by the phloem. Most roots originate from the pericycle.

Periplasm The space (periplasmic space) between the cell (cytoplasmic) membrane of a bacterium or fungus and the outer membrane or cell wall.

Permanent wilting point (PWP) The moisture content of soil at which plants wilt to such an extent that they fail to recover even when placed in a humid atmosphere.

Permeability Degree of leakiness of membranes.

Permeable Used of a membrane, cell or cell system through which substances may diffuse.

Peroxisome Small membrane-bounded organelle that uses molecular oxygen to oxidize organic molecules. Contains some enzymes that produce and others that degrade hydrogen peroxide (H_2O_2).

Persistent 1. Continues to exist or to remain attached.

2. Chemicals with a long inactivation time, such as some pesticides, which may accumulate in the food chain.

Pest resistant plants Same as insect resistant plants and are resistant to certain insect pests through a built-in or naturally occurring insecticide. They are mainly engineered to produce toxin proteins through genes, which come from a soil bacterium, Bacillus thuringiensis (Bt). In their natural form, Bt toxins are lethal to the larvae of certain pests but not to other living things.

Pesticide A toxic chemical product that kills harmful organisms (e.g., insecticides, fungicide, weedicides, rodenticides).

Pesticide resistance The development or selection of heritable traits (genes) in a pest population that allow individuals expressing the trait to survive in the presence of levels of a pesticide (biological or chemical control agent) that would otherwise debilitate or kill this pest. The presence of such resistant pests makes the pesticide less useful for managing pest populations.

Pest-resistant crops Plants with the ability to withstand, deter or repel pests and thereby prevent them from damaging the plants. Plant pests may include insects, nematodes, fungi, viruses, bacteria, weeds, and other.

Petal One of the parts of the flower that make up the corolla.

Petite mutant A respiration-deficient yeast mutant that produces small colonies when grown on glucose-containing medium.

Petri dish Flat round dish with a matching lid, made of glass or plastic material, and used for culturing organisms.

PH Common measure of the acidity of a solution: "p" refers to power of 10, "H" to hydrogen. Defined as the negative logarithm of the hydrogen ion concentration in moles per liter (M). Thus pH 3 (10^{-3} M H^+) is acidic and pH 9 (10^{-9} M H^+) is alkaline.

PH electrode-based sensor Sensor in which a standard electrochemical pH electrode is coated with a biological material. Many biological processes raise or lower pH, and the changes can be detected by the pH electrode.

Phage Abbreviation for bacteriophage.

Phagemids Cloning vectors that contain components derived from both phage chromosomes and plasmids.

Phagocyte General term for a professional phagocytic cell - that is, a cell such as a macrophage or neutrophil that is specialized to take up particles and microorganisms by phagocytosis.

Phagocytosis Process by which particulate material is endocytosed ("eaten") by a cell. Prominent in carnivorous cells, such as *Amoeba proteus*, and in vertebrate macrophages and neutrophils.

Pharmacological Pertains to drug-like actions of chemical substances.

Pharming The process of farming genetically engineered plants or animals to be used as living pharmaceutical factories The practice has used cows, sheep, pigs, goats, rabbits and mice to produced large amounts of human proteins in their milk. Plants are being used to produce vaccines and diagnostic reagents.

Phase space A classical system of N particles can be described by its $3N$ position and $3N$ momentum coordinates. The phase space associated with the system is the $6N$ dimensional space defined by these coordinates.

Phase state The coupling or repulsion of two linked genes.

Phasic activity Activity which is demonstrated in various phases.

Phenocopy An organism whose phenotype (but not genotype) has been changed by the environment to resemble the phenotype usually associated with a mutant organism.

Phenolic oxidation Many plant species contain phenolic compounds that blacken through oxidation. The process is initiated after plants are wounded. Phenolic oxidation may lead to growth inhibition or, in severe cases, to tissue necrosis and death. Antioxidants are incorporated into the sterilizing solution or isolation medium to prevent or reduce oxidative browning.

Phenols; phenolics Compounds with hydroxyl group(s) attached to the benzene ring, forming esters, ethers and salts. Phenolic substances are produced from newly explanted tissues, oxidizing to form coloured compounds visible in nutrient media.

Phenomics Phenomics is the study of an overall organism and how the characteristics or traits of an organism that we can see (its phenotype) fits with the information we know about its genes (genomics) and proteins (proteomics).

Phenotype The characteristic of a species or individual of a species that is inherited from generation to generation. Each phenotype is the result of a genotype, i.e., the genetic information stored in DNA. Most phenotypes are morphological, i.e., they describe a particular structure, size, texture, or color of an organism or part of an organism. Examples are the shape and arrangement of leaves of plants, or the legs, wings, and body segmentation of insects. These morphological phenotypes are the bases of most taxonomic classification of an organism, i.e., the organization of the evolutionary relationship among all life on Earth thought to originate from a single ancestral cell type. Phenotypes can also be functional characteristics and can best be thought of as hereditary diseases or metabolic processes (photosynthesis, skin color pigmentation).

Pheromone A hormone-like substance that is secreted by an organism into the environment as a specific signal to another organism, usually of the same species.

Phloem Food-conducting tissue in plants, consisting of sieve tubes, companion cells, phloem parenchyma and fibres.

Phonon A quantum of acoustic energy, analogous to the quantum of electromagnetic radiation, the photon. Thermal excitations in a crystal or in an elastic continuum can be described as a population of phonons (analogous to blackbody electromagnetic radiation). In highly inhomogeneous solids, a description in terms of phonons breaks down and localized vibrational modes become important.

Phosphatase An enzyme that hydrolyses esters of phosphoric acid, removing a phosphate group from an organic compound.

Phospho-diester bond A bond in which a phosphate group joins adjacent carbons through ester linkages. A condensation reaction between adjacent nucleotides results in a phospho-diester bond between 3′ and 5′ carbons in DNA and RNA.

Phospholipase A2 An enzyme which degrades type A2 phospholipids.

Phospholipid Main lipid component of cell membranes. Phospholipids are a heterogeneous type of molecule composed of glycerol, phosphate, two fatty acid residues, and 'headgroups' with different chemical properties. The organization of phospholipids in cell membranes is known as phospholipid bilayer where the fatty acid residues face the center of the membrane (hydrophobic or water insoluble) and the headgroups forming the surface of the membrane. As such, bilayers separate water filled compartments and provide an electrically insulating barrier between these two compartments. This barrier is overcome by the placement of membrane proteins penetrating the fatty acid core of the membrane and forming channels and transport pathways for metabolites. The latter are usually water soluble and/or charged and would not diffuse across cell membranes without the help of these transport proteins.

Phosphorolysis The cleavage of a bond by orthophosphate; analogous to hydrolysis referring to cleavage by water.

Phosphorylation The addition of a phosphate group to a compound.

Photo bioreactor Bioreactor dependent on sunlight, which is taken up by its content of plant material, usually algae.

Photon A quantum of light; the energy of a photon is proportional to its frequency; E = hv, where E is energy; h is Planck's constant, 6.62 X 10^{-27} erg-second; and v is the frequency.

Photoperiod The length of day or period of daily illumination provided or required by plants for reaching the reproductive stage.

Photoperiodism The response of an organism to changes in day length.

Photophosphorylation The formation of ATP from ADP and inorganic phosphate using light energy in photosynthesis.

Photoreactivation A DNA repair process that is light dependent.

Photosynthesis The fundamental chemical process in which green plants utilize the energy of sunlight or other light to convert carbon dioxide and water into carbohydrates, with the green pigment chlorophyll acting as the energy converter. This process releases oxygen and is the chief source of atmospheric oxygen.

Photosynthetic efficiency Efficiency of converting light energy into organic compounds.

Photosynthetically active radiation (PAR) Radiant energy captured by the photosynthetic system in the light reactions, usually taken to be the wavelengths between 400 and 700 nm.

Phototropism A growth curvature in which light is the stimulus.

Phylogeny A diagram illustrating the deduced evolutionary history of populations of related organisms.

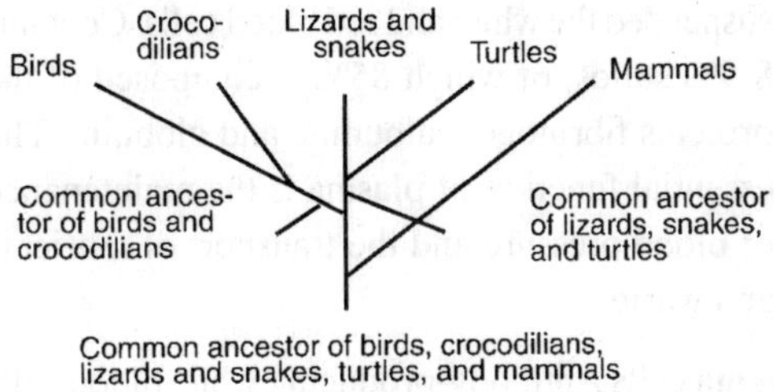

Fig. Phylogeny

Physical map A map showing physical locations on a DNA sequence, such as restriction sites and sequence-tagged sites. Also a diagram of a chromosome or a karyotype, showing the location of loci (genes and markers)

Phytochemicals Plant molecules with beneficial effects for our physiology such as antioxidant properties. They are also known as nutraceuticals or and are found in designer foods.

Phytohormone A substance that stimulates growth or other processes in plants. They include auxins, abscissic acid, cytokinins, gibberellins and ethylene. Phytohormones are chemical messengers that may pass through cells, tissues and organs and stimulate biochemical, physiological and morphological responses.

Phytoparasite Parasite on plants.

Phytopathogen An organism that causes disease in plants.

Phytoremediation Cleaning up waste or pollution using plants.

Phytosanitary Plant health, including quarantine.

Phytostat The name adopted by Tulecke in 1965 for an apparatus designed for the semi-continuous chemostat culture of plant cells.

Pi bond A covalent bond formed by overlap between two *p* orbitals on different atoms. Pi bonds are superimposed on sigma bonds, forming double or triple bonds.

Pigments Molecules that are coloured by the light they absorb. Some plant pigments are water soluble and are found mainly in the cell vacuole.

Pinocytosis The process by which a living cell engulfs a minute droplet of liquid.

Pipette A slender graduated tube into which small amounts of liquids are taken up by suction, for measuring and transferring.

Pistil Central organ of the flower, typically consisting of ovary, style and stigma. The pistil is usually referred to as the female part of a perfect flower.

Pituitary gland A small gland attached to the hypothalamus. It is responsible for the secretion of a range of hormones, including growth hormones. Its secretions control the functioning of other glands in the body and is therefore sometimes called the "master" gland.

Placebo A harmless and pharmacologically inactive substance, usually disguised, given to compare its effect with that of an active material.

Placenta The organ which links a developing foetus to the wall of the womb and the maternal blood supply.

Plant breeders' rights (PBR) Legal protection of a new plant variety granted to the breeder or his successor in title. The effect of PBR is that prior authorization is required before the material can be used for commercial purposes.

Plant breeding Is use of techniques involving crossing plants to produce varieties with particular characteristics (traits), which are carried in the genes of the plants and passed on to future plant generations.

Plant cell culture Growth of plant cells or roots of plants in bioreactors.

Plant cell immobilization Entrapment of plant cells in gel matrices so that they are protected from physical damage. The cells are suspended in liquified droplets which are then allowed to harden. Commonly used matrices are alginates, agar or polyacrylamide.

Plant genetic resources (PGR) Defined in the International Undertaking on Plant to mean the reproductive or vegetative propagating material of the following categories of plants: (i) cultivated varieties (cultivars) in current use and newly developed varieties;

(ii) obsolete cultivars;

(iii) primitive cultivars (landraces);

(iv) wild and weed species, near relatives of cultivated varieties; and

(v) special genetic stocks (including elite and current breeder's lines and mutants).

Plant growth regulator An organic compound, either natural or synthetic, and other than a nutrient, that modifies or controls one or more specific physiological processes within a plant.

Plant pests Organisms that may directly or indirectly cause disease, spoilage, or damage to plants, plant parts or processed plant materials. Common examples include certain insects, mites, nematodes, fungi, molds, viruses, and bacteria.

Plant-incorporated protectants (PIPs) Pesticidal substances introduced into plants by genetic engineering that are produced and used by the plant to protect it from pests. The protein toxins of Bt are often used as PIPs in the formation of Bt crops.

Plantlet A small rooted shoot developed from seed or from cultured cells either by embryogenesis or organogenesis.

Plaque A clear spot on an otherwise opaque culture plate of bacteria or cultured bacteria cells, showing where cells have been lysed by viral infection.

Plasma The fluid portion of the blood in which is suspended the white and red blood cells. Contains 8-9% solids, of which 85% is composed of the proteins fibrinogen, albumin, and globulin. The essential function of plasma is the maintenance of blood pressure and the transport of nutrients and waste.

Plasma cells Antibody-producing white blood cells derived from B lymphocytes.

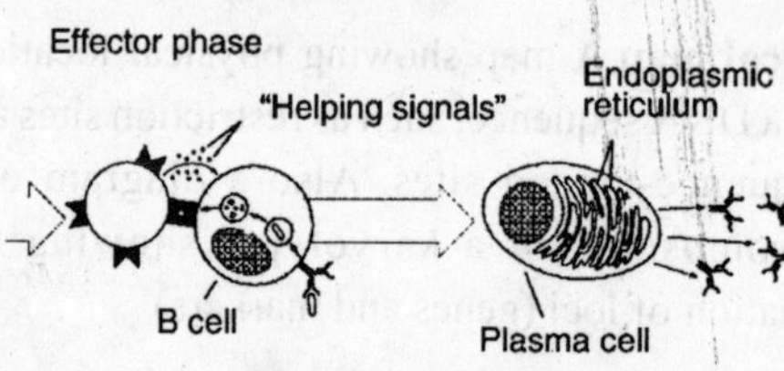

Fig. Plasma cells

Plasmapheresis A technique used to separate useful factors from blood.

Plasmid An extrachromosomal, autonomous circular DNA molecule found in certain bacteria, capable of autonomous replication. Plasmids can transfer genes between bacteria and are important tools of transformation in genetic engineering. They exist in an autonomous state and are transferred independently of chromosomes.

Plasmodesma Fine protoplasmic thread passing through the plant cell wall that separates two protoplasts.

Plasmolysis Separation of the cytoplasm from the cell wall, due to removal of water from the protoplast.

Plasticity The ability to be flexible. The ability of cells or tissue and their function to be influenced by an activity and how they respond to distinct environmental conditions.

Plastid A general term for a number of plant cell organelles which carry non-nuclear DNA.

Includes the pigment-carrying bodies: 1. chloroplasts in leaves,

2. chromoplasts in flowers, and

3. the starch-synthesizing amyloplasts in seeds.

Plastoquinone A quinone which is one of a group of compounds involved in the transport of electrons in photosynthesis in chloroplasts.

Plate 1. To distribute a thin film of something. Hence micro-organisms or plant cells are plated onto nutrient agar.

2. Refers to the two segments of a Petri dish or a similar-shaped item.

Platelet aggregation The naturally occurring process in the body during which platelets - disc shaped blood cells - clump together. The accumulation of platelets is essential in blood clotting at the site of an injury, for example. However, if there is platelet aggregation in a blood vessel, the result is a heart attack.

Platelet-activating factors Substances released by various cells, which cause platelets to clump together.

Plating efficiency The percentage of inoculated cells which give rise to cell colonies when seeded into culture vessels.

Pleiotrophy The effect of a particular gene on several different traits.

Pleiotropic response Multiple changes to an organism's appearance associated with a single change at the genetic level.

Ploidy The number of sets of chromosomes per cell, e.g., haploid, diploid, polyploid.

Plumule The first bud of an embryo or that portion of the young shoot above the cotyledons.

Pluralism The belief that there are multiple opinions about an issue, each of which contains part of the truth but none contain the whole truth.

Pluripotent The ability to be able to produce any cell in the body. Usually used when referring to embryonic stem cells.

Pluripotent cells Having the capacity to become any kind of cell or tissue in the body. Embryonic stem cells and cells of the inner cell mass are pluripotent. Adult stem cells are multipotent. The mammalian embryo (blastocyst trophoblast plus inner cell mass) is totipotent because it can become an entire organism. Fully differentiated cells from many plants are totipotent.

Point mutation A change in DNA at a specific site in a chromosome. Includes nucleotide substitutions and the insertion or deletion of one or a few nucleotide pairs.

Poisson's ratio A bar of an isotropic, elastic material ordinarily shrinks laterally when it is stretched longitudinally. The lateral contracting strain divided by the applied tensile strain is Poisson's ratio, which varies from material to material.

Polar bodies In female animals, the products of a meiotic division that do not develop into an ovum. The first polar body comprises one of the two products of meiosis I, and it may not go through meiosis II. The second polar body comprises one of the two products of meiosis II.

Polar mutation A mutation that influences the functioning of genes that are downstream in the same transcription unit.

Polar nuclei Two centrally located nuclei in the embryo sac that unite with a second sperm cell in a triple fusion. In certain seeds, the product of this fusion develops into the endosperm.

Polar transport The directed movement within plants of compounds (usually endogenous plant growth regulators) mostly in one direction; polar transport overcomes the tendency for diffusion in all directions.

Polarity The observed differentiation of an organism, tissue or cell into parts having opposed or contrasted properties or form.

Pole cells A group of cells in the posterior of *Drosophila* embryos that are precursors to the adult germ line.

Pollen The mass of germinated microspores or partially developed male gametophytes of seed plants.

Pollen culture The *in vitro* culture and germination of pollen grains. Callus cultures thus obtained will form shoots or embryoids which develop into monoploid plants.

Pollen grain A microspore produced in the pollen sac of angiosperms or the microsporangium of gymnosperms. Unicellular, with variable shape and size, and usually ovoid from 25 to 250mm.

Pollination Transfer of pollen from anther to stigma in the process of fertilization in angiosperms; transfer of pollen from male to female cone in the process of fertilization in gymnosperms.

Pollination control mechanisms A system developed to manage plant fertilization activities.

Poly (A) polymerase Enzyme that catalyses the addition of adenine residues to the 3′ end of pre-mRNAs to form the poly-(A) tail.

Polyacrylamide gel electrophoresis (PAGE) A method for separating nucleic acid or protein molecules according to their molecular size. The molecules migrate through the inert gel matrix under the influence of an electric field. In the case of protein PAGE, detergents such as sodium dodecyl sulphate (SDS) are often added to ensure that all molecules have a uniform charge. Secondary structure can often lead to the anomalous migration of molecules. Therefore it is common to denature protein samples by boiling them prior to PAGE. In the case of nucleic acids, denaturing agents such as formamide, urea or methyl mercuric hydroxide are often incorporated into the gel itself, which may also be run at high temperature. PAGE is used to separate the products of DNA-sequencing reactions and the gels employed are highly denaturing, since molecules differing in size by a single nucleotide must be resolved.

Polyacrylamide gels Often referred to incorrectly as acrylamide gels. These gels are made by cross-linking acrylamide with *N,N′*-methylene-*bis*-acrylamide. Polyacrylamide gels are used for the electrophoretic separation of proteins, DNA and RNA molecules. Polyacrylamide beads are also used as molecular sieves in gel chromatography, marketed as Bio-gelTM.

Polyadenylation Post-transcriptional addition of a polyadenylic acid tail to the 3′ end of eukaryotic mRNAs. Also called poly-(A) tailing. The adenine-rich 3′ terminal segments is called a poly (A) tail.

Polyclonal Derived from different types of cells.

Polyclonal antibody A serum sample that contains a mixture of immunoglobulin molecules secreted against a specific antigen, each recognizing a different epitope, some antibodies of which bind to different antigenic determinants of one antigen.

Polycyclic A cyclic structure contains rings of bonds; a structure having many such rings is termed polycyclic. In the polycyclic structures of interest in this volume, a large fraction of the atoms are members of multiple small rings, resulting in considerable rigidity.

Polyembryony In the ordinary course of events, one embryo is formed in each ovule and is derived from the fertilization of the ovum in the solitary embryo sac. However, two or more embryos could start development even though only one may reach maturity.

Polyethylene glycol (PEG); carbowax A polymer having the general formula: $HOCH_2(CH_2OCH_2)XCH_2OH$ and available in a range of molecular weights from *ca* 1000 to *ca* 6000. PEG 4000 and PEG 6000 are commonly used to promote cell or protoplast fusion, and to facilitate DNA uptake in the transformation of organisms such as yeast. PEG is also used to concentrate solutions by withdrawing water from them.

Polygene One of a number of genes, each of small effect, which together act to determine the phenotype of a quantitative trait. The result is continuous variation in the trait and a seemingly non-Mendelian mode of inheritance.

Polylinker A segment of DNA that contains a number of different restriction endonuclease sites.

Polymer A string of units covalently linked together either in linear form or with branching points. A macromolecule made of two or more units. The units can be identical forming a homo-polymer. Glycogen and starch, the glucose storage particles in animals and plants are examples of homopolymers. Polymers can be made of strings of different units. Proteins are made of 20 different amino acids, DNA and RNA are made of 4 different types of nucleotides. These polymers are called hetero-polymers Many polysaccharides are heteropolymers and found on cell surfaces where they serve as protective layer and receptors.

Polymerase An enzyme that catalyses the formation of polymeric molecules from monomers. A DNA polymerase synthesizes DNA from deoxynucleoside triphosphates using a complementary DNA strand and a primer. An RNA polymerase synthesizes RNA from monoribonucleoside triphosphates and a complementary DNA strand.

Polymerase chain reaction (PCR) A widespread molecular biology procedure that allows the production of multiple copies (amplification) of a specific DNA sequence, provided that the base pair sequence of each end of the target is known. It involves multiple cycles of DNA denaturation, primer annealing, and strand extension, and requires a thermostable DNA polymerase, deoxyribonucleotides, and specific oligonucleotides (primers).

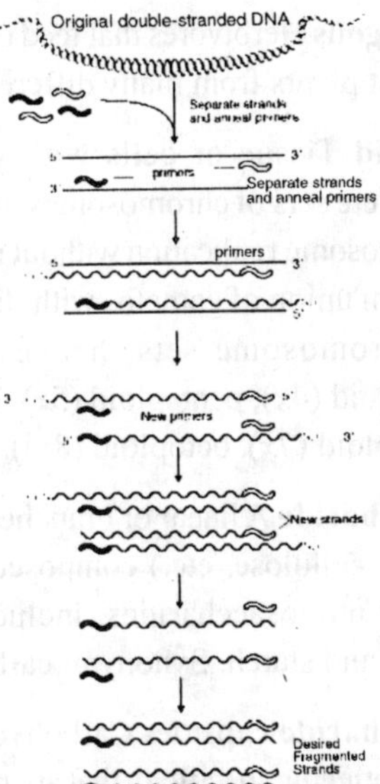

Fig. Polymerase chain reaction (PCR)

Polymerization Chemical union of two or more molecules of the same kind such as glucose or nucleotides to form a new compound (starch or nucleic acid) having the same elements in the same proportions but a higher molecular weight and different physical properties.

Polymery The phenomenon whereby a number of genes at different loci (which may be polygenes) can act together to produce a single effect.

Polymorphism 1. The occurrence of allelic variation at a locus. Polymorphism in nucleotide sequences has provided powerful diagnostic tools.

Polynucleotide A chain of nucleotides in which each nucleotide is linked by a single phosphodiester bond to the next nucleotide in the chain. They can be double- or single-stranded. The term is used to describe DNA or RNA.

Polypeptide A linear polymer composed of covalently linked amino acids. Each link is formed by a single peptide bond. Sometimes used as a synonym for protein, but also describes non-natural and low-molecular-weight polymers.

Polyphagous Herbivores that feed on a wide variety of host plants from many different families.

Polyploid Tissue or cells with more than two complete sets of chromosomes, that results from chromosome replication without nuclear division or from union of gametes with different number of chromosome sets, hence triploid (3*x*), tetraploid (4*x*), pentaploid (5*x*), hexaploid (6*x*), heptaploid (7*x*), octoploid (8*x*)).

Polysaccharide A linear or branched polymer (e.g. starch, cellulose, etc.) composed of covalently linked monosaccharides, including cellulose, pectin and starch. Synonym: carbohydrate.

Polysaccharide capsules Carbohydrate coverings with antigenic specificity that are present on some types of bacteria.

Polyspermy The entry of several sperm into the egg during fertilization, although only one sperm nucleus actually fuses with the egg nucleus.

Polytene chromosomes Giant chromosomes produced by interphase replication without division, and consisting of many identical chromatids arranged side by side in a cablelike pattern.

Polyunsaturated A molecule that contains two or more double or triple bonds.

Polyunsaturated fats & fatty acids Belonging to a class of animal or vegetable fats or oils containing multiple double or triple molecular bonds that are unsaturated by the incorporation of hydrogen; such fats and oils are associated with a low level of cholesterol in the blood.

Polyvalent vaccine A recombinant organism into which antigenic determinants have been cloned from a number of different disease-causing organisms, and used as a vaccine.

Polyvinylpyrrolidone (PVP) (C_6H_9NO)n) An occasional constituent of plant tissue culture isolation media. PVP is of variable molecular weight and has antioxidant properties, so is used to prevent oxidative browning of explanted tissues. PVP is less frequently used as an osmoticum in culture media.

Population The totality of a closely related number of individual organisms that belong to the same species and live in the same geographical area and interact with each other through sexual (or asexual for bacteria) reproduction.

Population density Number of cells or individuals per unit. The unit could be an area or volume of medium.

Population genetics The branch of genetics that deals with frequencies of alleles and genotypes in breeding populations.

Porcine endogenous retrovirus (PERV) The provirus of a porcine retrovirus. With the increasing interest in the use of pig organs for xenotransplantation to humans, there has been concern that PERVs could be activated after transplantation, creating an infection in the human recipient.

Positech™ The Positech™ gene is inserted into plant cells together with a gene of interest. When grown in the presence of mannose, which most plants can't utilise as food, only those cells with the Positech™ gene grow. These cells can be regenerated into plants which carry the new trait.

Position effect The situation in which a change in phenotype results from the change of the position of a gene or group of genes.

Positional candidate gene A gene known to be located in the same region as a DNA marker that has been shown to be linked to a single-locus trait or to a quantitative trait locus (QTL) and whose function suggests that it could be the source of genetic variation in the trait in question.

Positional cloning The process that commences with searching for markers linked to a particular inherited trait, then using those markers to identify the approximate location of the gene responsible for the trait, and then using various cloning strategies (including chromosome walking,) to identify, isolate and characterize the gene. Originally, the strategy was called *reverse genetics*, a term that some consider to be a misnomer and misleading.

Positional synthesis Control of chemical reactions by precisely positioning the reactive molecules, the basic principle of assemblers.

Positive control system A mechanism in which the regulatory protein(s) is required to turn on gene expression.

Post transcriptional changes Changes to a protein that occur after DNA has been copied to form messenger RNA and this has been translated to produce protein - for example, to facilitate excretion of the protein from the cell.

Postemergence herbicide Herbicides that are applied after the weeds have emerged from the soil.

Post-replication repair A recombination-dependent mechanism for repairing damaged DNA.

Post-translational modification The specific addition of phosphate groups, sugars (glycosylation), or other molecules to a protein after it has been synthesized.

Potential energy The energy associated with a configuration of particles, as distinct from their motions. In macroscopic experience, potential energy can be increased (for example) by stretching a spring or by lifting a mass against a gravitational force; in molecular systems, potential energy can be increased (for example) by stretching a bond or by separating molecules against a van der Waals attraction.

Potential energy surface The potential energy of a ground-state molecular system containing N atoms is a function of its geometry, defined by $3N$ spatial coordinates (a configuration space). If the energy is imagined as corresponding to a height in a $3N^{+}$ 1 dimensional space, the resulting landscape of hills, hollows, and valleys is the potential energy surface.

Potential well In a potential energy surface, the region surrounding a local energy minimum. Typically taken to include at least those points in configuration space such that a path of steadily declining energy can be found that leads to the minimum in question, and such that no similar path can be found to any other minimum. If the PES were a landscape, this would be the region around the minimum that could be filled with water without any flowing down and away toward another minimum.

Power The term power has quite a few different meanings. For Biotechnology Online, we are referring to an organization or individual's ability to act effectively according to their intentions, needs, or values.

Prebiotic Prebiotic refers to the state of matter before life existed, but was conducive to the formation of life (the origin of life). The prebiotic 'soup' is the mixture of organic molecules in bodies of water that are thought to have spontaneous, self-assembly property for the first formation of an organic complex with self-replicating qualities.

Precautionary principle The key principle of ecologically sustainable development. It basically states that in situations where there is lack of scientific certainty and possibility of serious or irreversible damage to the environment (both of which apply to gene technology). We should recognise this uncertainty; assess the possible impacts and our options; and put in place now whatever measures we can to avoid possible damage.

Preclinical Extensive research of a new health-care product conducted before a clinical trial. Testing in the pre-clinical stage is done on laboratory models and animals.

Preclinical studies Studies that test a drug on animals and in other nonhuman test systems. Safety information from such studies is used to support an investigational new drug application (IND).

Precocious germination Premature germination of the embryo, prior to completion of embryogenesis.

Predator Animal that kills another animal for food.

Pre-filter A coarse filter used to screen out large particles before air is forced through a much finer filter.

Pressure potential The pressure generated within a cell; it is the difference between the osmotic potential within the cell and the water potential of the external environment, provided the cell volume is constant.

Pre-transplant Stage III in tissue culture micropropagation; the rooting, hardening stage prior to transfer to soil.

Prevalence The number of existing cases of a disease in a defined population at a specified time.

Preventive immunization; vaccination Infection with an antigen to elicit an antibody response that will protect the organism against future infections.

Pribnow box Consensus sequence near the mRNA start-point of prokaryotic genes.

Primary First in order of time or development.

Primary antibody In an ELISA or other immunological assay, the antibody that binds to the target molecule.

Primary cell wall The cell wall layer formed during cell expansion. Plant cells possessing only primary walls may divide or undergo differentiation.

Primary culture A culture started from cells, tissues or organs taken directly from organisms. A primary culture may be regarded as such until it is sub-cultured for the first time. It then becomes a cell line.

Primary growth 1. Refers to apical meristem-derived growth; the tissues of a young plant.

2. Refers to explant growth during the initial culture period; such as primary callus growth.

Primary immune response The immune response that occurs during the first encounter of a mammal with a given antigen.

Primary meristem Meristem of the shoot or root tip giving rise to the primary plant body.

Primary tissue A tissue that has differentiated from a primary meristem.

Primary transcript The RNA molecule produced by transcription prior to any post-transcriptional modifications; also called a pre-mRNA in eukaryotes.

Primer A short DNA or RNA fragment annealed to a template of single-stranded DNA, providing a 3′ hydroxyl end from which DNA polymerase extends a new DNA strand to produce a duplex molecule.

Primer DNA polymerase DNA polymerase that provides primers for the DNA polymerization. Unlike RNA polymerase, DNA polymerase is unable to initiate the *de novo* synthesis of a polynucleotide chain. DNA polymerase can only add nucleotides to a free 3′ hydroxyl group at the end of a pre-existing chain. A short oligonucleotide, known as a primer, is therefore needed to supply such a hydroxyl group for the initiation of DNA synthesis.

Primer walking A method for sequencing long (>1 kb) cloned pieces of DNA. The initial sequencing reaction reveals the sequence of the first few hundred nucleotides of the cloned DNA. On the basis of these data, a primer containing about 20 nucleotides and complementary to a sequence near the end of sequenced DNA is synthesized, and is then used for sequencing the next few hundred nucleotides of the cloned DNA. This procedure is repeated until the complete nucleotide sequence of the cloned DNA is determined.

Primordial germ cells The precursors of reproductive cells within the embryo. They are detectable in an embryo after four weeks of development and will develop into either sperm or eggs.

Primordium A group of cells which gives rise to an organ.

Primosome A protein-replication complex that catalyses the initiation of synthesis of Okazaki fragments during discontinuous replication of DNA. It involves DNA primase and DNA helicase activities.

Prion Normal cell protein present on nerve cell membranes. It is found in most mammals, but its normal function is unclear. Abnormal prions are thought to cause certain diseases including mad cow disease.

Pro embryo A group of cells arising from the division of the fertilized egg cell or somatic embryo before those cells which are to become the embryo are recognizable.

Pro meristem The embryonic meristem that is the source of organ initials or foundation cells.

Pro nuclear micro injection The initial method of transgenesis in animals, involving injecting many copies of a particular gene into one of the two pro-nuclei of a fertilized ova. Characterised by a very low success rate. As animal cloning becomes more common, pro-nuclear micro-injection will be replaced by micro-injection of cloned genes into a culture of cloned embryos produced by nuclear transfer, which can be tested for expression of the transgene before transfer to recipient females.

Pro toxin A latent, non-active precursor form of a toxin.

Probability Statistics: The frequency of occurrence of an event.

Probability density function Consider an uncertain physical property and a corresponding space describing the range of values that the property can have (e.g., the configuration of a thermally excited *N* particle system and the corresponding 3*N* dimensional configuration space). The probability density function associated with a property is defined over the corresponding space; its value at a particular point is the probability per unit volume that the property has a value in an infinitesimal region around that point.

Proband The individual in a family in whom an inherited trait is first identified.

Probe 1. For diagnostic tests, the agent that is used to detect the presence of a molecule in a sample.

2. A DNA or RNA sequence labelled or marked with a radioactive isotope or that is used to detect the presence of a complementary sequence by hybridization with a nucleic acid sample.

Probe DNA A labelled DNA molecule used to detect complementary-sequence nucleic acid molecules by molecular hybridization. To localize the probe DNA sequence and reveal the complementary hybridization sequence, autoradiography or fluorescence is used.

Procambium A primary meristem that gives rise to primary vascular tissues and, in most woody plants, to the vascular cambium.

Processed food Any food product that has undergone physical or chemical treatment resulting in a substantial change in the original state of the food.

Processed pseudo-gene A copy of a functional gene which has no promoter, no introns and which, consequently, is not itself transcribed. Pseudo-genes are thought to originate from the integration into the genome of cDNA copies synthesized from mRNA molecules by reverse transcriptase. Pseudo-genes therefore have a poly (dA) sequence at their 5′ ends. Because they are not subject to any evolutionary pressure to maintain their coding potential, pseudo-genes accumulate mutations and often have stop codons in all three reading frames.

Product type (in microbial pest control products) Products fall into either of two product types: * End Use Products (EP); 1. and Manufacturing Products (MP).

2 End Use Products are the final products and must include labels and directions for use. Manufacturing Products are products that can either be used to manufacture an End Use Product or marketed alone as End Use Products with the correct labelling and approvals.

Production environment All input-output relationships, over time, at a particular location. The relationships will include biological, climatic, economic, social, cultural and political factors, which combine to determine the productive potential of a particular livestock enterprise. A production environment where all rate-limiting inputs to animal production can be managed to ensure high levels of survival, reproduction and output. Output and production risks are constrained primarily by managerial decisions. medium-input production environment A production environment where management of the available resources has the scope to overcome the negative effects of the environment on animal production, although it is common for one or more factors to limit output, survival or reproduction in a serious fashion. low-input production environment. A production environment where one or more rate-limiting inputs impose continuous or variable severe pressure on livestock, resulting in low survival, reproductive rate or output. Output and production risks are exposed to major influences which may go beyond human management capacity.

Production traits Characteristics of animals, such as the quantity or quality of the milk, meat, fibre, eggs, draught, etc., they (or their progeny) produce, which contribute directly to the value of the animals for the farmer, and that are identifiable or measurable at the individual level. Production traits of farm animals ate generally quantitatively inherited, i.e., they are influenced by many genes whose expression in a particular animal also reflects environmental influences.

Productivity The amount of product that is produced within a given period of time from a specified quantity of resource.

Progeny Offspring.

Progeny testing For a single locus: the practice of ascertaining the genotype of an individual from its offspring, such as by mating it to other individuals and examining the progeny. For a quantitative trait: the collection and use of progeny performance as a clue to the breeding value of the parent of those progeny.

Progesterone A hormone produced primarily by the corpus luteum of the ovary, but also by the placenta, that prepares the inner lining of the uterus for implantation of a fertilized egg cell.

Prokaryote A member of a large group of organisms, including bacteria and blue-green algae, which do not have the DNA separated from the cytoplasm by a membrane in their cells. The DNA is usually in one long strand. Prokaryotes do not undergo meiosis and do not have functional organelles such as mitochondria and chloroplasts.

Prolactin A hormone, produced by the anterior pituitary gland, that stimulates and controls lactation in mammals.

Proliferation Increase by frequent and repeated reproduction; growth by cell division.

Promoter 1. A short DNA sequence, usually upstream of (5' to) the relevant coding sequence, to which RNA polymerase binds before initiating transcription. This binding aligns the RNA polymerase so that transcription will initiate at a specific site. The nucleotide sequence of the promoter determines the nature of the enzyme that attaches to it and the rate of RNA synthesis.

2. A chemical substance that enhances the transformation of benign cells into cancerous cells.

Promoter sequence A regulatory DNA sequence that initiates the expression of a gene.

Pro-nucleus Either of the two haploid gamete nuclei just prior to their fusion in the fertilized ovum.

Proofreading The enzymatic scanning of newly-synthesized DNA for structural defects, such as mis-matched base pairs. It is effected by DNA polymerase.

Propagation The multiplication of plants by numerous types of vegetative material; an ancient practice dating from the dawn of agriculture, carried out in a nursery or directly in the field (vegetative propagation), and now in *in vitro* culture (micropropagation).

Propagule Any structure capable of giving rise to a new plant by asexual or sexual reproduction, including bulbils, leafbuds, etc.

Prophase An early stage in nuclear division, characterized by the shortening and thickening of the chromosomes and their movement to the metaphase plate. It occurs between interphase and metaphase. During this phase, the centriole divides and the two daughter centrioles move apart. Each sister DNA strand from interphase replication becomes coiled, and the chromosome is longitudinally double except in the region of the centromere. Each partially separated chromosome is called a chromatid. The two chromatids of a chromosome are sister chromatids.

Proposed use patterns The suggested uses of a new product in terms of location, target organism(s), quantities, and other criteria dictated by the specific regulations.

Prospective (cohort) study A study in which the study population is characterised at the start of the study and followed into the future. A population of people who do not (yet) have the disease under investigation is identified, and information is collected on the subjects' exposure to risk factors generally including nutritional factors. The frequency of the disease among subjects exposed to a particular risk factor during the follow up period is compared with the frequency among those who were not exposed.

Prostaglandins Fatty acids synthesised by various white blood cells and at the surface of mast cells, that can cause either contraction or relaxation of smooth muscles in the airways and in certain blood vessels.

Protamines Small basic proteins that replace the histones in the chromosomes of some sperm cells.

Protease An enzyme that catalyses the hydrolysis of proteins, cleaving the peptide bonds that link amino acids in protein molecules. Synonym: peptidase.

Protein A macromolecule composed of one to several polypeptides. Each polypeptide consists of a chain of amino acids linked together by covalent (peptide) bonds. They are naturally-occurring complex organic substances (egg albumen, meat) composed essentially of carbon, hydrogen, oxygen and nitrogen, plus sulphur or phosphorus, which are so associated as to form sub-microscopic chains, spirals or plates and to which are attached other atoms and groups of atoms in a variety of ways. The word was coined by Jφns J. Berzelius (1838) to emphasize the importance of this group of molecules.

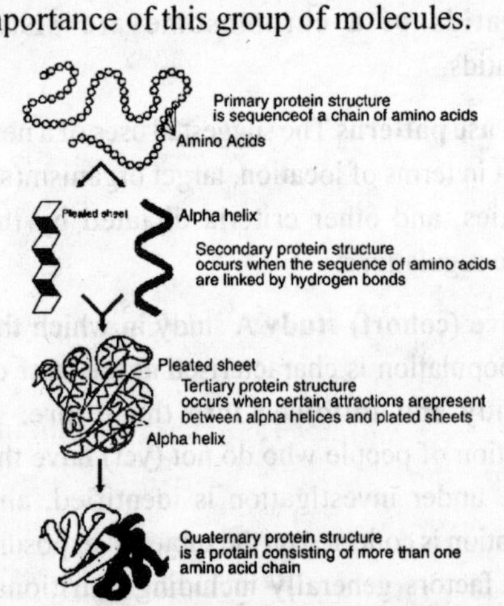

Fig. Protein

Protein coat Refers to the protein coat or shell that surrounds the genome or protein core of the viral body. It is a major structural feature of many viruses.

Protein crystallization Making crystals of a protein, as a key part of most methods used for determining a protein's three-dimensional structure.

Protein design / protein engineering The design and construction of new proteins; an enabling technology for nanotechnology.

Protein engineering Generating proteins with modified structures that confer properties such as higher catalytic specificity or thermal stability.

Protein kinase An enzyme that adds phosphate groups to a protein molecule at serine, threonine, or tyrosine residues.

Protein metabolic step One step in the chain of reactions that take place in an organism and dictate the composition of that organism.

Protein sequencing The process of determining the amino acid sequence of a protein. Usually achieved following initially partial hydrolysis of the protein into smaller peptides by enzymatic digestion.

Protein synthesis The creation of proteins from their constituent amino acids, in accordance with the genetic information carried in the DNA of the chromosomes.

Proteinaceous infectious particle; prion An abnormal form of a normal cell protein, with no detectable nucleic acid, found in the brain of mammals and believed to be the agent responsible for the class of diseases called spongiform encephalopathies, including scrapie in sheep and bovine spongiform encephalopathy (BSE; mad cow disease) in cattle.

Proteinase inhibitors A class of proteins capable of inhibiting insect feeding.

Proteolytic The ability to break down protein molecules.

Proteomics Each cell produces thousands of proteins, each with a specific function. This collection of proteins in a cell is known as the proteome, and, unlike the genome, which is constant irrespective of cell type, the proteome varies from one cell type to the next. The science of proteomics attempts to identify the protein profile of each cell type, assess protein differences between healthy and diseased cells, and uncover not only each protein's specific function but also how it interacts with other proteins.

Protoclone Regenerated plant derived from protoplast culture or a single colony derived from protoplasts in culture.

Protocol The step-by-step experiments proposed to describe or solve a scientific problem, or the defined steps of a specific procedure.

Protocorm In orchids, seeds contain an unorganized embryo comprising only a few hundred cells. During seed germination this embryo forms a tuberous structure, called a protocorm, from which develops a complete plant. In cultures, vegetative parts of several orchids form round, smooth protocorms which may be multiplied indefinitely or induced to regenerate a whole plant.

Protoderm A primary meristem tissue that gives rise to epidermis.

Protogyny The condition in which the female reproductive organs (carpels) of a flower mature before the male ones (stamens), thereby ensuring that self-fertilization does not occur.

Proto-oncogene A normal cellular gene that can be changed to an oncogene by mutation or by being incorporated into a retrovirus and then being transcribed at inappropriate times and/or in inappropriate tissues.

Protoplasm The essential, complex living substance of cells, upon which all vital functions of nutrition, secretion, growth and reproduction depend.

Protoplast A bacterial or plant cell for which the relatively rigid wall has been removed either chemically or enzymatically, leaving its cytoplasm enveloped by only a delicate peripheral membrane. Protoplasts are spherical and smaller than the elongate, angular shaped and often vacuolated cells from which they have been released.

Protoplast culture The isolation and culture of plant protoplasts by mechanical means or by enzymatic digestion of plant tissues or organs, or cultures derived from these. Protoplasts are utilized for selection or hybridization at the cellular level and for a variety of other purposes.

Protoplast fusion The coalescence of the plasmalemma and cytoplasm of two or more protoplasts in contact with one another. Initial adhesion of two protoplasts is a random process but coalescence can be induced.

Prototroph An organism such as a bacterium that will grow on a minimal medium.

Protozoa Any of a large group of single-celled, usually microscopic, organisms such as amoeba.

Provirus A retrovirus in which the single RNA strand has been converted into a double-strand DNA, which has been integrated into a host genome.

Pseudo-autosomal region A section at one end of the X and Y chromosomes for which there is sufficient homology that there is pairing (synapsis) during meiosis I.

Pseudocarp; false fruit A fruit that incorporates, in addition to the ovary wall, other parts of the flower, such as the receptacle (e.g., strawberry).

Pseudogenes A sequence of DNA that resembles a gene but is non functional and cannot be transcribed. It could be the remnant of a once-functional gene that has accumulated mutations.

Pseudomonas spp A common Gram-negative bacterial genus that is widely distributed. Many of the soil forms produce a pigment that fluoresces under ultraviolet light, hence the descriptive term fluorescent pseudomonas.

Psychophysical Concerning the relationships between measurable stimuli and the corresponding responses.

Psychrophile A micro-organism that can grow at temperatures below 30°C and as low as 0°C.

Punctuated equilibrium The occurrence of speciation events in bursts, separated by long intervals of species stability.

Pulsed field gel electrophoresis (PFGE) A procedure used to separate very large DNA molecules by alternating the direction of electric current in a pulsed manner across a semisolid gel.

Purine A double-ring, nitrogen-containing base present in nucleic acids; adenine (A) and guanine (G) are the two purines present in most DNA and RNA molecules.

Pure line A strain in which all members have descended by self-fertilization or close inbreeding . A pure line is genetically uniform.

Pyrethrins Active constituents of pyrethrum (*Tanacetum cinerariifolium*) flowers, used as insecticides.

Pyrimidine A single-ring, nitrogen-containing base present in nucleic acids. Cytosine (C) and thymine (T) are present in DNA, whereas uracil (U) replaces T in RNA. Thymine is a synonym for 5-methyluracil.

Pyrogen Bacterial substance that causes fever in mammals.

Q

Qualia A property of self, of experiencing the environment such as colors, tastes, or pain. The qualia refers to a part of consciousness that is different from the physical nature of the stimulus that provokes it. For example, in describing a color - red, blue, or green - we talk about a qualia of an object. The physical existence of a color is a quantum mechanical state that emits energy in from of electromagnetic radiation that is decoded by our sensory organs - the eye and the visual cortex in the brain as red, blue, or green. We have no a priori knowledge if these colors exists independent of us. We can only assert that the radiation has a certain energy who's quanta (the single photon) can elicit quantum mechanical effect in the retinal cell layer of our eyes.

Qualitative trait A trait that shows discontinuous variation - i.e. individuals can be assigned to one of a small number of discrete classes.

Quality The totality of characteristics of an entity that bear on its ability to satisfy stated or implied needs.

Quality assurance All the planned and systematic activities implemented within the quality system, and demonstrated as needed, to provide adequate confidence that an entity will fulfil requirements for quality.

Quality control The operational techniques and activities used to fulfil requirements for quality.

Quality management All activities of the overall management function that determine the quality policy, objectives and responsibilities and that implement them by means such as quality planning, quality control, quality assurance and quality improvement within the quality system.

Quality system The organisational structure, procedures, processes and resources needed to implement quality management.

Quantitative genetics The area of genetics concerned with the inheritance of quantitative traits that show continuous variation, as opposed to qualitative traits. Since many of the critical targets in both plant and animal breeding are of this type, most practical improvement programs involve the application of quantitative genetics.

Quantitative inheritance Inheritance of measurable traits (height, weight, colour intensity, etc.) that depend on the cumulative action of many genes.

Quantitative structure-activity relationship (QSAR). A computer modelling technique that enables the prediction of the likely activity of a molecule before it is synthesized. QSAR analysis relies on recognizing associations of molecular structures and activity from historical data.

Quantitative trait A measurable trait that shows continuous variation (e.g. height, weight, colour intensity, etc.) - i.e. the population cannot be classified into a few discrete classes.

Quantitative trait locus (QTL) A locus that affects a quantitative trait. The plural form (quantitative trait loci) is also abbreviated as QTL.

Quantum An elemental unit of energy. Its energy value is hv, where h, Planck's constant, is 6.62 X 10^{-27} erg-second and v is the frequency of the vibrations or waves with which the energy is associated.

Quantum mechanics Quantum mechanics describes a system of particles in terms of a wave function defined over the configuration space of the system. Although the concept of particles having distinct locations is implicit in the potential energy function that determines the wave function (e.g., of a ground-state system), the observable dynamics of the system cannot be described in terms of the motion of such particles from point to point. In describing the energies, distributions, and behaviors of electrons in nanometer-scale structures, quantum mechanical methods are necessary. Electron wave functions help determine the potential energy surface of a molecular system, which in turn is the basis for classical descriptions of molecular motion. Nanomechanical systems can almost always be described in terms of classical mechanics, with occasional quantum mechanical corrections applied within the framework of a classical mod

Quantum speciation The rapid formation of new species, primarily by genetic drift.

Quarantine Originally, keeping a person or living organism in isolation for a period (originally 40 days) after arrival to allow disease symptoms to appear, if there was any disease present. Now used for regulations restricting the sale or shipment of living organisms, usually to prevent disease or pest invasion of an area.

Quaternary structure Three-dimensional relationship of the different polypeptide chains in a protein complex.

Quiescent Quiet, at rest, but not necessarily dormant, and having the potential for resumed activity; can apply to non-meristematic cells.

Quinone An enzymatic cofactor that plays an important role in photosynthesis and respiration. More specifically, it is part of the electron transport chain in mitochondria and chloroplast membranes.

R

R loops Single-stranded DNA regions in RNA-DNA hybrids formed *in vitro* under conditions where RNA-DNA duplexes are more stable than DNA-DNA duplexes.

Rabies A viral disease of wild animals that can be transmitted to humans through the bite of an infected animal. The disease has not yet been detected in Australia.

Race A distinguishable group of organisms of a particular species. Criteria for distinctness can be one or a combination of geographic, ecological, physiological, morphological, genetic and karyotypic factors.

Raceme An inflorescence in which the main axis is elongated but the flowers are borne on pedicels that are about equal in length.

Rachilla Shortened axis of a spikelet.

Rachis Main axis of a spike; axis of fern leaf (frond) from which pinnae arise; in compound leaves, the extension of the petiole corresponding to the midrib of an entire leaf.

Radicle That portion of the plant embryo which develops into the primary or seed root.

Radioactive isotope Form of an atom with an unstable nucleus that emits radiation as it decays.

Radioimmunoassay (RIA) A test combining radioisotopes and immunology to detect trace substances. Such tests are useful for studying antibody interactions with cell receptors, and can be developed into clinical diagnostics.

Random amplified polymorphic DNA (RAPD) A technique using single, short (usually 10-mer) synthetic oligonucleotide primers for PCR. The primer, whose sequence has been chosen at random, initiates replication at its complementary sites on the DNA, producing fragments up to about 2 kb long, which can be separated by electrophoresis and stained with ethidium bromide. A primer can exhibit polymorphism between individuals, and polymorphic fragments can be used as markers.

Random mutagenesis A non-directed change of one or more nucleotide pairs in a DNA molecule.

Random primer method A protocol for labelling DNA *in vitro*. A sample of random oligonucleotides (6 or 14 nucleotides long) containing all possible combinations of nucleotide sequences is hybridized to a DNA probe. Then, in the presence of a DNA polymerase and the four deoxyribonucleotides - one of which is labelled - the 3′ hydroxy ends of the hybridized oligonucleotides provide initiation sites for DNA synthesis that uses the separated strands of the probe DNA as a template. This reaction produces labelled copies of portions of the probe DNA.

Rapeseed (brassica napus) The seed of the rape plant which is a source of edible oil. The rape plant is a bright yellow flowering member of the Brassicaceae (mustard) family and a specific variety is known as canola. It is generally grown and cultivated for animal feed, vegetable oil and biodiesel.

Ras protein One of a large family of GTP-binding proteins that help relay signals from cell-surface receptors to the nucleus. Named for the *ras* gene, first identified in viruses that cause rat sarcomas.

Rate-limiting enzyme An enzyme whose activity controls the overall flow of activity through a sequence of reactions.

Ratio scaling Method for quantification of effects through magnitude estimation of how many times the effect of a sample is compared to a standard.

Rational drug design Using the known three-dimensional structure of a molecule, usually a protein, to design a drug molecule that will bind to it. Usually viewed as an alternative to drug discovery through screening many molecules for biological activity.

Re differentiation Cell or tissue reversal from one differentiated type to another differentiated type of cell or tissue.

Re naturation The re-association of two nucleic acid strands after denaturation. The restoration of a molecule to its native form. In nucleic acid biochemistry, this term usually refers to the formation of a double-stranded helix from complementary single-stranded molecules. Some simple proteins can also be re-natured and regain their function.

Reaction In chemistry, any process in which the arrangement of atoms into molecules is changed.

Reactive oxygen species (ROS) Forms of oxygen that have enhanced chemical reactivity compared with the normal oxygen molecule; many are free radicals.Risk factorPhysical condition or lifestyle that in epidemiological studies appears to increase susceptibility to a particular disease.

Reading frame A series of triplets beginning from a specific nucleotide. Each triplet is represented by a single amino acid in the protein synthesized. The reading frame defines which sets of three nucleotides are read as triplets in the DNA, and hence as codons in the corresponding mRNA; this is determined by the initiation codon, AUG. Thus the sequence

AUGGCAAAAUUUCCC would read as

AUG/GCA/AAA/UUU/CCC/ and not as

A/UGC/CAA/AAU/UUC/CC.

Depending on where one begins, each DNA strand contains three different reading frames.

Read-through Transcription or translation that proceeds beyond the normal stopping point because of the absence of the transcription or translation termination signal of a gene.

Reagent Substance used in a chemical reaction.

Reca A protein in most bacteria, and that is essential for DNA repair and DNA recombination.

Recalcitrant Of seeds: unable to survive drying and subsequent storage at low temperature.

Receptacle Enlarged end of the pedicel or peduncle, to which other flower parts are attached.

Receptor Protein that binds a specific extracellular signaling molecule (ligand) and initiates a response in the cell. Cell-surface receptors, such as the acetylcholine receptor and the insulin receptor, are located in the plasma membrane, with their ligand-binding site exposed to the external medium. Intracellular receptors, such as steroid hormone receptors, bind ligands that diffuse into the cell across the plasma membrane.

Receptor-binding screening One of the biotechnology-based methods for discovering conventional drugs. The method relies on the fact that many drugs act by binding to specific proteins (receptors) on or in cells: these proteins usually bind to hormones or to other cells, and control the cell's behaviour, although they may be enzymes or structural elements of the cell. The drug interferes with the normal role of the protein.

Recessive Refers to the member of a pair of alleles that fails to be expressed in the phenotype of the organism when the dominant member is present. Also refers to the phenotype of an individual that has only the recessive allele.

Recessive gene or allele A gene that exercises little or no outward effect unless it is present in both of a pair of chromosomes, and therefore has been inherited from both parents, is said to be recessive.

Recessive genetic disorder A disorder which occurs only if there are two copies of a defective gene present in the genome. Genes are present in pairs, one from each parent. Individuals with a recessive genetic disorder inherit two defective genes from their parents. Cystic fibrosis is an example of a recessive genetic disorder.

Recessive oncogene; recessive acting oncogene; anti oncogene A single copy of this gene is sufficient to suppress cell proliferation; the loss of both copies of the gene contributes to cancer formation.

Recessive trait The quality of a gene or allele regarding its ability to express a phenotype. A recessive allele can only express its phenotype if both copies in a diploid organisms are identical. Often, a recessive allele is matched with a dominant allele, which overrides the activity of the recessive one. In this case, the effect of the gene cannot be seen, but it can still be inherited. These are important considerations in genetic diseases that are often caused by defective copies of a gene (allele) that cause the disease only, if both copies of the gene are defective (homozygous for the recessive allele). Heterozygous individuals with a healthy allele do not suffer the disease but are carriers. Examples of recessive genetic diseases are cystic fibrosis and hypercholesterolemia.

Reciprocal crosses Crosses between the same two strains, but with the sexes reversed; e.g., female A X male B and male A X female B.

Reciprocating shaker A platform shaker used for agitating culture flasks, with a back and forth action at variable speeds.

Recognition site A nucleotide sequence - composed typically of 4, 6 or 8 nucleotides - that is recognized by and to which a restriction endonuclease (restriction enzyme) binds. For type II restriction enzymes (those used in gene-cloning experiments) it is also the sequence within which the enzyme specifically cuts (and their corresponding enzymes methylate) the DNA, i.e., for type II enzymes, the recognition site and the target site are the same sequence. Type I enzymes bind to their recognition site and then cleave the DNA at some more or less random position outside that recognition site.

Recombinant A term used in both classical and molecular genetics. 1. In classical genetics: An organism or cell that is the result of recombination (crossing-over), e.g., Parents: *AB/ab* and *ab/ab*; recombinant offspring: *Ab/ab*.

2. In molecular genetics: A molecule containing DNA from different sources.

Recombinant DNA technology A set of techniques which enable one to manipulate DNA. One of the main techniques is DNA cloning (because it produces an unlimited number of copies of a particular DNA segment), and the result is sometimes called a DNA clone or gene clone (if the segment is a gene), or simply a clone. An organism manipulated using recombinant DNA techniques is called a genetically modified organism (GMO).

Among other things, recombinant DNA technology involves:

- identifying genes;

- cloning genes;

- studying the expression of cloned genes; and - producing large quantities of the gene product.

Recombinant organism An organism in which the DNA has been made by joining together segments of DNA using the techniques of genetic modification described in Directive 90\220\EEC.

Recombinant protein A protein whose amino acid sequence is encoded by a cloned gene.

Recombinant RNA A term used to describe RNA molecules joined *in vitro* by T4 RNA ligase.

Recombinant toxin A single multifunctional toxic protein that has been created by combining the coding regions of various genes.

Recombinant vaccine A vaccine produced from a cloned gene.

Recombination The process of crossing over, which occurs during meiosis I. It involves breakage in the same position of each of a pair of non-sister chromatids from homologous chromosomes, followed by joining of non-sister fragments, resulting in a reciprocal exchange of DNA between non-sister chromatids within an homologous pair of chromosomes. recombination fraction; recombination frequency The proportion of gametes that have arisen from recombination between two loci. It is estimated as the number of recombinant individuals among a set of offspring of a particular mating, divided by the total number of offspring from that mating. Represented by the Greek letter theta (q). Linkage maps are created from estimates of recombination fraction between all pair-wise combinations of loci.

Recombination frequency The frequency at which crossing over occurs between two chromosomal loci—the probability that two loci will become unlinked during meiosis.

Reconstructed cell A viable transformed cell resulting from genetic engineering.

Reduction Addition of electron density to an atom, as occurs during the addition of hydrogen to a molecule or the removal of oxygen from it.

Reductionism The doctrine that complex systems can be completely understood in terms of its simplest parts. For example, an organism is to be completely understood in terms of its genes, a society in terms of its individuals, and so on.

Refuge A stand of non-genetically modified crop in a GM crop designed to minimise the build up of pesticide resistance genes in the pest population by interbreeding of the two pest populations, i.e. exposed and unexposed to the pesticide concerned, thereby diluting the resistance genes.

Regeneration The growth of new tissues or organs to replace those injured or lost. In tissue culture, regeneration is used to define the development of organs or plantlets from a tissue, callus culture or from a bud.

Regenerative medicine A term applied to new medical advances in which damaged body parts or body tissue is replaced or the body is encouraged to heal itself. From our understanding of our genes and how they work to control the growth, building and repair of our body, regenerative medicine is studying how to create new tissues for transplant, transplant stem cells into the body or how to induce the body to regenerate from the body's own cells.

Regulations Regulations are a form of law, often referred to as delegated or subordinate legislation. They have the same binding legal effect as Acts and usually state rules that apply generally, rather than to specific persons or things. However, regulations are not made by Parliament. Rather, they are made by persons or bodies to whom Parliament has delegated the authority to make them, such as the Governor in Council, a minister or an administrative agency. Authority to make regulations must be expressly delegated by an Act. Acts that authorize the making of regulations are called enabling Acts.

Regulatory agency The agency or government department that has responsibility over the legislation (acts and regulations) for a given sector of the government.

Regulatory framework The due process of regulation surrounding a single topic, such as seeds or fertilizers, that entails all of the relevant legislative documents (acts, regulations, annexes) and describes the federal agency responsible for administering the framework.

Regulatory gene A gene with the primary function of controlling the rate of synthesis of the products of one or several other genes or pathways.

Rejuvenation Reversion from adult to juvenile stage.

Relative risk The ratio of the outcome rate among persons exposed to a certain factor divided by the outcome rate among persons not exposed.

Relaxed plasmid A plasmid that replicates independently of the main bacterial chromosome and is present in 10-500 copies per cell.

Release factors 1. Soluble protein that recognizes termination codons in mRNAs and terminate translation in response to these codons.

2. A hormone that is produced by the hypothalamus and stimulates the release of a hormone from the anterior pituitary gland into the bloodstream.

Renature The reannealing (hydrogen bonding) of single- stranded DNA and/or RNA to form a duplex molecule.

Renewable resource A natural resource that can be replenished by natural means at rates comparable to its rate of consumption.

Rennin An enzyme secreted by cells lining the stomach in mammals, and that is responsible for clotting milk.

Repeat unit A sequence of bases that occurs repeatedly in the genome, often end-on-end, i.e., tandemly.

Reperfusion The reoxygenation of tissue that has been deprived of adequate oxygen (ischaemia) as a result of either surgical procedures or physiological dysfunction. Vital organs can tolerate only a brief period of oxygen deprivation before cell injury and death occur. Subsequent reperfusion, however, can also cause tissue damage. Ischaemia reperfusion damage can be prevented or decreased in the presence of antioxidants.

Replacement; gene replacement A method of substituting a cloned gene, or part of a gene, which may have been mutated *in vitro*, for the wild-type copy of the gene within the host's chromosome.

Replacement therapy The administration of metabolites, co-factors or hormones that are deficient as the result of a genetic disease.

Replica plating A procedure for duplicating the bacterial colonies growing on agar medium in one Petri plate to agar medium in another Petri plate.

Replication, of DNA The replication of DNA is a fundamental process during cell division that provides the two daughter cells (offspring) with identical sets of genes. The replication requires the temporary separation of the DNA double helix into single strand forms where each of the two strands serves as a molecular template for the synthesis of one new complementary strand to form two new double helical copies of the original DNA. The accuracy of replication is based on the specific molecular interaction between two of the four existing nucleotides (A adenine; T thymine; G guanine; C cytosine) to form only AT or GC base pairs. Any other pairing can cause mutations in the daughter DNA, a process that occurs very rarely, and is one of the processes leading to genetic diversity of life.

Replicative form (RF) The molecular configuration of viral nucleic acid that is the template for replication in the host cell.

Replicator A system able to build copies of itself when provided with raw materials and energy.

Replicon The portion of a DNA molecule which is replicable from a single origin. Plasmids and the chromosomes of bacteria, phages and other viruses usually have a single origin of replication and, in these cases, the entire DNA molecule constitutes a single replicon. Eukaryotic chromosomes have multiple internal origins and thus contain several replicons. The word is often used in the sense of a DNA molecule capable of independent replication, e.g., "The shuttle vector pJDB219 is a replicon in both yeast and *E. coli*."

Replisome The complete replication apparatus present at a replication fork that carries out the semi-conservative replication of DNA.

Reporter gene A gene that encodes a product that can readily be assayed. Thus reporter genes are used to determinate whether a particular DNA construct has been successfully introduced into a cell, organ or tissue.

Repressible enzyme An enzyme whose synthesis is diminished by a regulatory molecule.

Repressible gene A gene whose expression can be diminished or extinguished by the presence of a regulatory molecule.

Repression Inhibition of transcription by preventing RNA polymerase from binding to the transcription initiation site: a repressed gene is "turned off."

Repressor A protein which binds to a specific DNA sequence (the operator) upstream from the transcription initiation site of a gene or operon and prevents RNA polymerase from commencing mRNA synthesis. Examples of repressors are the C1 protein of bacteriophage and the *lac*1 protein of the *lac* operon.

Reproduction The production of an organism, cell or organelle like itself (self propagation). 1. Sexual reproduction: the regular alternation (in the life-cycle of haplontic, diplontic and diplohaplontic organisms) of meiosis and fertilization (karyogamy) which provides for the production of offspring. The main biological significance of sexual reproduction lies in the fact that it achieves genetic recombination.

2. Asexual or agamic reproduction: the development of a new individual from either a single cell (agamospermy) or from a group of cells (vegetative reproduction) in the absence of any sexual process.

Reproductive cloning Making a full living copy of an organism. Currently illegal in Australia and many other countries around the world.

Repulsion The phase state in which a dominant (or wild-type) allele at one locus and a recessive (or mutant) allele at a second locus occur on the same chromosome. Also called trans configuration.

Residues The components of macromolecules, e.g., amino acids, nucleotides.

Resistance factor A plasmid that confers antibiotic resistance to a bacterium.

Resistance Term commonly used to describe the ability of an organism to withstand a stress, a force or an effect of a disease, or its agent or a toxic substance.

Resorption The break down of minerals from structures in the body. For example, the dissolution of calcium from bones.

Rest period An endogenous physiological condition of viable seeds, buds or bulbs that prevents growth even in the presence of otherwise favourable environmental conditions. By some seed physiologists, this is referred to as dormancy.

Restenosis Recurrent narrowing of a heart valve or artery after corrective surgery.

Restitution nucleus A nucleus with unreduced or doubled chromosome number that results from the failure of a meiotic or mitotic division.

Restriction endonuclease (enzyme) A class of endonucleases that cleaves DNA after recognizing a specific sequence, e.g., *Bam*H1 (5′GGATCC3′), *Eco*RI (5′GAATTC3′), and *Hin*dIII (5′AAGCTT3′). There are three types of restriction endonuclease enzymes: Type I: Cuts non-specifically a distance greater than 1000 bp from its recognition sequence and contains both restriction and methylation activities.

Type II: Cuts at or near a short, and often palindromic, recognition sequence. A separate enzyme methylates the same recognition sequence. They may make the cuts in the two DNA strands exactly opposite one another and generate blunt ends, or they may make staggered cuts to generate sticky ends. The type II restriction enzymes are the ones commonly exploited in recombinant DNA technology.

Type III: Cuts 24-26 bp downstream from a short, asymmetrical recognition sequence. Requires ATP and contains both restriction and methylation activities.

Restriction fragment A fragment of DNA produced by cleaving (digesting, cutting) a DNA molecule with one or more restriction endonucleases.

Restriction fragment length polymorphism (RFLP) The occurrence of variation in the length of DNA fragments that are produced after cleavage with a type II restriction endonuclease. The differences in DNA lengths are due to the presence or absence of recognition site(s) for that particular restriction enzyme. RFLPs were initially detected using hybridization with DNA probes after separation of digested genomic DNA by gel electrophoresis (Southern analysis). Now they are typically detected by electrophoresis of digested PCR product.

Restriction map The linear array of restriction endonuclease sites on a DNA molecule. .

Restriction nuclease A bacterial enzyme that cuts DNA at a specific site.

Restriction site The specific nucleotide sequence in DNA that is recognized by a type II restriction endonuclease and within which it makes a double-stranded cut. Restriction sites usually comprise four or six base pairs that typically are palindromic, e.g., 5′GGCC3′3′CCGG5′

The two strands may be cut either opposite to one another, to create blunt ends, or in a staggered manner, giving sticky ends, depending on the enzyme involved.

Reticulocyte A young red blood cell.

Reticuloendothelial system The system of macrophages, which serves as an important defense system against disease.

Retinoblastoma Eye cancer

Retro-element Any of the integrated retroviruses or the transposable elements that resemble them.

Retro-poson; retro-transposon A transposable element that moves via reverse transcription (i.e., from DNA to RNA to DNA) but lacks the long terminal repeat sequences.

Retroviral vectors Gene transfer systems based on viruses that have RNA as their genetic material.

Retrovirus A class of eukaryotic RNA viruses that can form double-stranded DNA copies of their genomes by using reverse transcription; the double-stranded forms integrate into chromosomes of an infected cell. Many naturally occurring cancers of vertebrate animals are caused by retroviruses. Also, the AIDS virus is a retrovirus.

Retting A biological process that removes gums and stem tissue from soft fibers, in the processing of natural fibers. It is carried out in water, dew, or snow through the action of certain groups of bacteria and fungi.

Reversal transfer Transfer of a culture from a callus-supporting medium to a shoot-inducing medium.

Reverse genetics Using linkage analysis and polymorphic markers to isolate a disease gene in the absence of a known metabolic defect, then using the DNA sequence of the cloned gene to predict the amino acid sequence of its encoded protein.

Reverse transcriptase (RNA dependent DNA polymerase) An enzyme isolated from retrovirus-infected cells that synthesizes a complementary (c)DNA strand from an RNA template.

Reversion; reverse mutation Restitution of a mutant gene to the wild-type condition, or at least to a form that gives the wild phenotype; more generally, the appearance of a trait expressed by a remote ancestor.

Rheology Study of the flow of matter such as fermentation liquids.

Rheumatoid arthritis A form of arthritis that attacks the connective tissues lining the joints.

Rhizobacteria Bacteria found closely associated with plant roots.

Rhizobacterium A micro-organism whose natural habitat is near, on or in plant roots.

Rhizobium A class of microorganisms that converts atmospheric nitrogen into a form that plants can utilize for growth. Species of this microorganism grow symbiotically on the roots of certain legumes, such as peas, beans and alfalfa.

Rhizosphere The soil zone immediately surrounding a plant root system.

Ri plasmid A class of large conjugative plasmids found in the soil bacterium *Agrobacterium rhizogenes*. Ri plasmids are responsible for hairy root disease of certain plants. A segment of the Ri plasmid is found in the genome of tumour tissue from plants with hairy root disease.

Ribonucleic acid (RNA) An organic acid polymer composed of adenosine, guanosine, cytidine and uridine ribonucleotides. The genetic material of some viruses, but more generally is the molecule, derived from DNA by transcription, that either carries information (messenger RNA), provides sub-cellular structure (ribosomal RNA), transports amino acids (transfer RNA), or facilitates the biochemical modification of itself or other RNA molecules.

Ribose A monosaccharide found in all ribonucleosides, ribonucleotides and RNA. Its close analogue, 2-deoxyribose, is similarly found in all deoxyribonucleosides, deoxyribonucleotides and DNA.

Ribosomal binding site A sequence of nucleotides near the 5′ end of a bacterial mRNA molecule that facilitates the binding of the mRNA to the small ribosomal sub-unit. Also called the Shine-Delgarno sequence.

Ribosomal DNA The coding locus for ribosomal RNA. This is generally a large and complex locus, typically composed of a large number of repeat units, separated from one another by the intergenic spacer. A repeat unit comprises a gene copy for each individual ribosomal RNA component, separated from one another by the internal transcribed spacer.

Ribosomal RNA The RNA molecules that are essential structural and functional components of ribosomes, where protein synthesis occurs. Different classes of rRNA molecule are identified by their sedimentation (S) values. E. coli ribosomes contain one 16S rRNA molecule (1541 nucleotides long) in one (small) ribosomal sub-unit, and a 23S rRNA (2904 nucleotides) and a 5S rRNA (120 nucleotides) in the other (large) sub-unit. These three rRNA molecules are synthesized as part of a large precursor molecule which also contains the sequences of a number of tRNAs. Special processing enzymes cleave this large precursor to generate the functional molecules. Constitutes about 80% of total cellular RNA.

Ribosome The cellular particles made of protein and RNA subunits that catalyze the synthesis of proteins along a messenger RNA (mRNA) template. This process is called translation that converts a nucleic acid genetic code into an amino acid sequence. The result are proteins (enzymes) with each having a specific structure and function carrying out a particular metabolic reaction in a cell.

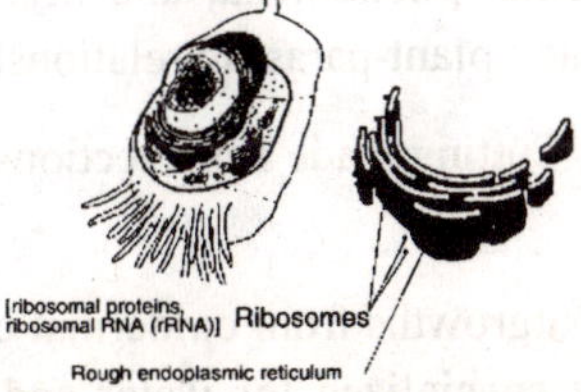

Fig. Ribosomes

Ribosome-binding site The region of an mRNA molecule that binds the ribosome to initiate translation.

Ribozyme; gene shears RNA molecule that can catalyse chemical reactions, often cutting other RNAs.

Ribulose A keto-pentose sugar ($C_5H_{11}O_5$) that is involved in carbon dioxide fixation in photosynthesis.

Ribulose biphosphate (RuBP) A five-carbon sugar that is combined with carbon dioxide to form a six-carbon intermediate in the first stage of the dark reaction of photosynthesis.

Rights These are entitlements. Some rights (human rights) belong to everyone by virtue of being human; some rights (legal rights) belong to people by virtue of their belonging to a particular political state.

Rinderpest Cattle plague; a viral infection of cattle, sheep and goats.

Risk Used as a term for a danger that arises unpredictably, such as being struck by a car.

Risk analysis A process consisting of three components: risk assessment, risk management and risk communication.

Risk assessment Approach taken by scientists, governments and companies to consider the safety of a new technology. On the basis of current knowledge and experience, it involves careful and systematic analysis and assessment of possible adverse impacts, the scale of them, and the likelihood of them happening. The aim is to estimate the benefits, and suggest ways to manage the risks, so as to make informed choices.

Risk characterisation The ,qualitative and\or quantitative estimation, including attendant uncertainties, of the probability of occurrence and severity of known or potential adverse health effects in a given population based on hazard characterisation and exposure assessment.

Risk communication The interactive exchange of information and opinions throughout the risk analysis process concerning hazards and risks, risk-related factors and risk perceptions, among risk assessors, risk managers, consumers, industry, the academic community and other interested parties, including the explanation of risk assessment findings and the basis of risk management decisions.

Risk management The process of weighing policy alternatives in light of the results of risk assessment and, if required, selecting and implementing appropriate control options, including regulatory measures.

RNA (ribonucleic acid) The most common form of nucleic acid used for storage of chemical energy, processing genetic information from genes (DNA) via messenger RNA (mRNA) into proteins. RNA is one of the most ancient form of molecular structures with enzymatic activity. As a matter of fact, protein biosynthesis is entirely controlled by RNA molecules including mRNA (genetic information), transfer RNA (tRNA) for translating the DNA code into amino acid code, and ribosomal RNA (rRNA) that provide the enzymatic linkage (chemical bond formation) of amino acids into proteins. The use of RNA probably precedes the use of proteins and most modern genomes (except for some viruses) are made of DNA instead of RNA but can only be read by proteins. DNA is the more chemically stable of the two forms of nucleic acids.

RNA editing The process in which the base sequence of the RNA transcript is changed by addition of bases to the RNA molecule or by chemical transformation of one base to another. This subverts the genetic information canied in the genes.

RNA interference A natural process used by organisms to block protein production.

RNA polymerase The enzyme that catalyses the synthesis of a complementary strand of RNA from either a DNA strand or an RNA strand in some viruses.

Roentgen (symbol: r) Obsolete unit of ionizing radiation. The SI unit is the sievert (symbol: Sv; 1 Sv » 8.4 r)

Root The descending axis of a plant, normally below ground, which serves to anchor the plant and to absorb and conduct water and mineral nutrients.

Root apex The apical meristem of a root; very similar to the shoot apical meristem in that it forms the three meristematic areas: the protoderm (developing into the epidermis); the procambium (which develops into the stele); and the growth meristem (which forms the cortex).

Root cap A thimblelike mass of cells covering and protecting the apical meristem of a root.

Root culture The culture of isolated root tips of apical or lateral origin to produce *in vitro* root systems with indeterminate growth habits. Root culture was among the first kinds of plant tissue cultures, and is still largely used in the study of developmental phenomena, and mycorrhizal, symbiotic and plant-parasitic relationships.

Root cutting Cutting made from sections of roots alone.

Root hairs Outgrowths from epidermal cell walls of the root specialized for water and nutrient absorption.

Root nodule A small round mass of cells that is located on the roots of plants and contains nitrogen-fixing bacteria.

Root stock The trunk or root material to which buds or scions are inserted in grafting.

Root tuber Thickened root that stores carbohydrates.

Root zone The volume of soil or growing medium containing the roots of a plant. In soil science, the depth of the soil profile in which roots are normally found.

Roslin technique A type of nuclear transfer used to clone mammals. It fuses a donor cell and an egg cell together and induces it to grow into an embryo by using electrical pulses.

Rotary shaker Rotating apparatus with a platform on which, containers can be shaken, such as Erlenmeyer flasks containing cells in liquid nutrient medium.

RRNA; ribosomal RNA The RNA molecules which are essential structural and functional components of ribosomes, the organelles responsible for protein synthesis. The different rRNA molecules are known by their sedimentation (Svedberg; symbol S) values. *E. coli* ribosomes contain one 16S rRNA molecule (1541 nucleotides long) in the same (small) sub-unit and a 23S rRNA (2904 nucleotides) and a 5S rRNA (120 nucleotides) in the large sub-unit. These three rRNA molecules are synthesized as part of a large precursor molecule which also contains the sequences of a number of tRNAs. Special processing enzymes cleave this large precursor to generate the functional moieties.

Ruminant animals; ruminants Animals having a rumen - a large digestive vat in which fibrous plant material is partially broken down by microbial fermentation, prior to digestion in a "true" stomach (the abomasum). There are also two other stomachs - the reticulum and the omasum. Typical ruminants are cattle and sheep.

Runner A lateral stem that grows horizontally along the ground surface and gives rise to new plants either from axillary or terminal buds.

Rust A generic descriptor for various plant diseases, especially those caused by a group of parasitic fungi of the phylum *Basidiomycota*, that attack the leaves and stems of crops.

S

S phase The phase in the cell cycle during which DNA synthesis occurs.

S1 nuclease An enzyme that specifically degrades RNA or single-stranded DNA to 5′ mononucleotides. Purified from the filamentous fungus *Aspergillus oryzae,* S1 nuclease is used in assessing the extent of a hybridization reaction by removing unpaired regions. It is also used to remove the sticky ends of restriction fragments. In S1 mapping, the coding region of a gene is detected by performing mRNA-DNA hybridization and removing unpaired DNA with S1 nuclease.

Saccharifaction Following liquefaction, the hydrolysis of polysaccharides by glucoamylase to maltose and glucose.

Saccharomyces Genus of yeasts that reproduce asexually by budding or sexually by conjugation. Economically important in brewing and baking, they are also widely used in genetic engineering and as simple model organisms in the study of eucaryotic cell biology.

Safety Freedom from danger.

Safety policy The overall intentions and direction of an organisation with regard to safety as formally expressed by top management.

Salmonella A genus of rod-shaped, Gram-negative bacteria that are a common cause of food poisoning.

Salmonids Members of the fish family "Salmonidae," including salmon, trout and chars.

Salt bridge An ionic bond between charged groups that are part of larger covalent structures; salt bridges occur in many proteins.

Salt tolerance; saline resistance The ability to withstand a concentration of sodium (Na^+ ion), or of any other salt, in the soil (or in culture), which is damaging or lethal to other plants. Breeding and selection for increased tolerance and resistance in crop plants is of great current interest.

Sap Fluid content of the xylem and phloem cells of plants. Fluid contents of the vacuole are referred to as cell sap.

Saprophyte A vegetable organism that derives its nutriment from decaying organic matter.

Sarcoma Cancer of connective tissue.

Sarcomere Repeating unit of a myofibril in a muscle cell, composed of an array of overlapping thick (myosin) and thin (actin) filaments between two adjacent Z discs.

Sarcoplasmic reticulum Network of internal membranes in the cytoplasm of a muscle cell that contains high concentrations of sequestered Ca2+ that is released into the cytosol during muscle excitation.

SARS (severe acute respiratory syndrome) It is caused by a virus thought to be a combination of the Coronavirus family, a virus that is often a cause of the common cold, and the paramyxovirus family which causes measles and mumps. The syndrome includes fever and coughing or difficulty breathing and can be fatal. It is thought to have originated in mainland China in 2003 and has spread to other countries.

Satellite DNA That portion of the DNA in plant and animal cells consisting of highly repetitious sequences (millions of copies) typically in the range from 5 to 500 bases. Thousands of copies occur tandemly (end-on-end) at each of many sites. It can be isolated from the rest of the DNA by density gradient centrifugation.

Satiation The term for the physiologic processes that bring eating a meal to an end and control meal size.

Saturated An organic molecule is described as saturated if it is a closed shell species lacking double or triple bonds; forming a new bond to a saturated molecule requires the cleavage of an existing bond.

Saturated fats/fatty acid A kind of fat often found in meat and dairy products, as well as some vegetable oils, such as coconut and palm oils. These fats contain single bonds and cannot incorporate any additional hydrogen atoms. They are associated with high cholesterol levels in the blood.

Saturated molecule Molecule containing carbon-carbon bonds that has only single covalent bonds.

SCA (single chain antigen) Antibody-binding domains in which the two chains are produced by a gene and linked by a short peptide.

Scaffold The central core structure of condensed chromosomes. The scaffold is composed of non-histone chromosomal proteins.

Scale up Conversion of a process, such as fermentation of a micro-organism, from a small scale to a larger scale.

Scanning electron microscope (SEM) An electron-beam-based microscope used to examine, in a three dimensional screen image, the surface structure of prepared specimens.

Scanning tunneling microscope A device in which a sharp conductive tip is moved across a conductive surface close enough to permit a substantial tunneling current (typically a nanometer or less). In a common mode of operation, the voltage is kept constant and the current is monitored and kept constant by controlling the height of the tip above the surface; the result, under faverable conditions, is an atomic-resolution map of the surface reflecting a combination of topography and electronic properties. The STM has been used to manipulate atoms and molecules on surfaces.

Scarification The chemical or physical treatment given to some seeds (where the seed coats are very hard or contain germination inhibitors) in order to break or weaken the seed coat sufficiently to permit germination.

Schwann cell Glial cell responsible for forming myelin sheaths in the peripheral nervous system.

Scientific name A unique identifier consisting of a genus and a species name (the specific epithet) in Latin, assigned to each recognized and described species of organism. Based on the Linnean system of classification. Susceptible to considerable blurring at the edges because, as Darwin came to realize, nature has not packaged living things into neatly discrete entities.

Scion The twig or bud to be grafted onto another plant, the root stock, in a budding or grafting operation.

Scion stock interaction The effect of a rootstock on a scion (and vice versa) in which a scion on one kind of rootstock performs differently than it would on its own roots or on a different rootstock.

Sclerenchyma A strengthening tissue in plants, composed of cells with heavily lignified cell walls.

Scrapie A disease of sheep; a spongiform encephalopathy.

Screen To separate by exclusion or collection on the basis of a set of criteria (biochemical, anatomical, physiological, etc.). Screening is often applied to the process of selection for specific purposes, such as disease resistance or improved agronomic qualities in plants, improved performance in animals, specific enzyme properties in micro-organisms, etc.

Sealed assembler lab A general purpose assembler system in a container permitting only energy and information to be exchanged with the environment.

Second messenger Small molecule that is formed in or released into the cytosol in response to an extracellular signal and helps to relay the signal to the interior of the cell. Examples include cAMP, IP3 and Ca2+.

Secondary antibody In an ELISA or other immunological assay system, the antibody that binds to the primary antibody. The secondary antibody is often conjugated with an enzyme such as alkaline phosphatase.

Secondary cell wall The innermost layer of cell wall, with a highly organized microfibrillar structure, which is formed in certain cells after cell elongation has ceased. It gives rigidity to the cells.

Secondary growth Type of growth characterized by an increase in thickness of stem and root and resulting from formation of secondary vascular tissues by the vascular cambium.

Secondary immune response The rapid immune response that occurs during the second (and subsequent) encounters of the immune system of a mammal with a specific antigen.

Secondary metabolite A compound that is not necessary for growth or maintenance of cellular functions but is synthesized, generally, for the protection of a cell or micro-organism, during the stationary phase of the growth cycle.

Secondary pests Those species within an ecosystem that are normally kept in check by natural enemies.

Secondary phloem Phloem tissue formed by the vascular cambium during secondary growth in a vascular plant.

Secondary plant products Metabolic products not having a known functional or structural use in plant cells. They have been extracted from plant tissue cultures for pharmaceutical and food processing purposes (e.g., essential oils, food additives, flavours).

Secondary root A branch or lateral root.

Secondary structure Regular local folding pattern of a polymeric molecule; in proteins, a helices and b-pleated sheets.

Secondary thickening Deposition of secondary cell wall materials which result in an increase in thickness in stems and roots.

Secondary vascular tissue Vascular tissue (xylem and phloem) formed by the vascular cambium during secondary growth in a vascular plant.

Secondary xylem Xylem tissue formed by the vascular cambium during secondary growth in a vascular plant.

Secretion The passage of a molecule from the inside of a cell through a membrane into the periplasmic space, or the extracellular medium.

Secretory vesicle Membrane-bounded organelle in which molecules destined for secretion are stored prior to release. Sometimes called secretory granule because darkly staining contents make the organelle visible as a small solid object.

Seed Botanically, a seed is the matured ovule without accessory parts. Colloquially, a seed is anything which may be sown; i.e., seed potatoes (which are vegetative tubers); seed of corn, sunflower, etc.

Seed shattering The spontaneous dispersal of mature seed from a plant following ripening.

Seed storage proteins Proteins accumulated in large amounts in seeds not because of their enzymatic or structural properties but simply as a convenient source of amino acid for use when the seed germinates. They are of interest to biotechnologists: 1. As a source of protein. Much of the world's food comes from plant seeds or fruits, and much of the protein in those seeds is storage protein. Thus a substantial amount of the world's food protein comes from plant storage protein. Any improvement of the nutritional content of those proteins could correspondingly improve human diet.

2. As expression systems. Storage proteins are produced in very large amounts relative to other proteins, and are stored in stable, compact bodies in the plant seed. Several workers are seeking to make the plants produce other proteins in similarly large amounts and in as convenient a form, by splicing the gene for a desired protein into the middle of a plant storage protein gene.

Segment-polarity gene A gene that functions to define the anterior and posterior components of body segments in *Drosophila*.

Segregation The separation of the two members of a chromosome pair from each other at meiosis; the result is seen as the separation of alleles from each other in the gametes of heterozygotes; the occurrence of different phenotypes among offspring, resulting from chromosome or allele separation in their heterozygous parents. Mendel's first principle of inheritance (the Law of Segregation) predicts that heterozygotes will produce equal numbers of gametes containing each allele.

Selectable Having a gene product that, when present, enables a researcher to identify and preferentially propagate a particular organism or cell type.

Selectable marker A gene, often encoding resistance to an antibiotic or an herbicide, introduced into a group of cells to allow identification of those cells that contain the gene of interest from the cells that do not. Selectable markers are used in genetic engineering to facilitate identification of cells that have incorporated another desirable trait that is not easy to identify in individual cells.

Selection 1. Differential survival and reproduction phenotypes.

2. A system for either isolating or identifying specific organisms in a mixed culture.

Selection coefficient, *s* The proportion by which the fitness of a genotype is less than the fitness of a standard genotype, which is usually the genotype with the highest fitness. In general, relative fitness = 1 - *s*.

Selection culture A selection based on difference(s) in environmental conditions or in culture medium composition, such that preferred variant cells or cell lines (presumptive or putative mutants) are favoured over other variants or the wild-type.

Selection differential, *S* The difference between the mean of the individuals selected to be parents and the mean of the overall population.

Selection pressure The intensity of selection acting on a population of organisms or on cells in culture. Its effectiveness is measured in terms of differential survival and reproduction, and consequently in changes in allele frequency in a population.

Selection response The difference between the mean of the individuals selected to be parents and the mean of their offspring. Predicted response = heritability (narrow-sense) X selection differential.

Selection unit The minimum number of organisms or cells effective in the screening process.

Selective agent An environmental or chemical agent characterized by its lethal or sub-lethal stress on growing plants, or portion thereof in culture. A selective agent is mainly used when selection of resistant or tolerant individuals is the research aim.

Selective breeding A process in which new or improved strains of plants or animals are developed, mainly through controlled mating or crossing and selection of progeny for desired traits.

Selective medium Nutrient material constituted such that it will support the growth of specific organisms while inhibiting the growth of others.

Self fertilization The process by which pollen of a given plant fertilizes the ovules of the same plant. Plants fertilized in this way are said to have been selfed. An analogous process occurs in some animals, such as nematodes and molluscs.

Self incompatibility In plants, the inability of the pollen to fertilize ovules (female gametes) of the same plant.

Self pollination Pollen of a plant is transferred to the female part of the same plant or another plant with the same genetic makeup.

Self-renewal Ability of stem cells to continue to divide and replenish themselves indefinitely.

Self replicating elements Extrachromosomal DNA elements that have origins of replication for the initiation of their own DNA synthesis.

Selfing Mating by a single individual containing both female and male reproductive systems.

Semi conservative replication During DNA duplication, each strand of a parent DNA molecule is a template for the synthesis of its new complementary strand. Thus, one half of a pre-existing DNA molecule is conserved during each round of replication.

Semi continuous culture Cells in an actively dividing state are maintained in culture by periodically draining off the medium and replenishing it with fresh medium.

Semi permeable membrane A cell or plasma membrane that is partially permeable; certain ions or molecules (water, solvents) can pass through it but others cannot (such as certain solutes).

Semi solid Gelled but not firmly so; small amounts of a gelling agent are used to obtain a semi-solid medium; called also semi-liquid.

Semi sterility A condition of only partial fertility in plant zygotes; usually associated with translocations.

Senescence The last stage in the post-embryonic development of multicellular organisms, during which loss of functions and degradation of biological components occur. A physiological ageing process in which cells and tissues deteriorate and finally die.

Sense RNA A primary transcript (RNA) that contains a coding region (contiguous sequence of codons) that is translated to produce a polypeptide.

Sensitivity For diagnostic tests, the smallest amount of the target molecule that the assay can detect.

Sepsis Destruction of tissue by pathogenic micro-organisms or their toxins, especially through infection of a wound.

Septum Any dividing wall or partition; frequently a cross wall in a fungal or algal filament.

Sequence The linear arrangement of building blocks in biological macromolecules like DNA, RNA, protein and polysaccharides. DNA and RNA macromolecules are linear polymers of nucleotides. Proteins are linear polymers of amino acids. Polysaccharides are linear and branched polymers of monosaccharides (sugars). While the sequence of RNA and proteins are encoded for by the nucleotide sequence in DNA (the genes and genomes), polysaccharides which play important roles in physiology are not encoded for by genetic information, but rather by the spatial and temporal activity of enzymes that synthesize these polysaccharides.

Sequence characterisation Determination of the order in which the nucleic acids making up a DNA molecule, or the amino acids making up a protein molecule, are linked together.

Sequence hypothesis Francis Crick's seminal concept that genetic information exists as a linear DNA code; DNA and protein sequence are colinear.

Sequence tagged site (STS) Short, unique DNA sequence (usually 200 to 500 bp) that, by being able to be amplified by PCR, is uniquely "tagged" to the site on the chromosome from which it was amplified.

Serial divisions Splitting at about monthly intervals of excised shoot-tip material growing on culture medium, in order to induce additional plantlets.

Serial float culture A technique of floating anthers on liquid medium developed by Sunderland. Anther dehiscence, pollen release and development occur at intervals of several days, and in different nutrient media.

Serine/threonine kinase Protein kinase that phosphorylates serines or threonines on its target protein.

Serology The study of serum reactions between an antigen and its antibody. Serology is mainly used to identify and distinguish between antigens, such as those specific to micro-organisms or viruses. Serology is also employed as an indicator technique to assay plants suspected of being virus-infected.

Sertoli cell Supporting cell of the mammalian testis that surrounds and nourishes developing sperm cells.

Serum Blood plasma that has had its clotting factor removed.

Serum albumin A globular protein obtained from blood and body fluids.

Sewage treatment Sewage treatment is one of the most widespread biotechnological processes in Western societies, to deal with the huge amounts of human and animal waste that such societies produce. Sewage treatment methods vary widely, but all have a biological basis to break down the organic material in sewage and convert it into something that can be safely discharged into the environment (usually rivers or seas).

Sex chromosome Differentiated chromosome which is responsible for the determination of sex of the individual. For all mammals, a small number of flowering plants and many insects, female individuals carry a pair of X chromosomes, and males carry one X and one Y. For birds, reptiles and most amphibians, male individuals carry a pair of W chromosomes, and females carry one W and one Z. In some insects there is only one sex chromosome, X, and sex is determined by the number of these present.

Sex determination The method by which the distinction between males and females is established in a species.

Sex duction The incorporation of bacterial genes into F factors and their subsequent transfer, by conjugation, to a recipient cell.

Sex factor A bacterial episome (e.g., the F plasmid in *E. coli*) that enables the cell to be a donor of genetic material. The sex factor may be propagated in the cytoplasm, or it may be integrated into the bacterial chromosome.

Sex hormones Steroid hormones that control sexual development.

Sex influenced dominance The tendency for the type of gene action to vary between the sexes within a species. Thus the presence of horns in some breeds of sheep appears to be dominant in males and recessive in females.

Sex limited Expression of a trait in only one sex; e.g., milk production in mammals; egg production in chickens.

Sex linkage The location of a gene on a sex chromosome, typically on the X chromosome.

Sexed embryos Embryos separated according to sex.

Sexual reproduction The process where two cells (gametes) fuse to form one fertilized cell or zygote.

Shake culture An agitated suspension in culture providing adequate aeration for cells in the liquid medium. Usually an Erlenmeyer flask containing the culture is attached to a horizontal or platform shaker, or agitated with a magnetic stirrer.

Shaker; platform shaker A platform fitted with clips for grasping Erlenmeyer flasks, with set or variable speed control. Shaking speed must be adjusted for gentle and even agitation of suspension cultures.

Shear 1. The sliding of one layer across another, with deformation and fracturing in the direction parallel to the movement. This term usually refers to the forces that cells are subjected to in a bioreactor or a mechanical device used for cell breakage.

2. To fragment DNA molecules into smaller pieces. DNA, as a very long and fairly stiff molecule, is very susceptible to hydrodynamic shear forces. Forcing a DNA solution through a hypodermic needle will fragment it into small pieces. The size of the fragments obtained is inversely proportional to the diameter of the needle's bore. The actual sites at which the shear force breaks a DNA molecule are approximately random. Therefore DNA fragments may be generated by random shear and then cloned (by either tailing their ends or using linkers) so as to create a complete gene library of an organism. This method is little used now, having been replaced by the use of partial digests with four-base-pair cutters, such as *Sau*3A, as a means of generating random DNA fragments.

Shear modulus Shear stress divided by shear strain; has the units of force per unit area.

Shine-dalgarno sequence A conserved sequence in prokaryotic mRNAs that is complementary to a sequence near the 5′ terminus of the 16S ribosomal RNA and is involved in the initiation of translation.

Shoot A young branch that grows out from the main stock of a tree, or the young main portion of a plant growing above ground.

Shoot differentiation The development of growing points, leaf primordia and finally shoots from a shoot tip, axial bud, or even a callus surface.

Shoot tip graft; micrograft. A shoot tip or meristem tip is grafted onto a prepared seedling or micropropagated rootstock in culture. Meristem tip grafting is mainly used for *in vitro* virus elimination with *Citrus* spp. and other plants.

Short day plant Plant that requires a night (or dark period) longer than its critical dark period to induce flower formation.

Short template A DNA strand that is synthesized during the polymerase chain reaction and has a primer sequence at one end and a sequence complementary to the second primer at the other end.

Short-chain fatty acids Fatty acids with chain lengths of two to six carbon atoms.

Shuttle vector; bifunctional vector A plasmid capable of replicating in two different host organisms because it carries two different origins of replication and can therefore be used to 'shuttle' genes from one to the other. For example, the YEp, pJDB219, is a shuttle vector able to replicate in *E. coli* from its pMB9 origin and in *Saccaromyces cerevisiae* from its 2 μm-plasmid origin.

Sib mating Crossing of siblings. Matings involving two individuals of the same parentage; brother-sister matings.

Sickle cell disease A single gene disorder caused by an altered polypeptide in the globin of haemoglobin. It was the first disorder in which a mutation in a structural gene was shown to produce an altered amino acid sequence. Inheritance is autosomal recessive. In homozygotes there is severe chronic anaemia, episodes of painful infarction (tissue death due to obstruction of the blood supply) especially of lungs, spleen and bones, and proneness to infection. It is connected with a resistance to malaria.

Siderophore A low molecular weight substance that binds very tightly to iron. Siderophores are synthesized by a variety of soil micro-organisms to ensure that the organism is able to obtain sufficient amounts of iron from the environment.

Sieve cell In vascular plants, a long and slender sieve element with relatively unspecialized sieve areas and with tapering end walls that lack sieve plates.

Sieve plate Perforated wall area in a sieve-tube element through which strands connecting sieve-tube protoplasts pass.

Sieve tube A tube within the phloem tissue of a plant, and composed of joined sieve elements.

Sievert (symbol: Sv) The SI unit of ionizing radiation.

Sigma bond A covalent bond in which overlap between two atomic orbitals (e.g., of *sp, sp2,* or *sp3* hybridization) produces a single bonding orbital in which the distribution of shared electrons has a roughly cylindrical symmetry about the axis linking the two atoms; see pi bond, single bond, double bond, triple bond. By themselves, sigma bonds present little barrier to rotation of one substructure with respect to another, although steric effects and cyclic structures may hinder or block rotation.

Sigma factor The sub-unit of prokaryotic RNA polymerases that is responsible for the initiation of transcription at specific initiation sequences.

Signal sequence A segment of about 15 to 30 amino acids at the N terminus of a protein, that enables the protein to be secreted (pass through a cell membrane). The signal sequence is removed as the protein is secreted. Also called signal peptide, leader peptide.

Signal to noise ratio A specifically produced response compared to the response level when no specific stimulus (activity) is present.

Signal transduction The biochemical events that conduct the signal of a hormone or growth factor from the cell exterior, through the cell membrane, and into the cytoplasm. This involves a number of molecules, including receptors, ligands and messengers.

Silencer A DNA sequence that helps to reduce or shut off the expression of a nearby gene.

Silencing A technique to stop or interrupt the expression of a particular gene, most commonly by the insertion of a reverse copy of all or part of that gene.

Simplicity For diagnostic tests, the ease with which an assay can be implemented.

SINEs (short interspersed nuclear elements) Families of short (150 to 300 bp), moderately repetitive elements of eukaryotes, occurring about 100,000 times in a genome. SINES appear to be DNA copies of certain tRNA molecules, created presumably by the unintended action of reverse transcriptase during retroviral infection.

Single bond A sigma bond having no associated pi bonds.

Single cell line; cell strain A culture initiated from a single cell, usually from suspension cultures of single cells or small aggregates plated on solidified medium. The latter may incorporate a selective agent, from which tolerant or resistant individual cell lines or cell clones can be selected.

Single cell protein (SCP) Protein produced by micro-organisms. The dried mass of a pure sample of a protein-rich-micro-organism, which may be used either as feed (for animals) or as a food (for humans).

Single copy A gene or DNA sequence which occurs only once per (haploid) genome. Most structural genes, those encoding functional proteins, are single-copy genes.

Single node culture Culture of separate lateral buds with each carrying a piece of stem tissue.

Single nucleotide polymorphism (SNP) A polymorphism at a particular base site in a coding sequence, e.g., at base 306 in a particular gene, one individual could be heterozygous for A and G: the maternal allele could have an A at this site, while the paternal allele has a G at this site. This type of polymorphism is extensive throughout the genome, and has the great advantage of being detectable without the need for gel electrophoresis, which opens the way for large-scale automation of genotyping.

Single stranded A term used to describe nucleic acid molecules consisting of only one polynucleotide chain. The genomes of certain phages, e.g., MI3, are single-stranded DNA molecules; rRNA, mRNA and tRNA are all single-stranded nucleic acids, but they all contain double-stranded regions formed by the intra-strand base-pairing of self-complementary sequences.

Single-stranded DNA (ssDNA) DNA molecules separated from their complementary strand, either by its absence or following denaturation.

Singlet An electronic state of a molecule in which all spins are paired; see doublet, triplet.

Sires Male animals used for breeding.

Site directed mutagenesis The introduction of base changes - mutations - into a piece of DNA at a specific site, using recombinant DNA methods.

Site specific A term used to describe any process or enzyme which acts at a defined sequence within a DNA or RNA molecule. Type II restriction enzymes are site-specific endonucleases and the recombination systems encoded by some transposons are site-specific, such as is the integration of phage into the *E. coli* chromosome.

Site specific mutagenesis A technique to change one or more specific nucleotides within a cloned gene in order to create an altered form of a protein with one or more specific amino acid changes.

Six base cutter A type II restriction endonuclease that binds (and subsequently cleaves) DNA at sites that contain a sequence of six nucleotide pairs that is uniquely recognized by that enzyme. Because any sequence of six bases occurs less frequently by chance than any sequence of four bases, six-base cutters cleave less frequently than do four-base cutters. Thus, six-base cutters create larger fragments than four-base cutters.

Small nuclear ribonucleoprotein (snRNP) A compound comprising small nuclear RNA and nuclear protein, that is heavily involved in the post-transcriptional processing of mRNA, especially the removal of introns. snRNPs are a major component of spliceosomes.

Smart materials and products Materials and products capable of relatively complex behaviour due to the incorporation of nanocomputers and nanomachines. Also used for products having some ability to respond to the environment.

Social hierarchy An arrangement within a group of animals, such as rabbits, where some individuals are dominant over others. The more dominant an animal, the more likely it is to have preferred access to mates and sources of food.

Soil amelioration The improvement of poor soils, usually using bacteria or fungi. This contrasts with bioremediation, which is the cleaning up of toxins, usually in soils. Amelioration includes breaking down organic matter; forming humus; by solubilizing them, making minerals - such as phosphates - in the soil available to plants; fixing nitrogen; and sometimes an element of bioremediation as well.

Soilless culture; soil free culture Tissue culture and hydroponics. Growing plants in nutrient solution without soil.

Solid media Nutrient media that has been solidified, such as by addition of agar.

Soluble fibre Soluble fibre dissolves in the gut and forms a viscous gel that slows down the release of some nutrients. It can also reduce the risk of heart disease, by reducing blood cholesterol levels. Soluble fibre is present in fruit, vegetables, pulses (e.g. kidney beans, baked beans, lentils) and foods containing oats, barley or rye.

Somaclonal Variation Genetic variations of plant cells arising in cell culture, due to enhanced genetic instability.

Somaclonal variation Altered characteristics in plant tissues by extended growth in a laboratory test tube (possibly a form of mutation).

Somatic Used to describe body tissues which do not include the sex cells (sperm & ova and their precursors)

Somatic cell Any nongerm cell that composes the body of an organism and which possesses a set of multiploid chromosomes (diploid in most organisms).

Somatic cell embryogenesis Embryos are produced either from somatic cells of explants (direct embryogenesis) or by induction on callus formed by explants (indirect embryogenesis).

Somatic cell gene therapy Somatic cell gene therapy involves the insertion of genes into cells for therapeutic purposes; for example, to induce the treated cells to produce a protein that the body is missing. It does not affect genetic makeup of a patient's offspring and generally does not change all, or even most, cells in the recipient. Somatic cell gene therapy is only one way of applying the science of genomics to improve health care.

Somatic cell nuclear transfer The transfer of a nucleus from a fully differentiated cell into an egg that has had its nucleus removed.

Somatic cell variant A somatic cell with unique characters not present in the other cells, such as might be selected for in a screening trial that following a mutation event.

Somatic embryo An organized embryo-like structure. Although morphologically similar to a zygotic embryo it is initiated from somatic plant cells. Under in vitro conditions, somatic embryos go through developmental processes similar to embryos of zygotic origin. Each somatic embryo is potentially capable of developing into a normal plantlet.

Somatic hybridization Naturally occurring or induced fusion of somatic protoplasts or cells of two genetically different parents. The difference may be as wide as interspecific. Wide synthetic hybrids formed in this way (i.e. not via gametic fusion) are known as cybrids. Not all cybrids contain the full genetic information (nuclear and non-nuclear) of both parents.

Somatic reduction Halving of the chromosomal number of somatic cells; a possible method of producing "haploids" from somatic cells and calluses by artificial means.

Somatostatin Growth-hormone-inhibiting hormone.

Sonication Disruption of cells or DNA molecules by high frequency sound waves.

SOS response The synthesis of a whole set of DNA repair, recombination and replication proteins in bacteria containing severely damaged DNA (e.g., following exposure to UV light).

Source DNA The DNA from an organism that contains a target gene; this DNA is used as starting material in a cloning experiment.

Southern blotting A technique for transferring denatured DNA molecules that have been separated electrophoretically, from a gel to a matrix (such as a nitrocellulose membrane) on which a hybridization assay can be performed.

Southern hybridization (southern blotting) A procedure in which DNA restriction fragments are transferred from an agarose gel to a nitrocellulose filter, where the denatured DNA is then hybridized to a radioactive probe (blotting).

Sp, sp2, sp3 An isolated carbon atom has four valence orbitals: three mutually perpendicular *p* orbitals, each with a single nodal plane, and one spherically symmetric *s* orbital. A carbon atom in a typical molecule can be regarded as bonding with four orbitals consisting of weighted sums (termed hybrids) of these *s* and *p* orbitals. One common pattern has four equivalent orbitals, each formed by combining the three *p* orbitals with the *s* orbital; this is *sp3* hybridization. An *sp3* carbon atom forms four sigma bonds, usually in a roughly tetrahedral arrangement. Another common pattern has three equivalent orbitals formed by combining two *p* orbitals with the *s* orbital; this is termed *sp2* hybridization. An *sp2* carbon atom forms three roughly coplanar sigma bonds, usually separated by ~120 , and one pi bond (or several fractional pi bonds). If a single *p* orbital is combined with the *s* orbital, the result is *sp* hybridization, forming two sigma bonds and two pi bonds (usually in a straight line). Atoms of other kinds (e.g., N and O) can hybridize in an analogous manner.

Sparger A device that introduces into a bioreactor air in the form of separate fine streams.

Specialized Anatomically or physiologically adapted for particular functions or habitats.

Species A group of organisms that can interbreed and reproduce. Used to distinguish sexually reproducing organisms into groups. Individuals from two different species cannot have offspring. They are said to be reproductively isolated. The biologist Ernst Mayr formulated this definition of a species advancing our understanding of the mechanism of evolution of higher organisms. For microbes, the species definition does not properly apply, because they do not reproduce sexually, but have an efficient mechanism to exchange genetic material even between evolutionarily distant forms. This exchange of genes is known as horizontal gene transfer. Unlike sexual reproduction, it usually involves only a fraction of an organisms genome that is being transferred and is a mechanisms of increasing genetic variability among microorganisms that does not depend on cellular reproduction (cell division).

Species barrier A hypothetical boundary believed to exist between one species and another (not a scientific concept!).

Species specific Pertaining to individuals of only one species. For example, a pesticide that is species specific affects only one species.

Specificity For diagnostic tests, the ability of a probe to react precisely with a specific target molecule.

Specificity of gene expression A measure of the extent to which the ability of an organism to produce a particular gene product is determined by factors affecting the cell, for example, its function, phase of growth or environmental pressures.

Spent medium After each sub-culture, the medium is discarded because it has been depleted of nutrients, dehydrated or accumulated toxic metabolic products.

Sperm Abbreviation of spermatozoon.

Sperm competition Competition between different spermatozoa to reach and fertilize the egg cell of a single female.

Sperm sexing The separation of sperm into those bearing an X chromosome and those bearing a Y chromosome, in order to be able to produce, via artificial insemination or *in vitro* fertilization, animals of a specified sex. Achieved by means of inactivation of X-bearing or Y-bearing sperm via antibodies directed against sex-specific peptides on the surface of sperm cells, or fluorescence-activated cell sorting (FACS) in which sperm that have been pre-treated with a fluorescent dye that binds to DNA are separated according to the quantity of fluorescence detected by a laser beam, based on the principle that X-bearing sperm contain more DNA than Y-bearing sperm.

Spermatid One of the four cells formed by the meiotic divisions in spermatogenesis. Spermatids become mature spermatozoa (sperm).

Spermatocyte Sperm mother cell. The cell that undergoes two meiotic divisions (spermatogenesis) to form four spermatids; the *primary* spermatocyte before completion of the first meiotic division; the *secondary* spermatocyte after completion of the first meiotic division.

Spermatogenesis The series of cell divisions in the testis by which maturation of the gametes (sperm) of the male takes place.

Spermatogonium (pl: spermatogonia) Primordial male germ cell that may divide by mitosis to produce more spermatogonia. A spermatogonium may enter a growth phase and give rise to a primary spermatocyte.

Spermatozoon The mature, mobile reproductive cell of male animals, produced by the testis.

Spheroplast (formerly also sphaeroplast) A microbial or plant cell from which most of the cell wall has been removed, usually by enzymic treatment. Strictly, in a spheroplast, some of the wall remains, while in a protoplast the wall has been completely removed. In practice, the two words are often used interchangeably.

Spike An inflorescence in which the main axis is elongated and the flowers are sessile.

Spindle In mitosis and meiosis, refers to the spindle-shaped intracellular structure in which the chromosomes move.

Spine Hard, sharp structure on the surface of a plant; usually a modified leaf.

Spirochaete A non-rigid, corkscrew-shaped bacterium that moves by means of muscular flexions of the cell.

Spliceosomes Organelles responsible for the removal of introns from mRNA by means of splicing.

Splicing During the maturation of eukaryotic mRNA, the process that eliminates intervening intron sequences and covalently joins exon sequences of RNA. In recombinant DNA technology, the term refers to the latter of the two processes just described, namely joining fragments of DNA together.

Split genes In eukaryotes, structural genes are typically divided up (split) by a number of non-coding regions called introns.

Sponsor(s) An individual, corporate body, institution, or organization that conducts a clinical trial.

Spore 1. A reproductive cell that develops into an individual without union with other cells; some spores such as meiospores are the product of the germ line, but others are asexual in nature.

2. A small, protected resting body, often synthesized by micro-organisms when nutrient levels are low.

Sporocyte A diploid cell that gives rise to four haploid spores by meiosis.

Sporophyll A leaf that bears spore producing structures (sporangia).

Sporophyte The diploid generation in the life cycle of a plant, and that produces haploid spores by meiosis.

Sport An individual or portion thereof distinguished by a spontaneous mutation. Sports are sometimes of great agricultural worth, but alternatively, they may be disadvantageous and may be rogued during agricultural production.

Spririllum A rigid, spiral-shaped bacterium.

Stable Strictly speaking, a system is termed stable if no rearrangement of its parts can form a system of lower free energy. In practice, the term is used with an implicit proviso regarding the transformations to be considered. Hydrogen is not considered unstable merely because it is subject to nuclear fusion at extreme temperatures. A system is usually regarded as stable (more precisely, as kinetically stable) if its rate of transformation to a state of lower free energy is negligible (by some standard) under the ambient conditions. In nanomechanical systems, a structure can commonly be regarded as stable if it has an extremely low rate of transformations when subjected to its intended operating conditions.

Staggered cuts Symmetrically cleaved phosphodiester bonds that lie on both strands of duplex DNA but are not opposite one another.

Stamen Flower structure made up of an anther (pollen-bearing portion) and a stalk or filament. The stamen is the male part of the flower.

Standards Standards are defined by the International Organization of Standardization (ISO) as "documented agreements containing technical specifications or other precise criteria to be used consistently as rules, guidelines or definitions of characteristics, to ensure that materials, products, processes and services are fit for their purpose."

Standard deviation A statistical measure of variability in a population of individuals or in a set of data; the square root of the variance.

Standard error A statistical measure of variation in a population of means, used to indicate how well sample estimates represent population parameters.

Staple length The length of the individual fibres of cotton. The length of the fibres affects the quality of the fabric that is made from it.

Starch (M.E. *strechen*, to stiffen) A complex insoluble carbohydrate, consisting of various proportions of two glucose polymers, amylose and amylopectin; the chief food storage substance of plants, which is composed of several hundred hexose sugar units and which easily breaks down on hydrolysis into these separate units.

Start codon; initiator codon The set of three nucleotides in an mRNA molecule with which the ribosome starts the process of translation. The start codon sets the reading frame for translation. The most commonly used start codon is AUG, which is decoded as methionine in eukaryotes and as *N*-formylmethionine in prokaryotes. AUG appears to be the only start codon used by eukaryotes, while in bacteria, GUG (valine) may sometimes be employed.

State A physical system is said to be in a particular state when its physical properties fall within some particular range; the boundaries of the range defining a state depend on the problem under consideration. In a classical world, each point in phase space could be said to correspond to a distinct state. In the real world, time-invariant systems in quantum mechanics have a set of discrete states, particular superpositions of which constitute complete descriptions of the system. In practice, broader boundaries are usually drawn. A molecule is often said to be in a particular excited electronic state, regardless of its state of mechanical vibration. In nanomechanical systems, the PES often corresponds to a set of distinct potential wells, and all points in configuration space within a particular well can be regarded as one state. Definitions of state in the thermodynamics of bulk matter are analogous, but extremely coarse by these standards.

Stationary culture A culture maintained in the growth chamber with no agitation movement. The antonym is shake culture.

Stationary phase The plateau of the growth curve after log growth, during which cell number remains constant. New cells are produced at the same rate as older cells die.

Statistic An estimate based on a sample or samples of a population, providing an indication of the true population parameter.

Statistical mechanics Statistical mechanics treats the detailed state of a system (its quantum state or, in classical models, its position in phase space) as unknown and subject to statistical uncertainties; entropy is a measure of this uncertainty. Statistical mechanics describes the distribution of states in an equilibrium system at a given temperature (describing either the distribution of probabilities of quantum states or the probability density function in phase space), and can be used to derive thermodynamic properties from properties at the molecular level. These equilibrium results are useful in nanomechanical design.

Steady state (equilibrium) A process in biochemistry that refers to situations of high activity even though the overall structure and composition of cells seems not to change. Steady state processes are used to explain metabolic homeostasis. The flow of water molecules in a river or free flowing traffic are examples of steady state processes where the overall movement of water molecules (cars) does not appear to change, although the composition of particular molecules (cars) is constantly changing.

Stele The central cylinder, inside the cortex, of roots and stems of vascular plants.

Stem The main body of the above-ground portion of a tree, shrub, herb or other plant; the ascending axis, whether above or below ground, of a plant.

Stem cell Stem cells are specialized cells of animals and plants that have the ability to grow and divide by mitosis. Cell division results in self-regeneration as well as differentiation. This means that stem cells can maintain their characteristic as stem cells (self-regeneration), but also change into a different cell type (differentiation).

Steric Pertaining to the spatial relationships of atoms in a molecular structure, and in particular, to the space-filling properties of a molecule. If molecules were rigid and had hard surfaces, steric properties would merely be an opaque way of saying "shape"; a flexible side-chain, however, has definite steric properties but no fixed shape. Nanomechanical systems make extensive use of the steric properties of relatively rigid molecules, for which the term "shape" has essentially its conventional meaning so long as one remembers that the surface interactions are soft on small length-scales.

Steric hindrance Slowing of the rate of a chemical reaction owing to the presence of structures on the reagents that mechanically interfere with the motions associated with the reaction, typically by obstructing the reaction site.

Sterile Incapable of reproduction. Not able to germinate or bear fruit.

Sterile room Operation room for performing inoculations under aseptic conditions; usually now replaced by use of laminar air-flow cabinets, in which filtered air is blown from the inside to the outside.

Sterility Inability to produce offspring.

Sterilize 1. The process of elimination of micro-organisms, such as by chemicals, heat, irradiation or filtration.

2. The operation of making an animal incapable of producing offspring.

Steward bottle Flask developed by Steward for the growth of cells and tissues in a liquid medium, in which they can be periodically submerged during rotation.

Sticky ends; cohesive ends The single-stranded nucleotide sequence left on a restriction fragment by type II restriction enzymes that cut each strand at a separate location. These unpaired regions are available for hybridization with complementary ends on other fragments during the creation of recombinant DNA.

Stiffness The stiffness of a system with respect to a deformation (e.g., the stiffness of a spring with respect to stretching) is the second derivative of the energy with respect to the corresponding displacement; this measures the curvature of the potential energy surface along a particular direction. Positive stiffness is associated with stability, and a large stiffness can result in a small positional uncertainty in the presence of thermal excitation. Negative stiffnesses correspond to unstable locations on the potential energy surface. Alternative terms for stiffness include force gradient and rigidity.

Stigma Receptive portion of the style, to which pollen adheres.

Stirred-tank fermenter A growth vessel in which cells or micro-organisms are mixed by mechanically-driven impellers.

Stochastic A stochastic event is based on random behavior. The occurrence of individual events cannot be predicted, although measuring the distribution of all observations usually follows a predictable pattern. These patterns can be described by statistical means. An example is the decay of radio active material, where a clump of matter has a measurable and thus predictable half-life time. It is impossible, however, to mark an individual atom and predict when it will decay and emit radiation. The latter process is a stochastic event.

Stock The lower portion of a graft.

Stock plant The source plant from which cuttings or explants are made. Stock plants are usually maintained carefully in an optimum state for (sometimes prolonged) explant use. Preferably they are certified-pathogen-free plants.

Stock solutions Pre-prepared solutions of individual components and used to prepare many different types of media. Certain substances, including Ca and Mg sulphates and phosphates must not be combined until actual medium preparation because insoluble combinations are formed and precipitate.

Stolon A lateral stem that grows horizontally along the ground surface. The runners of white clover, strawberry and bermuda grass are examples of stolons.

Stoma 1. Any of various small openings or pores in an animal body, especially an opening resembling a mouth in various invertebrates.

2. Botany: A minute pore in the epidermis of the leaf or stem of a plant, forming a slit of variable width between two specialized cells (guard-cells), which allows movement of gases, including water vapour, to and from the intercellular spaces. Also, the whole pore with its associated guard-cells.

Stomatal complex Includes the stoma, together with its guard cells and, when present, the subsidiary cells.

Stomatal index = (number of stomata per mm^2 X 100)/(number of stomata per mm^2 + number of epidermal cells per mm^2) This value has been found useful in comparing leaves of different sizes. Relative humidity and light intensity during leaf development affect the value of stomata index.

Stop codon; termination codon A set of three nulecotides for which there is no corresponding tRNA molecule to insert an amino acid into the polypeptide chain. Protein synthesis is hence terminated and the completed polypeptide released from the ribosome. Three stop codons are found: UAA (ochre), UAG (amber) and UGA (opal). Mutations which generate any of these three codons in a position which normally contains a codon specifying an amino acid are known as *nonsense mutations*. Stop codons can also be called nonsense codons.

STR (short tandem repeats) STRs are short DNA sequences that are repeated in a head-tail manner. They are useful in DNA profiling.

Strain In mechanical engineering, strain is a measure of the deformation resulting from stress (that is, force per unit area); the displacement of one point with respect to another, divided by their equilibrium separation in the absence of stress. In chemistry, a molecular fragment generally has some equilibrium geometry (bond lengths, interbond angles, etc.) when the rest of the molecular structure does not impose special constraints (e.g., bending bonds to form a small ring). Deviations from this equilibrium geometry are described as strain, and increase the energy of the molecule. Strain in the mechanical engineering sense causes strain in the chemical sense.

Strain isolation Isolation of any bacterium, animal or plant from the outside world.

Stratification Treatment of moist seeds at low temperature (+2° - +4°C) to break dormancy.

Stress Force per unit area applied by one part of an object to another. Pressure is an isotropic compressive stress. Suspending a mass from a fiber places it in tensile stress. Gluing a layer of rubber between two plates and then sliding one over the other (while holding their separation constant) places the rubber in shear stress.

Stringency Reaction conditions - notably temperature, salt concentration(s) and pH - that dictate the annealing of single-stranded DNA/DNA, DNA/RNA and RNA/RNA hybrids. At high stringency, duplexes form only between strands with perfect one-to-one complementarity; lower stringency allows annealing between strands with some degree of mismatch between bases.

Stringent plasmid A plasmid that only replicates along with the main bacterial chromosome and is present as a single copy, or at most several copies, per cell.

Stroke Interference with blood circulation in the brain, starving one or more parts of the brain of oxygen. The medical term is cerebrovascular accident.SubstrateThe substance on which a specific enzyme exerts its effects.

Structural gene A DNA sequence that forms the blueprint for the synthesis of a polypeptide.

Structural volume The interior of a diamondoid structure typically consists of a dense network of covalent bonds; a larger excluded volume, however, is determined by nonbonded repulsions at the surface. The structural volume corresponds to a region smaller than the excluded volume, chosen to make properties such as the strength and modulus nearly size independent by correcting for surface effects.

Structure, high-resolution The high resolution structure of a molecule refers to its atomic organization in three-dimensional space. It is either obtained from analysis of diffraction patterns of high energy radiation (X-rays, electron waves) or nuclear magnetic resonance spectra (NMR). Structural information has an important place in biological studies at the molecular-level, because structures can be used to elucidate the detailed mechanism of a chemical reaction, a biological binding events such as hormone signaling or immunological defenses, or nutrient transport (absorption) across intestinal epithelial cell layers and cell membranes. The structural analysis of DNA in 1953 has helped understand the mechanism of replication of genetic information during reproduction as well as the mechanism of genetic encoding, reading (transcription), and synthesis (translation)of amino acid sequences in proteins and enzymes.

Structure-functionalism The scientific tradition that stresses the relationship between a physical structure and its function, for example, the related disciplines of anatomy and physiology.

Style Slender column of tissue that arises from the top of the ovary and through which the pollen tube grows.

Sub cloning 1. Splicing part of a cloned DNA molecule into a different cloning vector.

2. The process of transferring a cloned DNA fragment from one vector to another.

Sub culture Division and transfer of a portion or inoculum of a culture to fresh medium. Sometimes used to denote the adding of fresh liquid to a suspension culture.

Sub culture interval The time between subsequent sub-cultures of cells. Sub-culture interval has no relationship to the term *cell generation time*.

Sub culture number The number of times cells, etc., have been sub-cultured, i.e., transplanted by inoculation from one culture vessel to another.

Subspecies Population(s) of organisms sharing certain characteristics that are not present in other populations of the same species.

Substantially equivalent The equivalence of a particular organism, in terms of its specific use and safety to the environment and human health, to those of that same species that are in use and generally considered as safe in Canada, based on valid scientific rationale.

Substitution A point mutation in which one base pair in the DNA sequence is replaced by another.

Substrate 1. A compound that is altered by an enzyme.

2. Food source for growing cells or micro-organisms.

3. Material on which a sedentary organism lives and grows.

Sub unit vaccine One or more immunogenic proteins either purified from the disease-causing organism or produced from a cloned gene. A vaccine composed of a purified antigenic determinant that is separated from the virulent organism.

Sub unit preparation The preparation of a subunit vaccine - a vaccine composed primarily of virus or other microorganism proteins that cannot cause disease.

Sucker A shoot that arises from an underground root or stem and grows at the expense of the parent plant.

Sucrose A disaccharide of glucose and fructose.

Sucrose density gradient centrifugation A procedure used to fractionate mRNAs or DNA fragments on the basis of size.

Sugars Simple carbohydrates, generally with a sweet taste and soluble in water. The term is usually reserved for mono- and disaccharides. In non-scientific English, the disaccharide sucrose is referred to as sugar.

Suicide gene A gene that codes for an antibiotic that can kill the host bacterial cell. It is genetically modified into the bacterium along with a molecular switch that is controlled by a nutrient in the environment. When the nutrient disappears, the suicide gene is switched on and the bacterium dies.

Sulphites Preservatives commonly used in, for example, wine.

Summer-fallowing (to fallow) To leave a field unseeded for a period of time after plowing, in order to recover natural fertility.

Superbug Jargon for the bacterial strain of *Pseudomonas* developed by Chakrabarty, who combined hydrocarbon-degrading genes carried on different plasmids into one organism. Although this genetically engineered micro-organism is neither "super" nor a "bug", it represents a landmark example because it showed how genetically modified microbial strains could be used in a novel way and because it was the basis for the precedent-setting legal decision that declared that genetically engineered organisms were patentable.

Supercoil A DNA molecule that contains extra twists as a result of overwinding (positive supercoils) or underwinding (negative supercoils).

Supercoiled plasmid The predominant in vivo form of plasmid, in which the plasmid is coiled around histone-like proteins. Supporting proteins are stripped away during extraction from the bacterial cell, causing the plasmid molecule to supercoil around itself in vitro.

Supergene A group of neighboring genes on a chromosome that tend to be inherited together and sometimes are functionally related.

Supernatant The soluble liquid &action of a sample after centrifugation or precipitation of insoluble solids.

Superpest A insect pest that has developed a resistance to an insecticide (that has been genetically engineered into a plant, or externally applied to a crop) it was once intolerant of.

Superweed A weed that has developed a resistance to a herbicide/weedkiller that once destroyed it.

Suppressor Mutations in suppressor genes are able to overcome (suppress) the effects of mutations in other, unlinked, genes. A common, and very useful, kind of suppressor mutation occurs within the gene encoding a tRNA molecule and results in a change in the tRNA's anticodon. Such a mutant tRNA can reverse the effect of chain-terminating mutations, such as amber or ochre, in protein-encoding genes.

Suppressor gene A gene that can reverse the effect of a mutation in other genes.

Suppressor mutation A mutation that reverses the effect of an earlier mutation, e.g. a mutation in a gene for a tRNA that permits it to read and override an amber mutation.

Suppressor-sensitive mutant An organism that can grow when a second genetic factor - a suppressor - is present, but not in the absence of this factor.

Surfactant A surface-active agent or wetting agent, such as Tween 20TM or Tween 80TM, TeepolTM, Lissapol FTM, AlconoxTM, etc. Surfactants act by lowering the surface tension and are common addenda to solutions used to surface sterilize materials prior to aseptic excision of explants.

Surrogate A person or animal that functions as a substitute for another. In the case of a surrogate mother, it is where a woman or female animal carries an embryo and ultimately gives birth to a baby that was formed from the egg of another female.

Susceptible The characteristic of a host organism such that it is incapable of suppressing or retarding an injurious pathogen or other factor.

Suspension culture A type of culture in which (single) cells and/or clumps of cells grow and multiply while suspended in a liquid medium..

Sustainability Has many definitions, but refers to those that encompass a holistic approach to using production systems while maintaining natural and non-renewable resources. Sustainable agricultural systems are considered ecologically sound, economically viable, socially just and humane.

Sustainable agriculture Agriculture which minimises inputs in the form of fuel and chemicals (pesticides, pertilisers) and outputs in the form of air, groundwater and soil pollutants whilst producing soils that retain fertility for future generations (e.g. biodynamic, organic and permaculture systems).

Sustainable development Or sustainable development has many definitions. One definition has it as 'using, conserving and enhancing the community's resources so that the ecological processes, on which life depends, are maintained and the total quality of life, now and in the future, can be increased'. Sustainable development closely incorporates environmental factors into economic growth.

Sustainable intensification of animal production systems The manipulation of inputs to, and outputs from, livestock production systems aimed at increasing production and/or productivity and/or changing product quality, while maintaining the long-term integrity of the systems and their surrounding environment, so as to meet the needs of both present and future human generations. Sustainable agricultural intensification respects the needs and aspirations of local and indigenous people, takes into account the roles and values of their locally adapted genetic resources, and considers the need to achieve long-term environmental sustainability within and beyond the agro-ecosystem.

Sweeteners Substances which taste sweet.

Symbiont An organism living in symbiosis with another, dissimilar organism.

Symbiosis The close association of two different kinds of living organisms where there is benefit to both or where both receive an advantage from the association. An example is the association of the mycelium of mycorrhizal fungi with roots of seed plants.

Symbiotic association An intimate partnership between two organisms, in which the mutual advantages normally outweigh the disadvantages.

Sympatric speciation The formation of new species by populations that inhabit the same or overlapping geographic regions.

Symplast The system of protoplasts in plants, that are interconnected by plasmodesmata.

Sympodial A type of plant development in which the terminal bud of the stem stops growing due to either its abortion or its development into a flower or an inflorescence, and the uppermost lateral bud takes over the further axial growth of the stem.

Synapse The synapse is a specialized portion of a neuron or nerve cell that is used for cell to cell communication with other neurons and muscle cells. The chemical synapse contains packaged neurotransmitters that can be released upon an electrical signal (action potential) reaching the synapse from the dendrites and cell body of the neuron, where action potentials originate (where synapses of other neurons interact with the signaling cell). A neuron can have multiple synapses, often with different signaling properties being excitatory or inhibitory synapses. Multiple synapses signaling to a receiving neuron or muscle can strengthen a stimulus or inhibition by activating some or all of the synapses through addition of signaling strength. In addition to chemical synapses, electrical synapses are propagating an action potential signal without a neurotransmitter, but directly by coupling membranes of adjacent cells using gap junctions. The feature of both chemical and electrical synapses allows the signal to propagate unidirectional. The signal cannot reverse. However, feedback signals between the signaling and receiving cell to strengthen or weaken the synaptic interaction, a process called synaptic plasticity.

Synaptonemal complex A ribbonlike protein structure formed between synapsed homologues at the end of the first meiotic prophase, binding the chromatids along their length and facilitating chromatid exchange.

Synchronous culture A culture in which the majority of the cells are dividing at the same time or are at a specific phase of the cell cycle.

Syndrome A group of symptoms that occur together and represent a particular disease.

Synergids The two nuclei within the embryo sac at the upper end in the ovule of the flower, which, with the third (the egg), constitute the egg apparatus.

Syngamy Fertilization.

Synkaryon A nucleus formed by the fusion of nuclei from two different somatic cells during somatic-cell hybridization.

Synteny The occurrence of two or more loci on the same chromosome, without regard to the distance between them.

Synthase One of the six primary classes of enzymes, composed of those that catalyze the division of carbon or other bonds without a hydrolysis or oxidation reaction.

Synthesis/synthesizing (in DNA molecules) The building of a known sequence of nucleotides into a chain called an oligonucleotide (of which genes are made) or DNA (In proteins) Chemical synthesis (manufacture) of a known protein molecule.

System In scientific usage, usually equivalent to "a collection of matter and energy being analyzed as a unit." In engineering usage, usually equivalent to "a set of components working together to serve a set of purposes."

Systems biology A hypothesis-driven field of research that creates predictive mathematical models of complex biological processes or organ systems.

T

T The thymine residue in DNA.

T cell mediated (cellular) immune response The synthesis of antigen-specific T cell receptors and the development of killer T cells in response to an encounter of immune system cells with a foreign immunogen.

T cell receptor An antigen-binding protein that is located on the surfaces of killer T cells and mediates the cellular immune response of mammals. The genes that encode T cell antigens are assembled from gene segments by somatic recombination processes that occur during T lymphocyte differentiation.

T4 DNA ligase An enzyme from bacteriophage-T4-infected cells, and that catalyses the joining of duplex DNA molecules and repairs nicks in DNA molecules. The enzyme requires that one of the DNA molecules has a 5′-phosphate group and that the other has a free 3′-hydroxyl group.

T lymphocytes (T cells) White blood cells that are produced in the bone marrow but mature in the thymus. They are important in the body's defense against certain bacteria and fungi, help B lymphocytes make antibodies and help in the recognition and rejection of foreign tissues. T lymphocytes may also be important in the body's defense against cancers.

Tailing The *in vitro* addition of the same nucleotide by the enzyme terminal transferase, to the 3′-hydroxyl ends of a duplex DNA molecule.

Tandem array The existence of two or more identical DNA sequences in series, i.e., end -to-end.

Tank bioreactor Vessel in which fermentation takes place. A tank bioreactor is a vessel in which a micro-organism is grown in a large volume of liquid. This contrasts with fibre or membrane bioreactors and immobilized cell reactors. The large majority of bioreactors used in biotechnology are tank bioreactors, and most tank bioreactors are stirred-tank bioreactors, because stirring helps to distribute effectively gas and nutrients to the growing organism.

Tap root Root system in which the primary root has a much larger diameter than the lateral roots.

Taq polymerase A heat-stable DNA polymerase isolated from the thermophilic bacterium *Thermus aquaticus,* and used in PCR.

Target For diagnostic tests, the molecule or nucleic acid sequence that is being sought in a sample.

Target site duplication A sequence of DNA that is duplicated when a transposable element inserts; usually found at each end the insertion.

Targeted drug delivery A method of delivering a drug to the site in the body where it is needed, rather than allowing it to diffuse into many sites.

Targeting vector A cloning vector carrying a DNA sequence capable of participating in a crossing-over event at a specified chromosomal location in the host cell.

TATA box Consensus sequence in the promoter region of many eucaryotic genes that binds a general transcription factor and hence specifies the position where transcription is initiated.

Tautomeric shift The transfer of a hydrogen atom from one position in an organic molecule to another position.

Tautomerism A type of isomerism in which the two isomes are in equilibrium.

Taxonomy The theories and techniques of describing, naming, and classifying organisms.

Tay sachs disease A lethal hereditary disease. The progressive accumulation of a substance called ganglioside in the brain causes paralysis, mental deterioration and blindness. Death usually occurs before the age of four.

Technology transfer The process of transferring discoveries made by basic research institutions, such as universities and government laboratories, to the commercial sector for development into useful products and services.

Telemeter The unique structure found at the end of eukaryotic chromosomes containing specialized sequences of DNA that assures the completion of a cycle of DNA replication.

Telomerase An enzyme that adds telomeric sequences to the ends of eukaryotic chromosomes.

Telomere The structure found at the end of eukaryotic chromosomes containing specialized repetitive (and widely conserved across species) DNA sequences, which are necessary to assure the completion of a cycle of DNA replication.

Telophase The last stage in each mitotic or meiotic division, in which the chromosomes are assembled at the poles of the division spindle.

Telos That which makes a thing (organism) what it is.

Temperate phage A phage (virus) that invades but may not destroy (lyse) the host (bacterial cell). However, it may subsequently enter the lytic cycle.

Temperature A system in which internal vibrational modes have equilibrated with one another can be said to have a particular temperature. Two systems A and B are said to be at different temperatures if, when brought into contact, heat flows from (say) A to B, increasing the thermal energy of B at the expense of the thermal energy of A.

Temperature sensitive (TS) mutant Organism or cell carrying a genetically altered protein (or RNA molecule) that performs normally at one temperature but is abnormal at another (usually higher) temperature.

Temperature-sensitive protein A protein that is functional at one temperature but loses function at another (usually higher) temperature.

Template An RNA or single-stranded DNA molecule upon which a complementary nucleotide strand is synthesized. A pattern or mould. DNA stores coded information and acts as a model or template from which information is copied into complementary strands of DNA or transcribed into mRNA.

Template strand The polynucleotide strand that a polymerase uses for determining the sequence of nucleotides during the synthesis of a new nucleic acid strand.

Term finalization Repelling movement of the centromeres of bivalents in the diplotene stage of the meiotic prophase, that tends to move the visible chiasmata toward the ends of the bivalents.

Terminal bud A branch tip, an undeveloped shoot containing rudimentary floral buds or leaves, enclosed within protective bud scales.

Terminal transferase An enzyme that adds nucleotides to the 3′ terminus of DNA molecules.

Termination codon Any of three mRNA sequences (UGA, UAG, UAA) that do not code for an amino acid and thus signal the end of protein synthesis. Also known as stop codon.

Termination signal In transcription, a nucleotide sequence that specifies RNA chain termination.

Terminator (of transcription) 1. A DNA sequence just downstream of the coding segment of a gene, which is recognized by RNA polymerase as a signal to stop synthesizing mRNA. In prokaryotes, terminators usually have an inverted repeat followed by a short stretch of Us at the very end of the transcribed portion. There may also be sequences beyond the transcribed part of the gene which influence the termination of transcription.

2. A name given to antisense DNA inserted in plants to make impossible the use of a second generation of seed by a farmer.

Terminator gene A patented gene technology that modifies plants so that they become sterile, such that the seed produced by the parent crop does not grow. This means the seed cannot be saved and grown by the farmer next year, but neither can the crop establish by itself in the environment.

Terminator region A DNA sequence that signals the end of transcription.

Terminator technology A biotechnological 'technology protection system' that renders sterile the seeds which are saved the first generation after the first sowing.

Tertiary structure The total three-dimensional shape of a protein that is essential to protein function.

Testcross Backcross to the recessive parental type, or a cross between genetically unknown individuals and a fully recessive tester to determine whether an individual in question is heterozygous or homozygous for a certain allele. It is also used as a test for linkage, i.e., to estimate recombination fraction.

Test tube Tube in which cells, tissues, etc., can be cultured.

Tetracycline An antibiotic that interferes with protein synthesis in prokaryotes.

Tetrad 1. The four cells arising from the second meiotic division in plants (pollen tetrads) or fungi (ascospores).

2. The quadruple group of chromatids that is formed by the association of duplicated homologous chromosomes during synapsis in meiosis I.

Tetrahymena Genus of ciliated protozoa used in studies of ciliary axonemes, self-splicing RNA, and telomere reproduction.

Tetraploid An organism whose cells contain four haploid ($4x$) sets of chromosomes.

Tetrasomic (tetrasome) Pertaining to a nucleus or an organism with four members of one of its chromosomes, whereas the remainder of its chromosome complement is diploid. Chromosome formula: $2n + 2$.

Tetratype In fungi, a tetrad of spores that contains four different types; e.g., *AB, aB, Ab* and *ab*.

Thalassaemias A group of disorders in which there is reduced production of one or more of the globin chains which constitute normal haemoglobin. This imbalance of globin chains results in chronic anaemia, which is a common cause of premature death and much suffering. Inheritance of each of the many genes which may be responsible is recessive.

Thallus Plant body without true roots, stems, or leaves.

Theory A scientific theory is an established and experimentally verified fact or collection of facts about the world. Unlike the everyday use of the word theory, it is not an unproved idea, or just some theoretical speculation. The latter meaning of a 'theory' in science is called a hypothesis.

Therapeutics Compounds that are used to treat specific diseases or medical conditions.

Therapeutic agent A compound that is used for the treatment of a disease or for improving the well-being of an organism.

Therapeutic cloning Generally referred to as somatic cell nuclear transfer technology. It involves replacing the nucleus of an egg cell with the nucleus from a cell from a patient's body and allowing it to develop to form a blastocyst. The embryonic stem cells from the inner cell mass are then harvested and used to establish a cell line that has the genetic makeup of the patient. These cells can then be directed to develop into the tissue needed for transplant.

Thermal energy The internal energy present in a system as a result of the energy of thermally equilibrated vibrational modes and other motions (including both kinetic energy and molecular potential energy). The mean thermal energy of a classical harmonic oscillator is kT.

Thermal expansion coefficient The rate of change of length with respect to temperature for a particular material.

Thermal fluctuations The thermal energy of a system (or of a particular part or mode of motion in a system) has a mean value determined by the temperature and by the structure of the system. Statistical deviations about that mean are termed thermal fluctuations; these are of great importance in determining both rates of chemical reactions and error rates in nanomechanical systems.

Thermic shock Exposure to reduced or increased temperature for several days.

Thermodynamics The physical theory of heat and energy distribution in the universe. The two important laws of thermodynamics state that the total energy of the universe is constant and energy can neither be made nor destroyed, and that the distribution of energy in the universe over time proceeds from a state of order to a state of disorder (entropy increase).

Thermoelastic Both stress and temperature changes alter the dimensions of an object having a finite stiffness and a nonzero thermal expansion coefficient. Applying a stress then produces a temperature change; this can result in a heat flow which then changes the stress: these are thermoelastic effects, and result in losses of free energy.

Thermolabile Not heatproof, e.g., a substance which disintegrates or is unstable upon heating.

Thermophile An organism which grows at a higher temperature than most other organisms. In general, a wide range of bacteria, fungi and simple plants and animals can grow at temperature up to 50°C; thermophiles are considered to be organisms which can grow at above 50°C. They can be classified according to their optimal growth temperature, into simple thermophiles (50-65°C), thermophiles (65-85°C), and extreme thermophiles (>85°C). Thermophiles and extreme thermophiles are usually found growing in very hot places, such as hot springs and geysers, smoker vents on the sea floor, and domestic hot water pipes.

Thermosensitivity Loss of activity of a protein at high temperature.

Thermostability Retention of activity at high temperature.

Thermotherapy Technique mainly used for virus or mycoplasma elimination. Plants are exposed to elevated temperatures as a treatment. Thermotherapy is used alone or in combination with meristem culture or meristem tip culture.

Thinning 1. Removal of older stems to promote new growth.

2. Removal of excess fruits to improve the size and quality of the remaining fruits.

3. Removal of seedlings spaced too closely for optimum growth.

Thioester bond High-energy bond formed by a condensation reaction between an acid (acyl) group and a thiol group (- SH); seen, for example, in acetyl CoA and in many enzyme-substrate complexes.

Thiol An SH group, or a molecule containing one. Also known as a sulfhydryl or mercapto group.

Thrombosis Blockage of an artery resulting from aggregation of platelets and fibrin, a fibrous protein.

Thylakoid Flattened sac of membrane in a chloroplast that contains pigment and carries out the light-gathering reactions of photosynthesis. Stacks of thylakoids form the grana of chloroplasts.

Thymidine The deoxyribonucleoside resulting from the combination of the base thymine (T) and the sugar 2-deoxy-D-ribose.

Thymidine kinase (TK) An enzyme that allows a cell to utilize an alternate metabolic pathway for incorporating thymidine into DNA. Used as a selectable marker to identify transfected eukaryotic cells.

Thymine (T) A chemical that is a basic structural unit of a living thing's DNA. It is one of 4 key bases or nucleotides (cytosine (C), guanine (G) and adenine (A)) that repeat throughout the length of a DNA chain. The order of their sequence defines the arrangement of genes.

Thymus (gland) An organ at the base of the neck which plays an important role in immune system responses. It produces T-cells which are crucial in identifying self and non-self antigens.

TI (tumor-inducing) plasmid A giant plasmid of Agrobac- terium tumefaciens that is responsible for tumor formation in infected plants. Ti plasmids are used as vectors to introduce foreign DNA into plant cells.

Ti plasmid Tumour-inducing plasmid. A giant plasmid of *Agrobacterium tumefaciens* that is responsible for the induction of tumours in infected plants. Ti plasmids are used as vectors to introduce foreign DNA into plant cells.

Tight junction Cell-cell junction that seals adjacent epithelial cells together, preventing the passage of most dissolved molecules from one side of the epithelial sheet to the other.

Tight-receptor structures A receptor structure in which a bound ligand of a particular kind is confined on all sides by repulsive interactions (note that favorable binding energies are compatible with repulsive forces). A tight-receptor structure discriminates strongly against all molecules larger than the target.

Tillage Cultivation, including hoeing and plowing. In agriculture, tillage is used to fight weeds.

Tissue A collection of cells specialised to perform a particular function. Tissues are usually composed of several cell types. Organs, eg. heart, are made up of such tissues.

Tissue culture A general term used to describe the culture of cells, tissues or organs in a nutrient medium under sterile conditions.

Tocopherols, tocotrienols Forms of vitamin E.

Tolerance A form of genetic resistance in which an organism attacked or affected by a disease pathogen (or pest) exhibits less reduction in yield or performance in comparison with members of other affected cultivars or breeds.

Tonoplast The cytoplasmic membrane bordering the vacuole, with a role in regulating the pressure exerted by the cell sap.

Topoisomerase (DNA topoisomerase) Enzyme that makes reversible cuts in a double helical DNA molecule for the purpose of removing knots or unwinding excessive twists.

Torr An obsolete unit of pressure equal to that exerted by a column of mercury 1mm high at 0°C and standard gravity (1mm Hg); named after Evangelista Torricelli (1608-1647), the inventor of the mercury barometer. 1 Torr = 1/760 atm = 133.322 Pa.

Total quality management An organisation's management approach centred on quality, based on the participation of all its members and aimed at long-term success through customer satisfaction and benefits to the members of the organisation and to society.

Totipotency Having the potentiality of forming all the types of cells in the body. The property of somatic cells to be induced to undergo regeneration. The diploid zygote formed at fertilization is a single cell which is capable of division and differentiation to give rise to the total range of cell types found in the adult organism.

Totipotent cell; totipotent nucleus An undifferentiated cell (or nucleus), such as a blastomere, that, when isolated or suitably transplanted, can develop into a complete embryo.

Toxic Poisonous

Toxicity Negative effect of a compound, as shown by altered morphology or physiology. It is meaningful only when the effect itself is also described, such as changes in the rate of cell growth, cell death, etc.

Toxicological effects The actions of foreign substances in the body, and the quality or degree of these substances being poisonous.

Toxin A compound produced by one organism, which is deleterious to the growth and/or survival of another organism of the same or different species.

TPA (tissue-type plasminogen activator) A protein in the blood which has an enzyme like function in helping the enzyme plasminogen to dissolve blood clots.

Tracer An added or injected substance that can be followed within a reaction or an organism, such as radioactive isotopes and certain dyes.

Tracheid An elongated, tapering xylem cell, with lignified pitted walls, and adapted for conduction and support. Found in conifers, ferns and related plants.

Tracker genes Same as marker genes.

Traditional breeding Modification of plants and animals through selective breeding. Practices used in traditional plant breeding may include aspects of biotechnology such as tissue culture and mutational breeding.

Trait The physical or metabolic phenotype of an organism such as red flower color and length of stem in plants and black fur or pink eye in mice.

Trans-acting A term describing substances that are diffusable and that can affect spatially separated entities within cells.

Transcapsidation The partial of full coating of the nucleic acid of one virus with a coat protein of a differing virus.

Transcription The process of making a complementary sequence of the gene sequence in the genome, which is either used directly, as in case of Ribosomal RNAs (rRNAs) and Transfer RNAs (tRNAs), or is further processed into the messenger RNA, which is translated into protein. The process is catalyzed by the enzyme known as DNA-dependent RNA polymerase.

Transcription factors Proteins in eukaryotes that regulate the transcription of other genes by binding to regulatory sequences of the gene, interaction with one another and with the RNA polymerase.

Transcription unit A segment of DNA that contains signals for the initiation and termination of transcription, and is transcribed into one RNA molecule.

Transcription vector A cloning vector that allows the foreign gene or DNA sequence to be transcribed *in vitro*.

Transcriptional anti-terminator A protein that prevents RNA polymerase from terminating transcription at specific transcription termination sequences.

Transdifferentiation The process whereby a specialized cell de-differentiates and re-differentiates into a different cell type; or the process whereby an adult stem cell from a specific tissue type becomes a cell type from a very different tissue (for example a nerve stem cell differentiates into a kidney cell).

Trans fatty acid A form of unsaturated fatty acid that is straight (rather than bent, i.e cis) at the double bond; not abundant in natural edible oils but occurring in ruminant fats and formed during some manufacturing processes.

Trans heterozygote A double heterozygote that contains two mutations arranged in the *trans* configuration.

Transduction 1. Genetic: the transfer by means of a viral vector of a DNA sequence from one cell to another.

2. Signal: any process that helps to produce biological responses to events in the environment (e.g. transduction of hormone binding into cellular events by hormone receptors).

Transfection Infection of a cell with nucleic acid from a virus, resulting in replication of the complete virus.

Transfer The process of moving cultured tissue or cells to a fresh medium.

Transfer RNA RNA molecules which transfer specific amino acids to the messenger RNA so that the polypeptide it encodes can be synthesized.

Transformation 1. The uptake and establishment of DNA in a bacterium or yeast cell, in which the introduced DNA often changes the phenotype of the recipient organism.

2. Conversion by various means of animal cells in tissue culture from controlled to uncontrolled cell growth. Typically through infection by a tumour virus or transfection with an oncogene.

Transformation efficiency The number of bacterial cells that uptake and express plasmid DNA divided by the mass of plasmid used (in transformants/microgram).

Transformation frequency The fraction of a cell population that takes up foreign DNA; expressed as the number of transformed cells divided by the total number of cells in a population.

Transforming oncogene A gene that upon transfection converts a previously immortalized cell to the malignant phenotype.

Transgenic Refers to an organism with one or more genes that have been transferred to it from another organism using recombinant DNA techniques. An organism whose gamete cells contain genetic material from another organism, or contain genetic material that has been altered in some other way for example using gene silencing techniques.

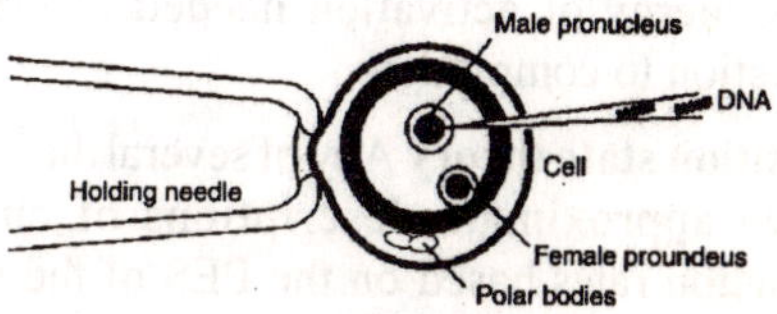

Fig. Transgenic

Transgenic animal Genetically enginnered animal or offspring of genetically engineered animals. The transgenic animal usually contains material from at lease one unrelated organism, such as from a virus, plant, or other animal.

Transgenic organism An organism formed by the insertion of foreign genetic material into the germ line cells of organisms. Recombinant DNA techniques are commonly used to produce transgenic organisms.

Transgenic plant Genetically engineered plant or offspring of genetically engineered plants. The transgenic plant usually contains material from at least one unrelated organisms, such as from a virus, animal, or other plant.

Transgressive variation The appearance in the F2 (or later) generation of individuals showing more extreme development of a trait than either of the original parents.

Transient Of short duration.

Transition The substitution in DNA or RNA of one purine by another purine, or of one pyrimidine by another pyrimidine.

Transition state At the saddle point of a col linking two potential wells, the direction of maximum negative curvature defines the reaction coordinate; the transition state is a hypothetical system of reduced dimensionality, free to move only on a hypersurface perpendicular to the reaction coordinate at its point of maximum energy.

Transition state intermediate In a chemical reaction, an unstable and high-energy configuration assumed by reactants on the way to making products. Enzymes are thought to bind and stabilize the transition state, thus lowering the energy of activation needed to drive the reaction to completion.

Transition state theory Any of several theories that give approximate descriptions of chemical reaction rates based on the PES of the system, and in particular, on the properties of two potential wells and a transition state between them.

Translation The process of converting the genetic infor- mation of an mRNA on ribosomes into a polypeptide. Transfer RNA molecules carry the appropriate amino acids to the ribosome, where they are joined by peptide bonds.

Translocation 1. The movement of nutrients or products of metabolism from one location to another.

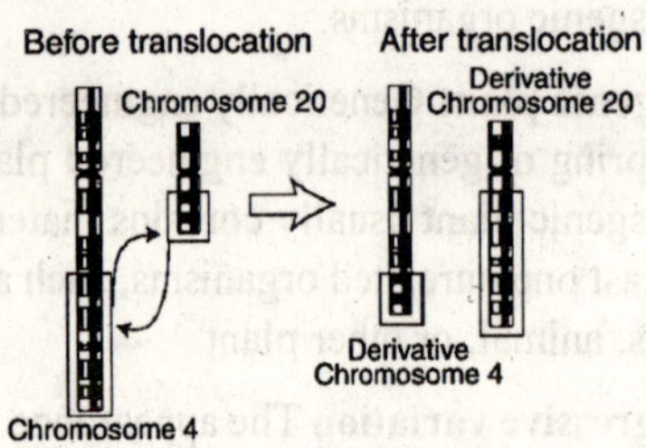

Fig. Translocation

2. Change in position of a segment of a chromosome to another, non-homologous chromosome.

Transmission electron microscope (TEM) A microscope which uses an electron beam to obtain images of objects, with a much greater resolving power than a light microscope.

Transplant 1. A plant grown in a cold frame, greenhouse, tissue culture or indoors for later planting outdoors.

2. To dig up and move a plant to another location.

Transplant rejection The rejection of the donor organ, tissue or cell by the transplant recipient's immune system.

Transporter A membrane protein that functions as an ion channel, solute transporter, facilitator or pump to move molecules across cell membranes.

Transposable genetic element A DNA element that can move from one location in the genome to another.

Transposase An enzyme encoded by a transposon gene and that catalyses the movement of a DNA sequence to a different site in a DNA molecule, by catalysing the excision of the transposon from one site and its insertion into a new chromosomal site.

Transposition The process whereby a transposon or insertion sequence inserts itself into a new site on the same or another DNA molecule. The exact mechanism is not fully understood and different transposons may transpose by different mechanisms. Transposition in bacteria does not require extensive DNA homology between the transposon and the target DNA. The phenomenon is therefore described as illegitimate recombination.

Transposon A transposable or movable genetic element. A relatively small DNA segment that has the ability to move (mobile genetic element) from one chromosomal position to another, e.g., Tn 5 is a bacterial transposon that carries the genes for resistance to the antibiotics neomycin and kanamycin and the genetic information for insertion and excision of the transposon.

Transposon tagging The insertion of a transposable element into or nearby a gene, thereby marking that gene with a known DNA sequence.

Transversion The substitution of a purine for a pyrimidine, or a pyrimidine for a purine.

Tri hybrid The offspring from homozygous parents differing in three pairs of genes.

Tri nucleotide repeats Tandem repeats of three nucleotides that are present in many genes. In several cases, these trinucleotide repeats have undergone expansions in copy number, and that has resulted in inherited diseases.

Triacylglycerols Compounds of glycerol and three fatty acids also called triglycerides; the main constituents of dietary and body fat.

Tribrid protein A fusion protein that has three segments, each encoded by parts of different genes.

Trichome A short filament of cells.

Triglycerides The constituent of fat composed of glycerol and three fatty acids. Most triglycerides are obtained from food as fats or synthesized by the liver and stored in liver and fat cells (adipocytes). Triglycerides are one of three major sources of metabolic fuel (the other sources are carbohydrates and proteins) providing energy for the physiological processes of the body.

Tripartite mating A process in which conjugation is used to transfer a plasmid vector to a target cell when the plasmid vector is not self-mobilizable. When 1. cells that have a plasmid with conjugative and mobilizing functions are mixed with

2. cells that carry the plasmid vector and

3. target cells, mobilizing plasmids enter the cells with the plasmid vector and mobilize the plasmid vector to enter into the target cells. Following tripartite mating, the target cells with the plasmid vector are separated from the other cell types in the mixture by various selection procedures.

Triple bond A double bond is formed when a pi bond is superimposed on a single bond; adding a second pi bond results in a triple bond. The two pi bonds have perpendicular nodal planes, and their sum has roughly cylindrical symmetry, permitting rotation in much the same manner as a single bond.

Triplet An electronic state of a molecule in which two spins are aligned. This term is derived from spectroscopy: a system of two aligned spins has three possible orientations with respect to a magnetic field; each has a different energy, resulting in sets of three field-dependent spectral lines.

Triploidy Three copies of genetic information in each cell rather than the normal two copies found in most plants and animals.

Trisomy Having three copies of a particular chromosome in each somatic (body) cell instead of the normal two copies. This leads to certain conditions for example Down syndrome (trisomy 21) or Edwards syndrome (trisomy 18)

Triticale The species formed by crossing wheat and rye.

TRNA; transfer RNA The class of small RNA molecules that transfer amino acids to the ribosome during protein synthesis. Transfer RNA molecules are folded into a 'clover-leaf' secondary structure by intrastrand base pairing. The anticodon loop contains a nucleotide-triplet complementary to a specific codon within the mRNA molecule. Each tRNA is 'charged' with the correct amino acid molecule, via its 3′ adenosine moiety, by an enzyme called aminoacyl-tRNA synthetase.

Tropism An involuntary plant response to a stimulus, in which a bending, turning or growth occurs, such as phototropism, geotropism or hydrotropism. The response may be positive (towards) or negative (away from) to the stimulus.

True to type Applied to a plant or propagation source, this term denotes correct cultivar identification and lack of variation in productivity or performance. Verification is determined visually by an expert or through biochemical, serological or other means.

Trypsin A proteolytic enzyme that hydrolyzes peptide bonds on the carboxyl side of the amino acids arginine and lysine.

Trypsin inhibitors Substances inactivating the enzyme trypsin, which is needed for digestion of peptides.

Tuber Food-storing modified roots in plants like potato.

Tumble tube A glass tube mainly used *in vitro* to agitate and consequently aerate suspension cultures. The tube, which is commonly attached to a slowly revolving platform, is closed at both ends, with a side-neck opening.

Tumor necrosis factors (TNFs) Rare proteins of the immune system that appear to destroy some types of tumor cells without affecting healthy cells.

Tumor The growth of abnormal tissue with no useful function to the organism. Tumors are known either as malignant or benign. Malignant tumors are known to grow continuously and have the ability to invade healthy tissues, eventually leading to death. Benign tumors tend to stop growing once they reach a certain point. And normally don't result in death.

Tumor virus A virus capable of transforming a cell to a malignant phenotype.

Tunica An outer one- to four-cell layered region of the apical meristem, where cell division is anticlinal, i.e., perpendicular to the surface.

Tunneling A classical particle or system could not penetrate regions in which its energy would be negative, that is, barrier regions in which the potential energy is greater than the system energy. In the real world, however, a wave function of significant amplitude may extend into and beyond such a region. If the wave function extends into another region of positive energy, the barrier is crossed with some probability; this process is termed tunneling (since the barrier is penetrated rather than climbed).

Turbidostat An open continuous culture in which a pre-selected biomass density is uniformly maintained by automatic removal of excess cells. The fresh medium flows in response to an increase in the turbidity (usually cell density) of the culture.

Turgid Swollen, distended; referring to a cell that is firm due to water uptake.

Turgor pressure The pressure within the cell resulting from the absorption of water into the vacuole and the imbibition of water by the protoplasm.

Turion An underground bud or shoot from which an aerial stem arises.

Twins Two individuals originating from the same zygote.

Twin studies Refers to the large-scale study of both fraternal and identical twins, raised together or apart, to determine the environmental and genetic factors involved in a specific development.

Type 1 diabetes This form of diabetes used to be known as juvenile or insulin-dependent diabetes. It is the form of diabetes disease that usually strikes young people, where the body's immune system destroys the insulin-producing cells of the pancreas.

U

U The uracil residue in RNA.

Ubiquinol An antioxidant produced by normal metabolism. Ubiquinol is the only known fat-soluble antioxidant synthesised by animal cells. It is believed to play an important role in cellular defence against oxidative damage. Ubiquinol can also contribute to the antioxidant defence system by regenerating the active form of vitamin E after that vitamin has reacted with a free radical. (The water-soluble vitamin C can also perform this function.).

Ubiquitin Small, highly conserved protein present in all eucaryotic cells that becomes covalently attached to lysines of other proteins. Attachment of a chain of ubiquitins tags a protein for intracellular proteolytic destruction in a proteasome.

Ultraviolet light; ultraviolet radiation (UV) The portion of the electromagnetic spectrum with wavelengths from about 100 to 400 nm; between ionizing radiation (X-rays) and visible light. UV is absorbed by DNA and is highly mutagenic to unicellular organisms and to the epidermal cells of multicellular organisms. UV light is used in tissue culture for its mutagenic and bactericidal properties.

Umbilicus Organ connecting foetus to placenta.

Uncertainty Principle (heisenberg's) A necessary consequence of the quantum mechanical description of matter. Unlike in classical mechanics where the position of a large object in time and space (including its speed) can exactly be described and predicted based on measurements (Newton's laws of gravity, actio and reactio), the position and speed of individual subatomic particles like electrons, and thus the behavior of individual molecular interaction, can neither be measured nor predicted with certainty, yet perfectly described by statistical methods when studied in large numbers.

Unconfined release The use of a novel organism that is not subject to reproductive and physical isolation from the natural or agricultural environment, site inspections, land use restrictions and/or restricted use of seed or progeny.

Undefined A medium or substance added to medium in which not all of the constituents or their concentrations are chemically defined, such as media containing coconut milk, malt extract, casein hydrolysate, fish emulsion or other complex compounds.

Understock Host plant for a grafted scion, a branch or shoot from another plant; an understock may be a fully grown tree or a stump with a living root system.

Undifferentiated In a meristematic state or resembling a meristem; lacking the specialized or differential gene expression characteristic of specialized cells.

Unequal crossing over Crossing over between repeated DNA sequences that have paired out of register, creating duplicated and deficient products.

Unicellular Describing tissues, organs or organisms consisting of a single cell.

Unimolecular Occurring to or within a single molecule; like intramolecular, but can refer to fragmentation reactions.

Unisexual Describing animals and plants possessing either male or female reproductive organs, but not both.

Univalent An unpaired chromosome at the first division of meiosis.

Univariate analysis Techniques for studying the effects of a single factor on an outcome variable.

Universal donor cell Cells that, after introduction into a recipient, will not induce an immune response that leads to their rejection.

Universality Referring to the genetic code, the triplet codons are translated to the same amino acid, with minor exceptions, in virtually all species.

Unorganized growth *In vitro* formation of tissues with few differentiated cell types and no recognizable structure; many call are unorganized.

Unsaturated Describes a molecule that contains one or more double or triple carbon-carbon bonds, such as isoprene or benzene.

Unsaturated fatty acid Fatty acid whose hydrocarbon chain contains at least one double bond. Monounsaturated fatty acids contain one double bond; polyunsaturated fatty acids contain tow or more.

Unspecialised Having no specific function.

Upstream 1. In molecular biology, the stretch of DNA base pairs that lie in the 5′ direction from the site of initiation of transcription. Usually the first transcribed base is designated +1 and the upstream nucleotides are marked with minus signs, e.g., -1, -10. Also, to the 5′ side of a particular gene or sequence of nucleotides.

2. In chemical engineering, those phases of a manufacturing process that precede the biotransformation step. Refers to the preparation of raw materials for a fermentation process. Also called upstream processing.

Uracil A pyrimidine base found in RNA but not in DNA. In DNA, uracil is replaced by thymine.

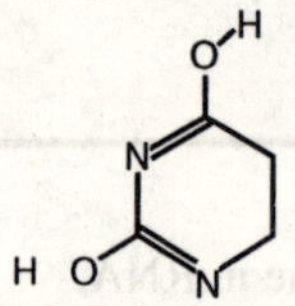

Fig. Uracil

Uridine The (ribo)nucleoside resulting from the combination of the base uracil (U) and the sugar D-ribose.

Urticaria A raised, itching rash, often with well-circumscribed areas (wheals).

Utilization of farm animal genetic resources The use and development of animal genetic resources for the production of food and agriculture. The use in production systems of AnGRs that already possess high levels of adaptive fitness to the environments concerned, and the deployment of sound genetic principles, will facilitate sustainable development of the AnGRs and the sustainable intensification of the production systems themselves. The wise use of AnGRs is possible without depleting domestic animal diversity. Development of AnGRs includes a broad mix of ongoing activities that must be well planned and executed for success, and compounded over time, hence with high value. It requires careful definition of breeding objectives, and the planning, establishment and maintenance of effective and efficient animal recording and breeding strategies.

UV radiation Ultraviolet radiation, an invisible, high energy component of sunlight can cause skin damage including cancer.

V

V regions Variable regions in antibodies.

V/V On a volume per volume basis; the percent of the volume of a constituent in 100 units of volume, e.g., (ml/100 ml) X 100.

Vaccine A preparation that contains an antigen, consisting of whole disease-causing organisms (killed or weakened) or parts of such organisms, that is used to confer immunity against the disease that the organisms cause. Vaccine preparations can be natural, synthetic or derived by recombinant DNA technology.

Vaccinia The cowpox virus used to vaccinate against smallpox and, experimentally, as a carrier of genes for antigenic determinants cloned from other disease organisms.

Vacuole A cavity in a plant cell, bounded by a membrane; in which various plant products and by-products are stored.

Vacuum Vacuum is created through use of a vacuum pump or aspirator pump, to facilitate specific biological preparations, such as inclusions or disinfection of material for *in vitro* culture, etc.

Valence In covalent compounds, the valence of an atom is the number of bonds it forms to other atoms.

Valence electrons Electrons that can participate in bonds and in chemical reactions; lone-pair electrons are valence electrons, although not participating in a bond.

Value-added traits Modified crops produced with traits such as improved taste, nutritional value, or utility - value for the consumer.

Van der waals force Any of several intermolecular attractive forces not resulting from ionic charges; in this volume, only London dispersion forces are described by this name. Descriptions of the potential energy of nonbonded interactions follow the convention of expressing van der Waals attractive forces and overlap repulsion forces as a single *van der Waals potential.*

Variable domains Regions of antibody chains that have different amino acid sequences in different antibody molecules. These regions are responsible for the antigen-binding specificity of the antibody molecule.

Variable expressivity Variation in the phenotype caused by different alleles of the same gene and/ or by the action of other genes and/or by the action of non-genetic factors.

Variable surface glycoprotein (VSG) One of a battery of antigenic determinants expressed by a microorganism to elude immune detection.

Variance In statistics: The sum of the squared deviations, divided by one less than the number of observations. A statistical measure of variation in a population.

Variation Differences in the frequency of genes and traits among individual organisms within a population.

Variegated Plants having both green and albino tissues. This difference in colour may result from viral infection, nutritional deficiency, or may be under genetic or physiological control.

Variety A naturally occurring subdivision of a species, with distinct morphological characters and given a Latin name according to the rules of the International Code of Nomenclature. A taxonomic variety is known by the first validly published name applied to it so that nomenclature tends to be stable.

Vascular Referring to any plant tissue or region consisting of or giving rise to conducting tissue, e.g., bundle, cambium, ray.

Vascular bundle; fascicle A strand of tissue containing primary xylem and primary phloem (and procambium if present) and frequently enclosed by a bundle sheath of parenchyma or fibres.

Vascular cambium In biennials and perennials, cambium giving rise to secondary phloem and secondary xylem.

Vascular plants Plants possessing organized vascular tissues.

Vascular system 1. A specialized network of vessels for the circulation of fluids throughout the body tissue of an animal.

2. The system of vascular tissue in plants.

Vascular tissue The tissue that conducts water and nutrients throughout the plant body in higher plants.

Vector 1. An organism, usually an insect, that carries and transmits pathogens.

2. A small DNA molecule (plasmid, virus, bacteriophage, artificial or cut DNA molecule) that can be used to deliver DNA into a cell. Vectors must be capable of being replicated and contain cloning sites for the introduction of foreign DNA.

Vegan Strict vegetarian who does not eat any food containing ingredients derived from animals.

Vegetal pole The end at which most of the yolk is located in an animal egg. The end opposite the animal pole.

Vegetative propagation Same as asexual or non-sexual propagation.

Vehicle The host organism used for the replication or expression of a cloned gene or other sequence The term is little used and is often confused with vector.

Velocity density gradient centrifugation A procedure used to separate macromolecules based on their rate of movement through a density gradient.

Velogenetics The combined use of marker-assisted selection and embryo technologies such as OPU, IVM and IVF, in order to increase the rate of genetic improvement in animal populations.

Ventral Situated toward the belly surface of an animal.

Vermiculite Coarse aggregate material made from expanded mica having a high cation exchange capacity and high water-holding capacity and used as a rooting medium and a soil additive.

Vernalization The process by which floral induction in some plants is promoted by exposing the plants to chilling for a certain duration.

Vesicle Small, membrane-bounded, spherical organelle in the cytoplasm of a eucaryotic cell.

Vessel 1. A series of xylem elements whose function is to conduct water and nutrients in plants.

2. A container, such as a Petri dish or test tube, used for tissue culture.

Vessel element A type of cell occurring within the xylem of flowering plants. Many are water-conducting vessels.

Viability The capability to live and develop normally.

Viability test Assay of the number or percent of living cells or plants in a population that have been given a specific treatment; such as ascertaining cell viability after cryopreservation treatment, or for seeds.

Viable Capable of germinating, living, growing and developing.

Vibrio Comma-shaped bacterium.

Vir genes A set of genes on a Ti plasmid that prepare the T-DNA segment for transfer into a plant cell.

Viral coat protein A protein present in the layer surrounding the nucleic acid core of a virus.

Viral oncogene A gene in the viral genome that contributes to malignancies in vertebrate hosts.

Viral pathogen A disease-causing virus.

Viral vaccines Vaccines consisting of live viruses rather than dead ones or separated parts of viruses. However, as the virus itself cannot be used, because that would simply give the patient the disease, the virus is genetically engineered so that it elicits the immune response to the viral pathogen without causing the disease itself.

Two genetic engineering methods can be used. The first is to make the disease virus harmless, but still able to replicate in cultured animal cells. This is similar to producing an 'attenuated' virus, i.e., one which has been grown in the laboratory until it loses its ability to cause disease. However, the genetic engineering route seeks to make sure that the attenuated virus has no chance of mutating back to a wild type, pathogenic virus, by deleting whole genes or replacing key regions of genes with completely different genetic material.

The second approach is to clone the gene for a protein from the pathogenic virus into another, harmless virus, so that the result 'looks' like the pathogenic virus but does not cause disease.

Virion An infectious virus particle. A plant pathogen that consists of a naked RNA molecule of approximately 250-350 nucleotides, whose extensive base pairing results in a nearly correct double helix.

Viroid A plant pathogen that consists of a naked RNA molecule of approximately 250-350 nucleotides, whose extensive base pairing results in a nearly correct double helix.

Virology Study of viruses.

Virtual reality system A combination of computer and interface devices (gloves, goggles, etc.) that present a user with the illusion of being in a three dimensional world of computer generated objects.

Virulence The degree of ability of an organism to cause disease. The relative infectiousness of a bacterium or virus, or its ability to overcome the resistance of the host metabolism.

Virulent phage A phage (virus) that destroys the host (bacterium).

Virus Smallest of all organisms and often not considered alive because they strictly depend on a cellular host organism (bacteria, plant, animal) to reproduce. Viruses have no metabolism of their own and depend on passive carriers to transport them around. Viruses are infectious particles with a DNA or RNA based small genome that can control the cellular mechanism of infected cells prompting the host cell to synthesize new viruses. Infections often cause mild to severe symptoms, yet some viruses do not cause any harm or not in all host organisms (called carriers). Because they have no metabolic activity they are not susceptible to antibiotics and most drugs, unless a drug can interfere during the infection and viral replication stages. Outside cells, viruses are passive and are easily destroyed by chemical intervention.

Virus free Plant, animal, cell, tissue or meristem which exhibits no viral symptoms or contains no identifiable virus-particles.

Virus resistant plants Plants that have been genetically modified so that they are resistant to a particular disease-causing virus.

Virus vector A virus used as a carrier for a gene when making a GMO.

Virus-tested Description of a organism or a cell stock certified as being free of certain specified viruses following recognized procedures of virus diagnosis.

Visible light The part of the electromagnetic spectrum with wavelengths between 380 nm and 750 nm and perceived by the human eye.

Vitamins Organic compounds that the body needs in very small amounts for normal growth and maintenance of good health to take place. They play important metabolic roles. There are two groups of vitamins, the fat soluble and the water soluble. Vitamins A, D, E, and K are fat soluble while vitamin C (ascorbic acid) and the vitamin B complex group are water soluble.

Vitamin B complex A large group of water soluble vitamins that function as co-enzymes, including thiamine (B1); riboflavin or vitamin G (B2); niacin or nicotinic acid (B3); pantothenic acid (B5); pyridoxine (B6); cyanocobalamin (B12); biotin or vitamin H; folic acid or vitamin M (Bc); inositol; choline; and others.

Vitrified; water soaked Cultured tissue having leaves and sometimes stems with a glassy, transparent or wet and often swollen appearance. The process of vitrification is a general term for a variety of physiological disorders that lead to shoot tip and leaf necrosis.

Vivipary 1. A form of reproduction in animals in which the developing embryo obtains its nourishment directly from the mother via a placenta or by other means.

2. A form of asexual reproduction in certain plants, in which the flower develops into a budlike structure that forms a new plant when detached from the parent.

3. The development of young plants in the inflorescence of the parent plant.

Vmax The maximal rate of an enzyme-catalysed reaction. *V*max is the product of *E*o (the total amount of enzyme) times the value of *K*eat (the catalytic rate constant).

VNTR (variable number tandem repeat) A short DNA sequence that is present as tandem repeats and in highly variable copy number.

Volatilization The conversion of a solid or liquid into a gas or vapour.

Volunteer plant Crop plants that persist for a few seasons without deliberate cultivation.

W

W/V Weight per volume; the weight of a constituent in 100 cm3 of solution, expressed as a percentage.

Walking A method for cloning large regions of a chromosome. Starting from a known site, a gene library is screened for clones that hybridize to DNA probes taken from the ends of the first clone. These clones are then isolated, and their ends used to screen the library again. These clones are then isolated and their ends used, and so on.

Wall pressure Pressure that a cell wall exerts against the turgor of the cell contents. Wall pressure is equal and opposite to the turgor potential.

Wash out The loss of the slower growing micro-organism when two organisms are being grown together.

Water potential Refers to the difference between the activity of water molecules in pure distilled water at atmospheric pressure and 30°C (standard conditions), and the activity of water molecules in any other system. The activity of these water molecules may be greater (positive) or less (negative) than the activity of the water molecules under standard conditions.

Water stress The condition when plants are unable to absorb enough water to replace that lost by transpiration. The results may be wilting, cessation of growth, or even death of the plant or plant parts.

Wave function In quantum mechanics, a complex function extending over the configuration space of a system; its complex conjugate yields the probability density function, and other mathematical operations yield other physical quantities.

Wavelength The distance between two corresponding points on any two consecutive waves. For visible light it is very small and is generally measured in nanometres.

Wax (A.S. *weax*, wax) Esters of alcohol higher than glycerol, which are insoluble in water and difficult to hydrolyze; wax forms protective waterproof layers on leaves, stems, fruits, animal fur and integuments of insects.

Weed Simply any plant growing where it is not wanted. In agriculture, used for a plant which has good colonizing capability in a disturbed environment, and can usually compete with a cultivated species therein. Weeds are typically considered as unwanted, economically useless or pest species.

Weediness In agriculture, the ability of a plant to colonize a disturbed habitat and compete with cultivated species.

Western blot A technique in which protein is transferred from an electrophoretic gel to a cellulose or nylon support membrane following electrophoresis. A particular protein molecule can then be identified by probing the blot with a radiolabelled antibody which binds only the specific protein to which the antibody was prepared. Useful, for example, for measurement of levels of production of a specific protein in a particular tissue or at particular developmental stage.

Wet weight The gross weight of a product with its full water content, or weight of fully hydrated tissue.

Wetting agent A substance that improves surface contact by reducing the surface tension of a liquid: e.g., Triton X-10TM added to disinfecting solutions promotes the disinfestation process.

White blood cell (leucocyte) Nucleated blood cell lacking hemoglobin; includes lymphocytes, neutrophils, eosinophils, basophils, and monocytes.

Wide cross A sexual cross between distantly related species that normally would not breed.

Wild type (WT) The wild type (WT) is a term referring to the natural genetic form of an organism. A wild type is distinguished from a mutant form (an organism with a genetic mutation). Usually, the distinction between wild type and mutant is based on a single mutation. It should be noted, that within a population of an organism, there is no such thing as a wild type. The term, however, is useful for geneticists because it allows a simple definition of a standard or control condition.

Wilt Drooping of stems and foliage due to loss of water and decreased turgidity of cells. May be caused by water stress or by disease.

Wilting point (WP) The moisture content of soil at which plants start to wilt, but not to the extent that they fail to recover when placed in a humid atmosphere.

Wobble hypothesis An explanation of how one tRNA may recognize more than one codon. The first two bases of the mRNA codon and anticodon pair properly, but the third base in the anticodon has some flexibility that permits it to pair with either the expected base or an alternative.

Work Energy transferred by applying a force over a distance; lifting a mass does work against gravity, and stores gravitational potential energy.

X

X The basic number of chromosomes in a polyploid series, monoploid = x; diploid = 2x; triploid = 3x; etc.

X chromosome The female sex chromosome. In species that have sex chromosomes that differ to one another this chromosome usually occurs as a pair in each female cell (i.e. XX in humans), and unpaired in each male cell (i.e. XY in humans).

X linked disease A genetic disease caused by an allele at a locus on the X-chromosome. In X-linked recessive conditions, a normal female "carrier" passes on the allele on her X chromosome to an affected son.

X ray crystallography The deduction of crystal structure from analysis of the diffraction pattern of X-rays passing through a pure crystal of a substance.

Xanthophyll A yellow oxygen-containing carotenoid, present in chloroplasts.

Xenia The immediate effect of pollen on some characters of the endosperm.

Xenobiotics Molecules entering an organism that has no physiological function and is not found in an organism if not taken up by eating, breathing, or injury. The term *xenobiotics* is used to describe a foreign particle or molecule that is potentially dangerous or toxic.

Xenopus laevis (South African clawed toad) Species of frog (not a toad) frequently used in studies of early vertebrate development.

Xenotransplantation The transplantation of tissue from one species to another species, typically from non-human mammals to humans. This technology has become very important because of a worldwide shortage of human organs for humans requiring a new organ. The most popular non-human species involved in this technology is the pig.

Xerophyte A plant very resistant to drought or that lives in very dry places.

X-linked: Describes the form of inheritance of diseases for which the gene responsible is carried on the X chromosome.

X-ray crystallography An essential technique for determining the three-dimensional structure of biological molecules. This information aids in the discovery of products that will interact with the biological molecule.

Xylem A complex tissue specialized for efficient conduction of water and mineral nutrients in solution. Xylem may also function as a supporting tissue, particularly secondary xylem.

Y

Y chromosome One of two sex chromosomes in higher organisms that defines the gender of the adult. In almost all sexually reproducing organisms, the Y-chromosome defines male characteristics.

YAC (yeast artificial chromosome) Used as a vector system for cloning DNA fragments that can be hundreds of kilobases long. Linear cloning vectors constructed from essential elements of yeast chromosomes. They can accommodate foreign DNA inserts of 200 to 500 kb.

Yeast A general term for single-celled fungi that reproduce by budding. Some yeasts can ferment carbohydrates (starches and sugars) and thus are important in brewing and baking.

Yeast cloning vectors The yeasts, and especially *Saccaromyces cerevisiae*, are favourite organisms in which to clone and express DNA. They are eukaryotes, and so can splice out introns, the non-coding sequences in the middle of many eukaryotic genes.

Yeast episomal vector (YEp) A cloning vector for the yeast *Saccaromyces cerevisiae* that uses 2mm plasmid as origin of replication and is maintained as an extrachromosomal nuclear DNA molecule.

Yeast extract A mixture of substances from yeast.

Young's modulus A modulus relating tensile (or compressive) stress to strain in a rod that is free to contract or expand transversely in accord with its Poisson's ratio. The relevant measure of strain is the elongation divided by the initial length.

Young's syndrome A genetic disorder that affects the hairs, called cilia, that are found in the tubes and ducts of the body. This syndrome results in an inability to produce sperm and chronic respiratory problems in men.

Z

Z disc (z line) Platelike region of a muscle sarcomere to which the plus ends of actin filaments are attached.

Z DNA; zig zag DNA A form of DNA duplex in which the double helix is wound in a left-hand, instead of a right-hand, manner. DNA adopts the Z configuration when purines and pyrimidines alternate on a single strand, e.g.,

5′CGCGCGCG3′ or 5′CACACACACA3′

3′GCGCGCGC5′ 3′GTGTGTGTGT5′.

Zebra fish The Zebra fish is one of many model organisms used in biomedical research to understand development of higher organisms, the functioning of nervous systems, and fundamental aspects of physiology and the cause of diseases.

Zinc finger Structural motif seen in many DNA-binding proteins, composed of a loop of polypeptide chain held in a hair-pin bend bound to a zinc atom.

Zone of elongation The section of the young root or shoot just behind the apical meristem, in which the cells are enlarging and elongating rapidly.

Zoo blot Hybridization of cloned DNA from one species to DNA from other organisms to determine the extent to which the cloned DNA is evolutionarily conserved.

Zoo FISH (fluorescent *in situ* hybridization) Of DNA from one species on metaphase chromosomes of another species. Typically, the hybridization is done separately for DNA libraries representing each chromosome. The result is a fascinating picture of the regions of chromosomal homology between species.

Zoospore A spore that possesses flagella and is therefore motile.

Zygonema Stage in meiosis during which synapsis occurs; coming after the leptotene stage and before the pachytene stage in the meiotic prophase.

Zygospore A thick-walled resistant spore developing from a zygote resulting from the fusion of isogametes.

Zygote A zygote is a fertilized egg containing two sets of chromosomes, one from the egg (oocyte) and one form the sperm. The zygote is a single cell and the result of a fusion between two gametes, an egg (female) and one sperm cell (male).

Zymogen Inactive enzyme precursor that after secretion is chemically altered to the active form of the enzyme.